Bernard

D1583512

Records

Records

BARRY C. ARNOLD

Department of Statistics
University of California, Riverside

N. BALAKRISHNAN

Department of Mathematics and Statistics
McMaster University, Hamilton, Ontario

H. N. NAGARAJA

Department of Statistics
The Ohio State University, Columbus

A Wiley-Interscience Publication
JOHN WILEY & SONS, INC.
New York • Chichester • Weinheim • Brisbane • Singapore • Toronto

Copyright © 1998 by John Wiley & Sons, Inc.

Library of Congress Cataloging in Publication Data:

Arnold, Barry C.
    Records / Barry C. Arnold, N. Balakrishnan, H.N. Nagaraja.
        p.   cm. — (Wiley series in probability and statistics.
    Probability and statistics section)
    "Wiley-Interscience publication."
    Includes bibliographical references and indexes.
    ISBN 0-471-08108-6 (cloth : alk. paper)
    1. Order statistics.   2. World records.   I. Balakrishnan, N.,
1956–   . II. Nagaraja, H. N. (Haikady Navada). 1954–   .
III. Title.   IV. Series: Wiley series in probability and statistics.
Probability and statistics.
QA278.7.A748   1998
                                              98-12671
                                              CIP

Printed in the United States of America

10 9 8 7 6 5 4 3 2 1

To Sabrina

BCA

To Sarah and Julia

NB

To Chaitra and Smitha

HNN

Contents

Notations and Abbreviations   xiii

Preface   xvii

1. Introduction   1

     1.1 Who Cares?, 1

     1.2 A Road Map, 2

2. Basic Distributional Results   7

     2.1 Introduction, 7

     2.2 Standard Record Values Processes, 7

     2.3 Record Values From the Classical Model, 9

     2.4 Record Values From Specific Distributions, 20
         2.4.1 Weibull Records, 20
         2.4.2 Power Function Distribution Records, 21
         2.4.3 Pareto Records, 21
         2.4.4 Extreme Value Records, 22

     2.5 Record Times and Related Statistics, 22

     2.6 Markov Chains, 28

     2.7 Moments of Record Values, 29
         2.7.1 Weibull Distribution, 30
         2.7.2 Power Function Distribution, 31
         2.7.3 Pareto Distribution, 31
         2.7.4 Extreme Value Distribution, 32
         2.7.5 Normal Distribution, 32
         2.7.6 Covariance and Correlations, 33

     2.8 A Discrete Interlude, 34

     2.9 Geometric Results, 38

2.10 Counting Process and k-Records, 41
    2.10.1 A Point Process View, 41
    2.10.2 k-Record Statistics, 42
    Exercises, 44

3. Moment Relations, Bounds and Approximations   **51**

3.1 Introduction, 51

3.2 Exponential Distribution, 52

3.3 Weibull Distribution, 54

3.4 Gumbel Distribution, 56

3.5 Lomax Distribution, 61

3.6 Normal Distribution, 63

3.7 Logistic Distribution, 65

3.8 Bounds and Approximations, 68

3.9 Results for k-Records, 81
    Exercises, 89

4. Characterizations   **93**

4.1 Introduction, 93

4.2 Characterizing Properties of Record Values, 94
    4.2.1 The Moment Sequence, 95
    4.2.2 Regression of Adjacent Record Values, 96

4.3 Families of Distributions, 97
    4.3.1 Families Defined by Reliability Properties, 97
    4.3.2 Linear Regressions of Adjacent Record Values, 99

4.4 The Exponential Distribution, 101
    4.4.1 The Incomplete Catalog, 101
    4.4.2 Integrated Cauchy Functional Equation, 103
    4.4.3 Characterizations Based on the ICFE and
          Other Functional Equations, 105
    4.4.4 Lower Record Statistics, 107

4.5 Other Continuous Distributions, 108

4.6 Geometric-Tail Distributions, 111

4.7 Dependence Structures of Record Values and
      Order Statistics, 114
    Exercises, 116

5. Inference **121**

   5.1 Introduction, 121

   5.2 Maximum Likelihood Estimation, 122

   5.3 Best Linear Unbiased Estimation, 127

   5.4 Best Linear Invariant Estimation, 143

   5.5 Interval Estimation and Tests of Hypotheses, 145

   5.6 Point Prediction, 150
      5.6.1 Best Linear Unbiased Prediction, 150
      5.6.2 Best Linear Invariant Prediction, 153
      5.6.3 Asymptotic Linear Prediction, 154

   5.7 Interval Prediction, 156
      5.7.1 Prediction Intervals Based on BLUE's, 156
      5.7.2 Conditional Prediction Intervals, 159
      5.7.3 Tolerance Region Prediction, 161
      5.7.4 Bayesian Prediction Intervals, 162

   5.8 Illustrative Examples, 163

   5.9 Inference with Records and Inter-Record Times, 169

   5.10 Distribution-Free Tests in Time-Series Using
       Records, 170

       Exercises, 175

6. General Record Models **183**

   6.1 Introduction, 183

   6.2 Geometrically Increasing Populations, 184

   6.3 The F^α Record Model, 187
      6.3.1 Finite-sample Properties, 188
      6.3.2 Asymptotic Properties, 190

   6.4 Linear Drift Record Model, 194

   6.5 The Pfeifer Model, 198

   6.6 Characterizations, 202
      6.6.1 F^α and Linear Drift Record Models, 202
      6.6.2 The Pfeifer Model, 207

   6.7 Records From Dependent Sequences, 208
      6.7.1 Markov Sequences, 209

6.7.2 Exchangeable Observations, 211

6.7.3 Dependent Models Based on Archimedean Copula, 213

6.7.4 A Random Power Record Model, 214

Exercises, 215

7. Random and Point Process Record Models   **223**

7.1 Introduction, 223

7.2 Basic Random Record Model, 224
    7.2.1 Joint Distribution of Record Values, 224
    7.2.2 Dependence Structure of Record Values and
        Record Counts, 229
    7.2.3 Number of Records, 231

7.3 Basic Point Process Model and Applications, 233

7.4 Records Over Poisson Processes, 236
    7.4.1 Homogeneous Poisson Pacing Process, 236
    7.4.2 Comparison with the Classical Model, 239
    7.4.3 Nonhomogeneous Poisson Pacing Process, 241

7.5 Records Over a Renewal Process, 243

7.6 Records Over Birth Processes, 246
    7.6.1 General Birth Process, 246
    7.6.2 Yule Process, 248
    7.6.3 Poisson Record Arrival Process, 249

7.7 Records Over Other Pacing Processes, 252

7.8 The Secretary Problem, 254

Exercises, 258

8. Higher Dimensional Problems   **265**

8.1 How Should We Define Multivariate Records?, 265

8.2 Bivariate Records With Independent Coordinates, 267

8.3 Concomitants of Records, 271

8.4 Lower and Upper Records and the Record Range, 274

8.5 Records in Partially Ordered Sets, 277

Exercises, 277

Bibliography   **279**

Author Index   **303**

Subject Index   **309**

Notations and Abbreviations

i.i.d.	independent and identically distributed
a.e.	almost everywhere
a.s.	almost surely
iff	if and only if
R.H.S.	right-hand side
L.H.S.	left-hand side
SE	standard error
MSE	mean square error
cdf, $F(x)$	cumulative distribution function
pdf, $f(x)$	probability density function *also* density function
$\bar{F} = 1 - F$	survival function
$F(x-)$	$= P(X < x)$
PGF	probability generating function
MLE	maximum likelihood estimator
MVUE	minimum variance unbiased estimator
BLUE	best linear unbiased estimator
BLIE	best linear invariant estimator
BLUP	best linear unbiased predictor
BLIP	best linear invariant predictor
IFR	increasing failure rate
DFR	decreasing failure rate
DMRL	decreasing mean residual life
IFRA	increasing failure rate on average
NBU	new better than used
NWU	new worse than used
NBUE	new better than used on expectation
NWUE	new worse than used on expectation
PPRM	point process record model
CFE	Cauchy functional equation
ICFE	integrated Cauchy functional equation
γ	Euler's constant

S_n^k	Stirling number of the first kind
$\Gamma(\cdot)$	complete gamma function
$B(a,b)$	$= \frac{\Gamma(a)\Gamma(b)}{\Gamma(a+b)}$, complete beta function (with real or complex argument)
$\zeta(\cdot)$	Riemann zeta function
$I_p(\alpha)$	$= \frac{1}{\Gamma(\alpha)} \int_0^p e^{-t} t^{\alpha-1} dt$, incomplete gamma ratio
$[x]$	integer part of x
$x^{(i)}$	$= x(x-1)\cdots(x-i+1)$, descending factorial
$\mathbf{R}^+$	positive real line
$L^p(0,1)$	class of p-integrable functions in $(0,1)$, $p > 0$
$h(x)$	hazard or failure rate function
$H(x)$	hazard function
$\psi_F(u) = H^{-1}(u)$	inverse of the hazard function
$\phi(x)$	standard normal pdf
$\Phi(x)$	standard normal cdf
$\psi(z)$	$= \frac{d}{dz} \log \Gamma(z)$, digamma function
$\psi^{(k)}(z)$	$= \frac{d^{k+1}}{dz^{k+1}} \log \Gamma(z)$, polygamma functions
$\bar{G}(u)$	greatest convex minorant of $G(u)$
$E(X)$	expected value of X
$\text{Var}(X)$	variance of X
$\text{Cov}(X,Y)$	covariance of X and Y
$\text{Corr}(X,Y)$	correlation coefficient between X and Y
$\text{Ber}(p)$	Bernoulli distribution with parameter p
$\text{Geo}(p)$	geometric distribution with parameter p
$\text{Geo}T(j,p)$	geometric-tail distribution with positive integer j and parameter p
$\text{NBIN}(r,p)$	negative binomial distribution with parameters r and p
$\text{Beta}(m,n)$	beta distribution with parameters (m,n)
$\text{Beta}_{\text{II}}(a,b,c)$	beta distribution of the second kind with parameters (a,b,c)
χ_n^2	chi-square distribution with n degrees of freedom
$\text{Exp}(\theta)$	exponential distribution with mean θ
$\text{Exp-Gamma}(a,b,c,d)$	exponential-gamma distribution with parameters (a,b,c,d)
$F_{a,b}$	F distribution with (a,b) degrees of freedom
$\text{Gamma}(p,\theta)$	gamma distribution with shape parameter p and scale parameter θ

Gumbel(μ, σ)	Gumbel distribution with location parameter μ and scale parameter σ
$L(\mu, \sigma^2)$	logistic distribution with location parameter μ and scale parameter σ
Log-gamma(n)	log-gamma distribution with shape parameter n
$N(\mu, \sigma^2)$	normal distribution with mean μ and variance σ^2
Pareto(σ, α)	Pareto distribution with shape parameter α and scale parameter σ
Uniform(a, b)	uniform distribution over (a, b)
Weibull(c)	Weibull distribution with shape parameter c
$\beta(m, n; c)$	cth quantile of Beta(m, n) distribution
$\gamma(p, \theta; c)$	cth quantile of Gamma(p, θ) distribution
$\chi^2_{m,\beta}$	βth percentile of χ^2_m distribution
$F_1 \underset{c}{<} F_2$	F_1 c-precedes F_2
$F_1 \underset{s}{<} F_2$	F_1 s-precedes F_2
$X_{i:n}$	ith order statistic in a sample of size n
$F_{i:n}(x)$	cdf of $X_{i:n}$
$\alpha_{i:n}$	expected value of $X_{i:n}$
$D_M(G)$	domain of maximal attraction of G
$D_R(G)$	domian of record attraction of G
$\boldsymbol{P}$	point process
R_n	upper record values
R'_n	lower record values
$\tilde{R}_n$	weak records
T_n	record times
J_n	record increments
Δ_n	inter-record times
N_n	record counts
I_n	record indicators
$M_{R_n}(t)$	moment generating function of R_n
$K_{R_n}(t)$	cumulant generating function of R_n
α_n	mean of R_n
$\alpha_n^{(k)}$	kth moment of R_n
$\kappa_n^{(k)}$	kth cumulant of R_n
$\alpha_{m,n}$	$E(R_m R_n)$
$\alpha_{m,n}^{(k_1, k_2)}$	(k_1, k_2)th moment of (R_m, R_n)
$\alpha_{n_1, n_2, \ldots, n_\ell}^{(k_1, k_2, \ldots, k_\ell)}$	$(k_1, k_2, \ldots, k_\ell)$th moment of $(R_{n_1}, R_{n_2}, \ldots, R_{n_\ell})$
$\sigma_{n,n}(\sigma_n^2)$	variance of R_n
$\sigma_{m,n}$	covariance of R_m and R_n

$\alpha_n'^{(k)}$	kth moment of R_n'
$\mathbf{R}$	vector of record values $(R_0, R_1, \ldots, R_n)$
$\boldsymbol{\alpha}$	mean vector $(\alpha_0, \alpha_1, \ldots, \alpha_n)$
$\boldsymbol{\Sigma}$	variance-covariance matrix of $(R_0, R_1, \ldots, R_n)$
$\mathbf{1}$	column vector of 1s
$\mathbf{0}$	vector of 0s
$(_iR_n(x),\ _iR_n(y))$	record of the ith kind in (X, Y) sequence, $i = 1, 2, 3, 4$
$_iT_n$	record times corresponding to $(_iR_n(x),\ _iR_n(y))$, $i = 1, 2, 3, 4$
$_1N_n$	number of records of the first kind
$_1I_n$	record indicator of the first kind
$R_{[n]}$	nth record concomitant
ρ_n	initial ranks
$R_n^{(k)}$	Type 1 k-records
$R_{n(k)}$	Type 2 k-records
$T_{n(k)}$	record times of $R_{n(k)}$
RV_α	regularly varying at infinity with index α
$\overset{d}{=}$	equals in distribution
$\overset{P}{\to}$	converges in probability
$\overset{d}{\to}$	converges in distribution

Preface

It was with some reluctance that we eschewed the title *The Wiley Book of Records* for our book. Confusion with the enormously popular *Guinness Book of World Records* would have guaranteed that the have stood as a record for Statistics books. But just as Statistics books do not consist of artfully arranged arrays of numbers, so our book of records does not address the listing of records. Our focus is the theory behind sequences of record values and times. We contend that this theory is truly beautiful; much of it is elegant *and* elementary; an unbeatable combination. If you came seeking the latest information on the tallest resident of Lithuania or on the most recent record of the renowned pole vaulter Sergei Bubka, we apologize for disappointing you, but we urge you to sample the material. Try it. We know you'll like it.

Our main aim is to provide a resource for a one-term course on *record statistics*, which could be offered in a mathematics/statistics graduate program. This book hopefully will also prove useful if you are considering an independent study for credit or curiosity. We present the essential topics related to record values and record counting processes in a text-book format. In many instances, the exercises expand on the material covered in the text. There we provide relevant references, in order to acknowledge the contribution of various researchers and also to direct you to the appropriate source(s). We assume that you have gone through a one-year course in introductory mathematical statistics. Additional basic knowledge of stochastic processes will make your expedition a more pleasurable one.

We are indebted to numerous researchers who have enriched this field and have shared their findings enthusiastically with us over the years. Dr. Glenn Hofmann read a substantial portion of the manuscript and made several useful suggestions. Dr. Ping-Shing Chan carried out some of the computations presented in Chapters 3 and 5. Professor F. Thomas Bruss was helpful in untangling some mysteries of the secretary problem considered in Chapter 6. Reviewers of the manuscript had several helpful suggestions. We appreciate the assistance of all these individuals.

We are thankful to Mr. Steve Quigley (Executive Editor) and Ms. Lisa Van Horn (Production Editor) at John Wiley & Sons for their excellent support of this project. We had the outstanding assistance of Ms. Debbie Iscoe (Hamilton, Ontario), who undertook the arduous task of preparing the final LaTeX version of the manuscript in camera-ready form. Ms. Peggy Franklin (UC Riverside) was also helpful in preparing portions of the manuscript. It is a pleasure to offer our sincere thanks to them. Thanks are also due to Dr. Carole Arnold, for her constructive linguistic suggestions, and to Ms. Jyothi Nagaraja, for checking reference citations.

Now, for some personal acknowledgements. BCA is grateful to Dr. Robert Houchens, for initially kindling his interest in records. NB thanks the Natural Sciences and Engineering Research Council of Canada, for providing financial support during the course of this project. HNN expresses his sincere thanks to Professors Herbert A. David and Thomas Santner, for their friendship and longstanding support. We are indeed appreciative of our family members, for bearing with us during the preparation of this book. Numerous moments were stolen from the treasured family time as the manuscript evolved at a record (!) slow pace.

We also thank the Royal Statistical Society and Elsevier Science B.V., for providing permission to reproduce some of the previously published tables.

As we conclude, we would like to recall that in the preface of the book *A First Course in Order Statistics*, published in 1992, we promised a sequel. Though the present book is not a sequel, in the strict sense, it is a natural continuation of that volume, as it ended with a chapter on *record values*. As a matter of fact, while a basic understanding of the theory of order statistics will provide additional insight into many results on record statistics, these two books can be read quite independently. We hope you try both and appreciate your input for further improvement in these presentations.

BARRY C. ARNOLD
N. BALAKRISHNAN
H. N. NAGARAJA

April 1998

Records

CHAPTER 1

INTRODUCTION

1.1   WHO CARES?

No one can resist being interested in record values. The hottest day ever, the longest winning streak in professional basketball, the lowest stock market figure, these we cannot resist. January and October are peak months for record reporting. End-of-year reviews highlight record numbers of robberies, and the like, for the last year and, in October, World Series record mania hits us and we learn of the longest streak of hits between second and third base by a left-hander in night games, off knuckleball pitching, hit when the team is trailing, and other such delectable fare. Indeed, every at bat seems to produce a new record of some sort (as, in retrospect, it obviously does). So we can't resist records and there are lots of probabilists and statisticians among us. So the literature on record values must be enormous, right? Wrong. Two reasons probably account for this. First, the basic record model involving i.i.d. observations introduced by Chandler (1952) is, in a sense, too easy to analyze and as soon as we add bells and whistles to better model reality, it suddenly becomes too hard. That's a sure way to scare away researchers. Though almost the same can be said for queueing theory, the literature there *is* enormous. The other problem is intrinsic to the nature of records. They are rare. So anyone developing inferential techniques involving record values will always be faced with small samples. That deters anyone who is dedicated to large sample properties of estimates. It does not and has not deterred probabilists. Just because records are rare, it doesn't stop us from generating beautiful asymptotic distribution theory for them. The mathematics is not concerned with potential utility. And the mathematics for small and for large samples is remarkably pretty. Up until now, the study of record values has

1

been carried out by a relatively small but highly talented group of individuals. Chandler was the founding father. Stuart was a pioneer in finding applications of record counting statistics in inference. Barton, Mallows and F.N. David were entranced with combinatorial aspects of record sequences. While Dwass established the independence of record indicators, Rényi developed some of the first limit theorems. The fine-tuning of the asymptotic theory can be attributed to Tata, Neuts, Galambos, Resnick, Goldie, Shorrock, de Haan and Deheuvels. Their work is scattered in the literature in a broad spectrum of publications.

Several books on order statistics and related processes have chapters on record values. Many of them emphasize the close parallel between record value theory and the theory of sample maxima. In particular, Chapter 6 of Galambos (1987) and Chapter 4 of Resnick (1987) provide extensive discussion of record values. A fine early introductory article by Glick (1978) also merits attention, as do the survey articles by Nevzorov (1987), Nagaraja (1988a) and Nevzorov and Balakrishnan (1998). Ahsanullah's (1995) *Record Statistics* and its precursor (Ahsanullah, 1988) provide a limited coverage of certain topics in the domain of record statistics. The coverage in the present volume is much more extensive and inevitably more up to date.

Record value theory justifiably has its own existence, apart from the study of sample extremes. This book will not deny the possible interrelationship between the two topics, but neither will it present the results as a special topic in order statistics. (We confess to having already committed that crime when we finished up our book *A First Course in Order Statistics* with a brief chapter on record values.) Our current presentation is not an exhaustive study of record values, but it is comprehensive. There are indeed many papers that have been written on record statistics. The number of references easily exceeds the number of pages in this book!

1.2   A ROAD MAP

The time honored recipe for a good technical talk – tell them what you are going to explain, explain it, and then tell them what you have explained – applies to books also. We'll just follow steps one and two. This section will sketch what the remaining chapters will try to explain.

The classical record value model will be introduced in Chapter 2. The emphasis will be on the continuous case. This permits a smooth monotone transformation to the case involving records from i.i.d. exponential sequences. The lack of memory property of the exponential distribution renders the analysis of exponential record sequences a trivial exercise. As much as possible, results from the general classical model (i.e., based on i.i.d. non-exponential observations) will be derived via transformation from the exponential case. Both exact and asymptotic distributions are discussed. The asymptotic theory is aesthetically pleasing, though, as we point out, the low density of records in any i.i.d. sequence makes it unlikely that practical applications of the asymptotic theory can be envisioned. The basic problem is that you die long before the 50th record is encountered and a theory describing the distribution of the nth record for large n will be necessarily of limited utility. In addition to studying the actual values of the records, we are interested in the distribution of times at which records occur. These phenomena in the continuous case will be unchanged by monotone transformations of the data and, again, a transformation to exponentiality or uniformity will clarify matters.

The study of record values in many ways parallels the study of order statistics, indeed they are inextricably related. This means in, general, that things that could be done relatively easily for order statistics are also feasible for records. Things that were difficult for order statistics are unfortunately equally or more difficult for records. For example, the exact distribution and exact expressions for moments of order statistics are rarely available. The exceptional cases correspond to friendly distributions, such as the exponential or uniform and positive or negative powers of those such as Pareto and Weibull. Analogously, exact distribution results for the nth record are also usually available only for friendly distributions. Fortunately, the parallel goes further. Interrelations, bounds and approximations abound for distributions and moments of functions of order statistics. See Arnold and Balakrishnan (1989) for an extensive, though by now, out-of-date survey. In Chapter 3, a spectrum of parallel results for record values are recorded. We may not be able to give exact distributional results, but we frequently can give approximate and/or iterative methods of evaluating distributional features of the record sequence.

Knowledge of certain distributional functions of the record value sequence is adequate to determine the common distribution of the underlying observations. For example, the sequence of expected records $E(R_n)$ is adequate for such a characterization. Chapter 4 provides a survey of such characterization results. As is to be expected, the catalog parallels the corresponding list of characterizations based on order statistics. The exponential distribution is the boundary case for many of the inequalities introduced in Chapter 3 and, as a consequence, several characterizations can be stated in terms of the attainment of equality in some of the inequalities listed in the earlier chapter. The curious features of exponential record sequences (for which record increments are i.i.d.) spawn an imposing array of exponential characterizations using record values. As in the case of many order statistics based characterizations of the exponential distribution, integrated lack of memory (equivalently, the Lau-Rao theorem) will play a central role. Parallel discrete distribution characterizations (highlighting the geometric distribution) are also surveyed in the chapter.

In Chapter 5 we address the problems associated with statistical inference based on observed record sequences. Estimation of parameters and prediction of future records are discussed, both from a classical and a Bayesian viewpoint. Hypothesis tests regarding the tenability of the basic record model are also discussed. Thus, given a presumed realization of a standard record value sequence we could (anticipating, in some sense, the general record models to be discussed in Chapter 6) test for trends and/or spurious observations.

The basic record model is essentially a straw man when we deal with real data sequences. The oftremarked on plethora of Olympic records is a fairly typical instance in which the classical record model has little hope of explaining the data. Assumptions regarding changing and/or improving populations are naturally invoked to better describe such phenomena. Chapter 6 provides a survey of the literature in this area. Yang's (1975) model involving increasing populations was an early entry in the field. Trends can be, and have been, built into the models in a variety of ways to pump up the record production process to better mirror reality. Pfeiffer led the way in considering models in which the basic distribution of the observations is affected directly by the number of records thus far observed. Another way to manipulate and enhance the record production process is to allow for dependence among the observations. Efforts in this direction are also surveyed.

Random record models in which the number of available observations is a random variable form the focus of Chapter 7. In these models, with a positive probability the current record will not be broken and this feature adds new twists to the distribution theory for record counting statistics. Further distributional possibilities arise when observations arrive at time points determined by an independent point process. Some common point processes are used to model these arrival times and distributions of record and inter-record times are obtained. Results for records driven by Poisson processes and certain birth processes are particularly attractive. Role of records in finding optimal solution to the Secretary Problem and some of its generalizations is discussed there.

The final chapter ventures into more treacherous waters. Even the definitions of record times and record values for multivariate observations are open to discussion. Several competing definitions are introduced. Very limited progress is documented in the development of appropriate multi-variate record theory. There are some nice results (due to Goldie and Resnick) regarding when the number of records will be almost surely fi-nite, but there is very little to report on the distribution of the multivariate records themselves. Further work is clearly needed in this area. With this roadmap in mind, we are ready to begin.

CHAPTER 2

BASIC DISTRIBUTIONAL RESULTS

2.1 INTRODUCTION

Chandler (1952) introduced the study of record values and documented many of the basic properties of records. The major results summarized in this chapter were obtained during the period 1952–1983. The standard record value process corresponding to an infinite sequence of independent identically distributed (i.i.d.) observations is our focus. Non-standard variations on the theme dealing with non-independent or non-identically distributed observations will be discussed in Chapter 6. Chapter 7 discusses models where the number of available observations is a random variable.

2.2 STANDARD RECORD VALUE PROCESSES

Let $X_1, X_2, \ldots$ be an infinite sequence of i.i.d. random variables having the same distribution as the (population) random variable X. Denote the cumulative distribution function (cdf) of X by F. For the moment we put no constraints on F save that it be non-degenerate. An assumption of continuity is often invoked. This avoids the possibility of ties and allows a cleaner theoretical development.

An observation X_j will be called an upper record value (or simply a record) if its value exceeds that of all previous observations. Thus X_j is a record if $X_j > X_i$ for every $i < j$. An analogous definition deals with

7

lower record values. The times at which records appear are of interest. For convenience, let us assume that X_j is observed at time j. Then the *record time* sequence $\{T_n, n \geq 0\}$ is defined in the following manner:

$$T_0 = 1 \text{ with probability } 1$$

and, for $n \geq 1$,

$$T_n = \min\{j : X_j > X_{T_{n-1}}\}. \tag{2.2.1}$$

The *record value* sequence $\{R_n\}$ is then defined by

$$R_n = X_{T_n}, \quad n = 0, 1, 2, \ldots . \tag{2.2.2}$$

Here R_0 is referred to as the reference value or the trivial record. The rest are treated as nontrivial records. The above definition of the record sequence implicitly assumes that the cdf F will not produce any unbreakable record. This will not be the case if there exists some value x_0 such that $F(x_0) - F(x_0-) > 0$ and $F(x_0) = 1$, that is, if there is a largest possible real value that can be achieved with positive probability by the X_j's. For example, if we are repeatedly rolling a die and X_j denotes the number obtained on the jth throw, the value of $X_j = 6$ is an unbreakable record. In such a case we must make minor modifications in our definition of records. Here $\{R_n\}$ may be thought of as being a Markov chain with "6" as an absorbing state. Though such processes may be of interest in particular cases, we will eliminate them from our discussion and assume that the cdf F cannot yield unbreakable records so that $\{R_n, n \geq 0\}$ will be a strictly increasing non-terminating sequence. Note that this does *not* imply that $\{R_n\}$ is unbounded. For example, if the cdf F is uniform on the interval $(0, 1)$, then $\{R_n\}$ will be an increasing sequence that is bounded above by 1.

We may define a *record increment* process $\{J_n, n \geq 0\}$ (J for jump) by

$$J_0 = R_0$$

and, for $n > 1$,

$$J_n = R_n - R_{n-1}. \tag{2.2.3}$$

An *inter-record time* sequence, Δ_n, is also of interest. Here

$$\Delta_n = T_n - T_{n-1}, \quad n = 1, 2, \ldots . \tag{2.2.4}$$

The number of records may be tracked by means of the record counting process, $\{N_n, n \geq 1\}$, where

$$N_n = \{\text{number of records among } X_1, \ldots, X_n\}. \tag{2.2.5}$$

Of course $N_1 = 1$, since X_1 is always a record.

We name the setup where the X_i's are i.i.d. continuous random variables the *classical record model*. Observe that for the classical model, the parent cdf F will not affect distributions of the record counting statistics – T_n, Δ_n or N_n. The distributions of the record values R_n are predictably affected by F. These observations suggest the desirability of carefully selecting a common distribution for the X_j's, to make derivations as simple as possible. As we shall see, for the classical model, a strong argument can be made in favor of studying i.i.d. exponentially distributed X_j's.

2.3 RECORD VALUES FROM THE CLASSICAL MODEL

As suggested in the last section, it is convenient to first consider exponential observations. Let $\{X_j^*, j \geq 1\}$ denote a sequence of i.i.d. Exp(1) random variables. The asterisk on the X_j^*'s and on the related record values will serve to remind us that we are focusing on the specially tractable standard exponential case. The exponential distribution has the lack of memory property and consequently the differences between successive records will be i.i.d. standard exponential random variables. Thus $\{J_n^*\} = \{R_n^* - R_{n-1}^*, n \geq 1\}$ are i.i.d. Exp(1) random variables. It follows that for the nth record, R_n^*, corresponding to an i.i.d. Exp(1) sequence, we have

$$R_n^* \sim \text{Gamma}(n+1, 1), \quad n = 0, 1, 2, \dots . \tag{2.3.1}$$

We may use this result to obtain the distribution of the nth record corresponding to an i.i.d. sequence of random variables $\{X_j\}$ with common *continuous* cdf F. (See Sections 2.8 and 2.9 for further discussion of records from discrete populations.) If X has a continuous cdf F, then

$$H(X) \equiv -\log(1 - F(X)) \tag{2.3.2}$$

has a standard exponential distribution. Consequently, $X \overset{d}{=} F^{-1}(1 - e^{-X^*})$ where, as usual, X^* is Exp(1). Since X is a monotone function of X^*, the nth record of the $\{X_j\}$ sequence is expressible as a simple function of the nth record of an $\{X_j^*\}$ (standard exponential) sequence. Specifically we have

$$R_n \overset{d}{=} F^{-1}(1 - e^{-R_n^*}), \quad n = 0, 1, 2 \dots . \tag{2.3.3}$$

Repeated integration by parts can be used to justify the following expression for the survival function of R_n^* (a Gamma$(n+1, 1)$ random variable):

$$P(R_n^* > r^*) = e^{-r^*} \sum_{k=0}^{n} (r^*)^k / k!, \quad r^* > 0. \tag{2.3.4}$$

We may then use the relation (2.3.3) to immediately derive the survival function of the nth record corresponding to an i.i.d. F sequence.

$$P(R_n > r) = [1 - F(r)] \sum_{k=0}^{n} [-\log(1 - F(r))]^k / k! \tag{2.3.5}$$

or equivalently

$$P(R_n \le r) = \int_0^{[-\log(1-F(r))]} w^n e^{-w} / n! \; dw. \tag{2.3.6}$$

If F is absolutely continuous with corresponding probability density function (pdf) f, we may differentiate either of the above expressions to obtain the pdf for R_n. In fact, we have

$$f_{R_n}(r) = f(r)[-\log(1 - F(r))]^n / n! \; . \tag{2.3.7}$$

The joint pdf of a set of Exp(1) records $(R_0^*, R_1^*, \ldots, R_n^*)$ is easily written down, since the record spacings $R_n^* - R_{n-1}^*$ are i.i.d. Exp(1). One finds

$$f_{R_0^*, R_1^*, \ldots, R_n^*}(r_0^*, r_1^*, \ldots, r_n^*) = e^{-r_n^*}, \quad 0 < r_0^* < r_1^* < \ldots < r_n^*. \tag{2.3.8}$$

Applying the transformation (2.3.3) coordinatewise we obtain the joint pdf of the set of records $R_0, R_1, \ldots, R_n$ corresponding to an i.i.d. F sequence (assuming F is absolutely continuous with pdf f). It is given by

$$
\begin{aligned}
f_{R_0, R_1, \ldots, R_n}(r_0, r_1, \ldots, r_n) &= \prod_{i=0}^{n} f(r_i) / \prod_{i=0}^{n-1} [1 - F(r_i)], \\
&= f(r_n) \prod_{i=0}^{n-1} h(r_i) \\
&\qquad -\infty < r_0 < r_1 < \ldots < r_n < \infty.
\end{aligned} \tag{2.3.9}
$$

where $h(r) = dH(r)/dr = f(r)/[1 - F(r)]$ represents the failure rate function.

The joint pdf of any pair of records (R_m, R_n) can of course be obtained from (2.3.9) by integration. It is perhaps easier to first find the joint distribution for two Exp(1) records (R_m^*, R_n^*) and then use the transformation (2.3.3). When $m < n$, since $(R_m^*, R_n^*) \stackrel{d}{=} (Y_1, Y_1 + Y_2)$ where $Y_1 \sim$ Gamma$(m + 1, 1)$ and $Y_2 \sim$ Gamma$(n - m, 1)$ are independent, we may use a simple Jacobian argument beginning with the joint pdf of (Y_1, Y_2) to obtain

$$f_{R_m^*, R_n^*}(r_m^*, r_n^*) = \frac{1}{m!(n - m - 1)!} \, r_m^{*m}(r_n^* - r_m^*)^{n-m-1} e^{-r_n^*},$$

$$0 < r_m^* < r_n^* < \infty. \qquad (2.3.10)$$

From this we obtain [with (2.3.3) applied to R_m^* and R_n^*]

$$f_{R_m, R_n}(r_m, r_n)$$
$$= \frac{[-\log(1 - F(r_m))]^m}{m!} \frac{[-\log(\frac{1-F(r_n)}{1-F(r_m)})]^{n-m-1}}{(n - m - 1)!} \frac{f(r_m)f(r_n)}{[1 - F(r_m)]},$$
$$-\infty < r_m < r_n < \infty. \qquad (2.3.11)$$

Conditional distributions of records given a particular record are also easily obtained by transforming the results for exponential records. For example, from well-known properties of Poisson processes, we conclude that, for any n, the conditional distribution of $R_0^*, \ldots, R_{n-1}^*$, given $R_n^* = r_n^*$, is the same as the distribution of the order statistics of a random sample of size n drawn from a Uniform$(0, r_n^*)$ distribution. For future records, we have the conditional distribution of $R_{n+1}^*, \ldots, R_{n+j}^*$ given $R_n^* = r_n^*$ is the same as the joint distribution of $r_n^* + R_0^*, r_n^* + R_1^*, \ldots, r_n^* + R_{j-1}^*$; that is,

$$f_{R_{n+1}^*, \ldots, R_{n+j}^* | R_n^*}(r_{n+1}^*, \ldots, r_{n+j}^* | r_n^*)$$
$$= e^{-(r_{n+j}^* - r_n^*)}, r_n^* < r_{n+1}^* < \cdots < r_{n+j}^* < \infty. \qquad (2.3.12)$$

Applying the transformation (2.3.3), we obtain the parallel results for records corresponding to a general absolutely continuous cdf F. Thus

$$f_{R_0, \ldots, R_{n-1} | R_n}(r_0, r_1, \ldots, r_{n-1} | r_n) = \frac{n! \prod_{i=0}^{n-1} [\frac{f(r_i)}{(1-F(r_i))}]}{[-\log(1 - F(r_n))]^n},$$

$$-\infty < r_0 < r_1 < \cdots < r_{n-1} < r_n$$

and

$$f_{R_{n+1}, R_{n+2}, \ldots, R_{n+j} | R_n}(r_{n+1}, \ldots, r_{n+j} | r_n) = \prod_{i=0}^{j-1} \left[\frac{f(r_{n+i+1})}{1 - F(r_{n+i})} \right],$$

$$r_n < r_{n+1} < \cdots < r_{n+j} < \infty.$$

The nth record corresponding to an i.i.d. F sequence is distributed as a smooth function of a sum of $(n+1)$ i.i.d. standard exponential variables, namely the X_i^*'s. This should allow us to obtain an appropriate limiting distribution for R_n relatively easily. From (2.3.3) we may write

$$R_n \overset{d}{=} \psi_F \left(\sum_{i=0}^{n} X_i^* \right), \tag{2.3.13}$$

where

$$\psi_F(u) = F^{-1}(1 - e^{-u}) = H^{-1}(u) \tag{2.3.14}$$

represents the inverse of the hazard function $H(x) = -\log(1 - F(x))$. The strong law of large numbers assures us that $\frac{1}{n+1}\sum_{i=0}^{n} X_i^* \overset{a.s.}{\longrightarrow} 1$ and that $\sum_{i=0}^{n} X_i^* \overset{a.s.}{\longrightarrow} \infty$ as $n \to \infty$. It is evident consequently that the crucial feature of the function ψ_F, with regard to the limiting distribution of R_n, is its behavior as $u \to \infty$. Note that ψ_F is non-decreasing with domain $(0, \infty)$.

Intuition might suggest that smooth functions of sums of i.i.d. random variables would be asymptotically normal. Smooth functions of averages of i.i.d. random variables usually are and so asymptotic normality of R_n is an appealing possibility. Sometimes this works, as the following theorem shows.

Theorem 2.3.1 (Resnick, 1973a) *If ψ_F is such that*

$$\lim_{s \to \infty} \frac{\psi_F(s + x\sqrt{s}) - \psi_F(s)}{\psi_F(s + \sqrt{s}) - \psi_F(s)} = x, \ \forall \ x \tag{2.3.15}$$

then

$$\frac{R_n - \psi_F(n)}{\psi_F(n + \sqrt{n}) - \psi_F(n)} \overset{d}{\longrightarrow} N(0, 1).$$

Proof. Recall that $R_n = \psi_F(S_{n+1}^*)$ where S_{n+1}^* is a sum of $n + 1$ i.i.d. exponential random variables. The central limit theorem applies and $(S_{n+1}^* - (n+1))/\sqrt{n+1} \overset{d}{\longrightarrow} N(0, 1)$ and, a bit awkwardly but conveniently for us, $(S_{n+1}^* - n)/\sqrt{n} \overset{d}{\longrightarrow} N(0, 1)$ also. We can argue as follows:

$$\frac{R_n - \psi_F(n)}{\psi_F(n + \sqrt{n}) - \psi_F(n)} = \frac{\psi_F(S_{n+1}^*) - \psi_F(n)}{\psi_F(n + \sqrt{n}) - \psi_F(n)}$$

$$= \frac{\psi_F(n + [(S_{n+1}^* - n)/\sqrt{n}]\sqrt{n}) - \psi_F(n)}{\psi_F(n + \sqrt{n}) - \psi_F(n)}.$$

This can be written as

$$g_n\left(\frac{S^*_{n+1} - n}{\sqrt{n}}\right)\left[\frac{S^*_{n+1} - n}{\sqrt{n}}\right]$$

where $g_n(x) \to 1 \ \forall \ x$ by (2.3.15). From this we conclude that $(R_n - \psi_F(n))/[\psi_F(n + \sqrt{n}) - \psi_F(n)]$ and $(S^*_{n+1} - n)/\sqrt{n}$ have the same $N(0,1)$ limiting distribution. $\bigcirc$

   The condition (2.3.15) may appear unusual but it holds in a broad spectrum of cases.

Example 2.3.1 (Weibull Distribution) If $F(x) = 1 - e^{-x^c}, 0 < x < \infty$, $c > 0$, i.e., the X_i's are Weibull(c) random variables, then

$$\psi_F(u) = u^{\frac{1}{c}}$$

and

$$\lim_{s \to \infty} \frac{\psi_F(s + x\sqrt{s}) - \psi_F(s)}{\psi_F(s + \sqrt{s}) - \psi_F(s)} = \lim_{s \to \infty} \frac{(s + x\sqrt{s})^{\frac{1}{c}} - s^{\frac{1}{c}}}{(s + \sqrt{s})^{\frac{1}{c}} - s^{\frac{1}{c}}}$$

$$= \lim_{s \to \infty} \frac{(1 + x/\sqrt{s})^{\frac{1}{c}} - 1}{(1 + \frac{1}{\sqrt{s}})^{\frac{1}{c}} - 1} = x.$$

Consequently, Weibull records are asymptotically normal. $\bigcirc$

Example 2.3.2 (Logistic Distribution) If $F(x) = e^x/(1+e^x), -\infty < x < \infty$, i.e., the X_i's have a standard logistic distribution, then

$$\psi_F(u) = \log(e^u - 1)$$

and again we have

$$\lim_{s \to \infty} \frac{\psi_F(s + x\sqrt{s}) - \psi_F(s)}{\psi_F(s + \sqrt{s}) - \psi_F(s)} = \lim_{s \to \infty} \frac{\log(\frac{e^{s+x\sqrt{s}}-1}{e^s-1})}{\log(\frac{e^{s+\sqrt{s}}-1}{e^s-1})} = x.$$

Thus, logistic records are asymptotically normal. $\bigcirc$

Example 2.3.3 (Extreme Value Distribution) Suppose $F(x) = 1 - e^{-e^x}$, $-\infty < x < \infty$, i.e., the X_i's have an extreme value distribution, then $\psi_F(u) = \log u$. It is possible, though a bit tiresome, to verify condition (2.3.15) and conclude asymptotic normality of extreme records. $\bigcirc$

An alternative, which is convenient in some cases, is to use the following sufficient condition for the asymptotic normality of records.

Theorem 2.3.2 (Tata, 1969) *If the hazard function $H(x) = \psi_F^{-1}(x)$ is convex, twice differentiable and satisfies*

$$H(x)/x \uparrow \infty$$

and

$$H(x)/x^2 \downarrow 0$$

as $x \to \infty$, then there exist constants a_n and $b_n > 0$ such that $(R_n - a_n)/b_n \overset{d}{\longrightarrow} N(0, 1)$.

Tata's condition allows a simple derivation of the asymptotic normality of Weibull records when $c \in (1, 2)$.

Of course, not every distribution satisfies condition (2.3.15). If $F(x) = 1 - x^{-\alpha}, x > 1$, i.e., a classical Pareto distribution, then

$$\psi_F(u) = e^{u/\alpha}$$

and it is readily verified that (2.3.15) does not hold. In fact, in this case, R_n cannot be normalized to obtain any non-trivial limiting distribution.

Non-normal limit laws are possible for R_n. For example, if (for some $\alpha > 0$),

$$F_0(x) = 1 - e^{-\frac{\alpha^2}{4}(\log x)^2}, \qquad x > 1, \tag{2.3.16}$$

then R_n is attracted to a limit law of the log-normal form

$$\Phi_\alpha(x) = \begin{cases} 0, & x < 0 \\ \Phi(\log x^\alpha), & x \geq 0, \end{cases} \tag{2.3.17}$$

where $\Phi(\cdot)$ is the standard normal cdf. In order for R_n to be attracted to the law (2.3.17), the common cdf F of the X_i's must be such that $F(x) < 1, \forall\, x$ and, in fact, must be similar to F_0 (in (2.3.16)) in tail behavior, in the sense that

$$\lim_{x \to \infty} \log(1 - F(x))/\log(1 - F_0(x)) = 1.$$

For such distributions it can be shown that the asymptotic distribution of $R_n/\psi_F(n)$ is given by $\Phi_\alpha(\cdot)$, as in (2.3.17).

Certain cdfs F with $F(x_0) = 1$ for some $x_0 < \infty$ have a third kind of limit law for their associated records. The typical example is provided by a distribution of the form

$$\tilde{F}_0(x) = 1 - e^{-\frac{\alpha^2}{4}[\log(x_0 - x)]^2}, \qquad x_0 - 1 \leq x \leq x_0, \tag{2.3.18}$$

for some $\alpha > 0$ and $x_0 < \infty$. In this case R_n will be attracted to the limit law

$$\tilde{\Phi}_\alpha(x) = \begin{cases} \Phi(\log(-x)^{-\alpha}), & x < 0 \\ 1, & x \geq 0. \end{cases} \qquad (2.3.19)$$

In order for R_n to be attracted to the distribution (2.3.19), F must be such that $F(x_0) = 1$ for some x_0 and the tail behavior as $x \uparrow x_0$ must be similar to that of $\tilde{F}_0$ (in (2.3.18)), in the sense that

$$\lim_{x \to x_0} \log(1 - F(x))/\log(1 - \tilde{F}_0(x)) = 1.$$

For such distributions, the appropriate normalization of R_n is again quite simple. One finds that $(R_n - x_0)/(x_0 - \psi_F(n))$ has a limiting distribution given by $\tilde{\Phi}_\alpha(x)$, as in (2.3.19).

In summary, for a relatively broad class of distributions, condition (2.3.15) will hold and records will be asymptotically normal. Under special circumstances distributions similar to (2.3.16) and (2.3.18) in their upper tail will have non-normal limiting distributions for records.

The first detailed derivation of the three kinds of limit laws which can arise in the study of record values was provided by Resnick (1973a). His development of the topic involved the identification of an associated cdf F_a for each cdf F defined by

$$F_a(x) = 1 - \exp\left[-\sqrt{H(x)}\right], \qquad (2.3.20)$$

where H is the hazard function of F. In other words, the hazard function of $F_a(x)$ is given by $\sqrt{H(x)}$. The limiting distribution of records corresponding to the cdf F turns out to be related to the limiting behavior of maxima of samples drawn from the associated cdf F_a. Recall that there are three possible limit laws for sample maxima:

(i) Frechét distributions:

$$G_{1,\alpha}(x) = \begin{cases} 0, & x \leq 0 \\ \exp[-x^{-\alpha}], & x > 0, \end{cases} \qquad (2.3.21)$$

where $\alpha > 0$,

(ii) Weibull distributions:

$$G_{2,\alpha}(x) = \begin{cases} \exp[-(-x)^\alpha], & x < 0 \\ 1, & x \geq 0, \end{cases} \qquad (2.3.22)$$

where $\alpha > 0$, and

(iii) Gumbel distribution:

$$G_3(x) = \exp[-e^{-x}], \qquad -\infty < x < \infty. \qquad (2.3.23)$$

No other distributions can arise as limiting distributions of sample maxima (Gnedenko, 1943). There has arisen considerable controversy regarding what names should be associated with the three kinds of limiting distributions for sample maxima. Many would attach Gumbel's name to G_3 as we have done here. There appears to be no unanimity. The choice of names used here is not definitive. The choice of symbols $G_{1\alpha}, G_{2\alpha}$, and G_3 is also idiosyncratic [for example, Resnick (1973a) and others use Φ_α, Ψ_α and Λ instead]. The notation here is consistent with that used in Chapter 8 of Arnold, Balakrishnan and Nagaraja (1992), which may serve as a convenient elementary reference for needed results on asymptotic distributions of sample maxima. Resnick (1973a) derived a parallel result for record values. He showed that there are only three kinds of distributions that could arise as limiting distributions of suitably normalized record values. Using, as usual, the symbol Φ to denote a standard normal distribution, the three possibilities are

(i) Log-normal distributions:

$$\Phi_\alpha(x) = \begin{cases} 0, & x < 0 \\ \Phi(\log x^\alpha), & x \geq 0, \end{cases} \qquad (2.3.24)$$

where $\alpha > 0$,

(ii) Negative-log-normal distributions:

$$\tilde{\Phi}_\alpha(x) = \begin{cases} \Phi(\log(-x)^{-\alpha}), & x < 0 \\ 1, & x \geq 0, \end{cases} \qquad (2.3.25)$$

where $\alpha > 0$, and

(iii) the normal distribution:

$$\Phi(x). \qquad (2.3.26)$$

As Resnick (1973a) observed, these limit laws are all of the form $\Phi(-\log(-\log G(x)))$, where G is some limit law for sample maxima. This observation was crucial in the development of Resnick's remarkable Duality theorem. It is convenient to introduce notation for maximal and record attraction as follows: Consider i.i.d. observations X with common cdf F. Denote as usual the maximum of n such observations by $X_{n:n}$ and the nth record in the sequence by R_n. Let G be one of the possible limit laws for maxima (i.e., of one of the forms (2.3.21), (2.3.22) or (2.3.23)). We will say that F is in the domain of maximal attraction of G and write $F \in D_M(G)$ if there exist normalizing constants such that

$$\lim_{n \to \infty} P((X_{n:n} - a_n^0)/b_n^0 \leq x) = G(x),$$

for all x that are continuity points of G. In parallel fashion, let $\hat{G}$ be one of the possible limit laws for records (i.e., of one of the forms (2.3.24), (2.3.25) or (2.3.26)). We will say that F is in the domain of record attraction of $\hat{G}$ and write $F \in D_R(\hat{G})$ if there exist normalizing constants such that

$$\lim_{n \to \infty} P((R_n - a_n)/b_n \leq x) = \hat{G}(x),$$

for all continuity points of $\hat{G}$. With these notational conventions in place the duality theorem is of the following form.

Theorem 2.3.3 (Duality Theorem) (Resnick, 1973b) *If F is a continuous cdf with associated cdf F_a (as defined in (2.3.20)), then*

(i) $F \in D_R(\Phi_\alpha)$ *iff* $F_a \in D_M(G_{1,\alpha/2})$ *and in this case one may use as normalizing constants*

$$a_n = 0, \qquad b_n = \psi_F(n),$$

to obtain

$$\lim_{n \to \infty} P(R_n/\psi_F(n) \leq x) = \Phi_\alpha(x)$$

[ψ_F is as defined in(2.3.14)].

(ii) $F \in D_R(\tilde{\Phi}_\alpha)$ *iff* $F_a \in D_M(G_{2,\alpha/2})$. *In this case $F^{-1}(1)$ is necessarily finite and we may use as normalizing constants*

$$\begin{aligned} a_n &= F^{-1}(1), \\ b_n &= F^{-1}(1) - \psi_F(n) \end{aligned}$$

to obtain

$$\lim_{n \to \infty} P((R_n - F^{-1}(1))/(F^{-1}(1) - \psi_F(n)) \leq x) = \tilde{\Phi}_\alpha(x).$$

(iii) $F \in D_R(\Phi)$ *iff* $F_a \in D_M(G_3)$ *and in this case one may take as normalizing constants*

$$a_n = \psi_F(n), \quad b_n = \psi_F(n + \sqrt{n}) - \psi_F(n)$$

to obtain

$$\lim_{n \to \infty} P((R_n - \psi_F(n))/(\psi_F(n + \sqrt{n}) - \psi_F(n)) \leq x) = \Phi(x).$$

For more details the reader is referred to Resnick (1973a) or to Chapter 4 of Resnick (1987).

Even more is possible. de Haan (1970) provided characterizations of distributions in the domains of maximal attraction of all of the extremal limiting laws (2.3.21), (2.3.22) and (2.3.23). These can, via the duality theorem, be translated into necessary and sufficient conditions for record attraction to the three possible kinds of limit laws [given in (2.3.24), (2.3.25) and (2.3.26)]. de Haan provides a spectrum of alternative specifications of necessary and sufficient conditions for maximal attraction derived from a spectrum of alternative descriptions of regularly varying functions. We report here the conditions which appear simplest. They may not be, in a particular application, the most convenient to check and if that proves to be the case, recourse to alternatives described in Resnick (1973a) and in de Haan (1970) may be helpful. The "simple" conditions are as described in the following theorem.

Theorem 2.3.4 (Resnick, 1973a)

(i) $F \in D_R(\Phi_\alpha)$ iff for all real x,

$$\lim_{s \to \infty} \frac{\psi_F^{-1}(sx) - \psi_F^{-1}(s)}{\sqrt{\psi_F^{-1}(s)}} = \alpha \log x.$$

(ii) $F \in D_R(\tilde{\Phi}_\alpha)$ iff for all $x > 0$,

$$\lim_{\epsilon \to 0+} \frac{\psi_F^{-1}(x_0 - \epsilon x) - \psi_F^{-1}(x_0 - \epsilon)}{\sqrt{\psi_F^{-1}(x_0 - \epsilon)}} = -\alpha \log x,$$

where $x_0 = F^{-1}(1)$ is finite.

(iii) $F \in D_R(\Phi)$ iff for all real x,

$$\lim_{s \to \infty} \frac{\psi_F(s + x\sqrt{s}) - \psi_F(s)}{\psi_F(s + \sqrt{s}) - \psi_F(s)} = x.$$

It will be now noticed that the condition for record attraction to the normal distribution used in Theorem 2.3.1 was in fact a necessary and sufficient condition. It is, of course, as a consequence of parts (i) and (ii) of Theorem 2.3.4 that the distributions (2.3.24) and (2.3.25) are, as claimed, essentially prototypical of those distributions in the domain of record attraction of the log-normal and negative log-normal distributions respectively. But see Exercise 2.30.

We will illustrate the utility of the Resnick approach by using it to analyze the asymptotic distribution of normal records.

Example 2.3.4 (Standard Normal Distribution) When $F = \Phi$ we have

$$
\begin{aligned}
\psi_F^{-1}(u) &= -\log(1 - \Phi(u)) \\
&\approx u^2/2 + \log u,
\end{aligned}
$$

since $1 - \Phi(u) \approx \phi(u)/u$ (see e.g., Feller, 1968; p. 175). Consequently, the associated cdf F_a satisfies

$$
1 - F_a(x) \approx e^{-x/\sqrt{2}} .
$$

Thus F_a is asymptotically equivalent to an exponential distribution. Now it is known that exponential maxima are attracted to the limit law $G_3(x) = \exp\{-e^{-x}\}$ and consequently so are maxima associated with F_a. But then, according to Resnick's duality theorem, since maxima from F_a are attracted to G_3, records associated with $F(= \Phi$ in this case) will be asymptotically normal with appropriate norming constants as given in Theorem 2.3.1. Thus, for records from a standard normal distribution we have

$$
\frac{R_n - \Phi^{-1}(1 - e^{-n})}{\Phi^{-1}(1 - e^{-(n+\sqrt{n})}) - \Phi^{-1}(1 - e^{-n})} \xrightarrow{d} N(0, 1).
$$

An application of the mean value theorem (see, e.g., Nayak and Inginshetty, 1995) gives an equivalent simpler statement of this limiting result for normal records, namely:

$$
\sqrt{2}(R_n - \Phi^{-1}(1 - e^{-n})) \xrightarrow{d} N(0, 1). \qquad (2.3.27)
$$

○

The representation (2.3.13) [i.e., $R_n = \psi_F(\sum_{i=0}^{n} X_i^*)$] permits development of strong laws, laws of the iterated logarithm and functional limit theorems for R_n, borrowing heavily on the parallel results available for sums of i.i.d. random variables (such as the X_i^*'s). Convenient references for these results are Resnick (1973b), de Haan and Resnick (1973) and Vervaat (1973).

The general expression (2.3.9) gives the joint pdf for the first n records from any cdf F with corresponding pdf f. On close inspection it will be seen that formula (2.3.9) is valid for distribution absolutely continuous with respect to any measure on the real line including counting measure concentrated on a countable set. So the equation remains valid for discrete distributions with a countable number of possible values where the supremum of the support is not an atom (to ensure no unbreakable records are encountered).

2.4 RECORD VALUES FROM SPECIFIC DISTRIBUTIONS

For certain choices of the cdf F, the record value sequence admits a simple stochastic description. We have already remarked on the simplicity encountered when F is Exp(1). In this case the nth record has a Gamma($n+1, 1$) distribution. This is included as a special case of the Weibull example described below. The list of parent distributions for which the record sequence admits a simple representation [cases in which (2.3.3) simplifies markedly] is not long, but includes distributions frequently referred to in the sequel, especially in the context of characterizations via record values.

2.4.1 Weibull Records

Suppose that the distribution of the population random variable X is Weibull with cdf

$$F(x) = \begin{cases} 0, & x < \mu \\ 1 - e^{-(\frac{x-\mu}{\sigma})^c}, & x \geq \mu, \end{cases} \tag{2.4.1}$$

where the parameters are μ real and $\sigma, c > 0$. Such a random variable admits the representation

$$X \overset{d}{=} \mu + \sigma X^{*\frac{1}{c}}, \tag{2.4.2}$$

where X^*, as usual, is an Exp(1) random variable. Consequently, for $n = 0, 1, 2, \ldots,$

$$R_n \overset{d}{=} \mu + \sigma \left(\sum_{i=0}^{n} X_i^* \right)^{\frac{1}{c}}, \tag{2.4.3}$$

where $\{X_i^*,\ i \geq 0\}$ is a sequence of i.i.d. Exp(1) random variables. If $\mu = 0$ in (2.4.3), then quotients of successive records are independent random variables. Specifically if $Q_m = R_m/R_{m-1}, m = 1, 2, \ldots$, then the Q_m's are independent random variables with $P(Q_m > q) = q^{-cm}, q > 0$.

In the particular case where $c = 1$, i.e., the exponential case, we have (i) J_n and R_{n-1} are independent random variables, (ii) $E(R_n|R_{n-1} = r_{n-1})$ is a linear function of r_{n-1} with unit slope, and (iii) $J_n \overset{d}{=} R_0 - \mu$. All of these are characteristic properties of exponential records.

2.4.2 Power Function Distribution Records

If the population cdf is of the form

$$F(x) = \begin{cases} 0, & x \leq 0 \\ x^\alpha, & 0 < x < 1 \\ 1, & x \geq 1, \end{cases} \tag{2.4.4}$$

where $\alpha > 0$, a simple description of the record value sequence is possible. For $n = 0, 1, 2, \ldots$ we have

$$R_n \overset{d}{=} \left[1 - \prod_{j=0}^{n} (1 - U_j) \right]^{\frac{1}{\alpha}}, \tag{2.4.5}$$

where $\{U_j, j \geq 0\}$ is a sequence of i.i.d. Uniform$(0, 1)$ random variables. The lower record sequence, $\tilde{R}_n$, admits a slightly simpler representation

$$\tilde{R}_n = \left[\prod_{j=0}^{n} V_j \right]^{\frac{1}{\alpha}}, \tag{2.4.6}$$

where $\{V_j, j \geq 0\}$ are i.i.d. Uniform$(0, 1)$ random variables. In the particular case when $\alpha = 1$, i.e., the uniform case, we have (i) $(1 - R_n)/(1 - R_{n-1})$ and $1 - R_{n-1}$ are independent, (ii) $E(R_n|R_{n-1} = r_{n-1}) = \frac{1}{2} + \frac{1}{2}r_{n-1}$, and (iii) $(1 - R_n)/(1 - R_{n-1}) \overset{d}{=} 1 - R_0$. All of these are characteristic properties of the uniform distribution.

2.4.3 Pareto Records

Suppose that the observations are taken from a cdf

$$\begin{aligned} F(x) &= 0, & x < \sigma \\ &= 1 - \left(\tfrac{x}{\sigma}\right)^{-\alpha}, & x \geq \sigma, \end{aligned} \tag{2.4.7}$$

where $\alpha > 0$ and $\sigma > 0$. If (2.4.7) holds, we say the population random variable, X, is Pareto(σ, α). Observe that for such a random variable we have

$$X \overset{d}{=} \sigma U^{-\frac{1}{\alpha}}, \tag{2.4.8}$$

where U is Uniform$(0, 1)$. Consequently for $n = 0, 1, 2, \ldots$

$$R_n \stackrel{d}{=} \sigma \left(\prod_{j=0}^{n} U_j \right)^{-\frac{1}{\alpha}}, \qquad (2.4.9)$$

where the U_j's are i.i.d. Uniform$(0, 1)$ random variables. Quotients of successive Pareto records are again independent Pareto random variables. Thus if $Q_m = R_m/R_{m-1}$, $m = 1, 2, \ldots$, then the Q_m's are i.i.d. random variables with

$$P(Q_m > q) = q^{-\alpha}.$$

It is evident from (2.4.9) that (i) R_n/R_{n-1} and R_{n-1} are independent random variables, (ii) $E(R_n | R_{n-1} = r_{n-1}) = \alpha r_{n-1}$ when $\alpha > 1$, and (iii) $R_n/R_{n-1} \stackrel{d}{=} R_0$. All of these are characteristic properties of Pareto records.

2.4.4   Extreme Value Records

Suppose that the population random variable X has an *extreme value* distribution with cdf

$$F(x) = 1 - \exp[-e^{(x-\mu)/\sigma}], \qquad (2.4.10)$$

where μ is real and $\sigma > 0$. For such a random variable we have

$$X \stackrel{d}{=} \mu + \sigma \log X^*, \qquad (2.4.11)$$

where $X^* \sim \mathrm{Exp}(1)$. The corresponding record value sequence can be described by

$$R_n \stackrel{d}{=} \mu + \sigma \log \left(\sum_{i=0}^{n} X_i^* \right) . \qquad (2.4.12)$$

In such cases, one can show that the spacings of successive records are independent exponential random variables. Specifically, if we define $J_i = R_i - R_{i-1}$, $i = 1, 2, \ldots$ then the J_i's are independent random variables with $P(J_i > x) = e^{-ix/\sigma}$, $x > 0$ (for details, see, for example, Houchens 1984; p. 45–46).

2.5   RECORD TIMES AND RELATED STATISTICS

Recall the definition of the record time sequence $\{T_n, n \geq 0\}$, namely

$$T_0 = 1 \text{ with probability } 1$$

and for $n \geq 1$

$$T_n = \min\{j : X_j > X_{T_{n-1}}\}.$$

In order to derive the joint distribution of the first n non-trivial record times [i.e., $(T_1, T_2, \ldots, T_n)$], we introduce a sequence of record indicator random variables as follows:

$$I_1 = 1 \quad \text{with probability 1}$$

and for $n > 1$,

$$I_n = I(X_n > \max\{X_1, X_2, \ldots, X_{n-1}\}). \tag{2.5.1}$$

Thus $I_n = 1$ if and only if (iff) X_n is a record value.

We will invoke the assumption that F, the common cdf of the X_i's, is continuous. This avoids ties and ensures that the distribution of record times and counts does not depend on F (i.e., the T_n's, N_n's etc. are distribution free).

To determine the joint distribution of the I_n's, consider, for any finite sequence of m integers $1 = j_1 < j_2 \ldots < j_m$, the probability that observations numbered $j_1, j_2, \ldots, j_m$ are records. To do this computation we assume without loss of generality, that the X_i's are Uniform$(0, 1)$ random variables. We argue as follows:

$$P(I_{j_1} = 1, I_{j_2} = 1, \ldots, I_{j_m} = 1)$$

$$= \int \cdots \int_{0 < x_1 < \ldots < x_m < 1} P(I_{j_1} = 1, \ldots, I_{j_m} = 1 | X_{j_1} = x_1, \ldots, X_{j_m} = x_m)$$

$$\cdot dx_1 \ldots dx_m$$

$$= \int \cdots \int_{0 < x_1 < \ldots < x_m < 1} x_1^{j_2 - 1} x_2^{j_3 - j_2 - 1} \cdots x_m^{j_m - j_{m-1} - 1} dx_1 \ldots dx_m$$

$$= \frac{1}{j_1} \frac{1}{j_2} \frac{1}{j_3} \cdots \frac{1}{j_m} \quad (\text{recall } j_1 = 1).$$

Since this holds for any sequence $1 = j_1 < j_2 < \ldots < j_m$ and any m, it is not difficult to verify that the I_n's are independent random variables with

$$P(I_n = 1) = \frac{1}{n}, \quad n \geq 1. \tag{2.5.2}$$

In other words, I_n is Ber$(1/n)$, where Ber(p) represents a Bernoulli random variable with success parameter p. An alternative argument that establishes the independence of the I_n's is sketched in Exercise 2.9. In his

pioneering paper on record times and related statistics, Rényi (1962) gives
two other arguments.

The joint pdf of the first m (non-trivial) record times can be easily
obtained using the I_n's. For integers $1 < n_1 < n_2 < \ldots < n_m$ we have

$$
\begin{aligned}
P(T_1 &= n_1, T_2 = n_2, \ldots, T_m = n_m) \\
&= P(I_2 = 0, \ldots, I_{n_1-1} = 0, I_{n_1} = 1, \ldots, I_{n_m} = 1) \\
&= [(n_1 - 1)(n_2 - 1) \cdots (n_m - 1)n_m]^{-1}.
\end{aligned}
\tag{2.5.3}
$$

The marginal pdf of T_1 can be obtained from (2.5.3) (by setting $m = 1$)
or, alternatively, can be obtained by observing that $T_1 > n$ iff X_1 exceeds
$\max(X_2, \ldots, X_n)$. One finds

$$
P(T_1 = n_1) = \frac{1}{n_1(n_1 - 1)}, \qquad n_1 = 2, 3, \ldots .
\tag{2.5.4}
$$

In order to get the marginal distribution of T_k for $k > 1$, it is convenient
to first review some facts about the record counting process $\{N_n, n \geq 1\}$
defined by

$$
\begin{aligned}
N_n &= \{\text{number of records among } X_1, X_2, \ldots, X_n\} \\
&= \sum_{j=1}^{n} I_j.
\end{aligned}
\tag{2.5.5}
$$

Since the $\{I_j\}$ are independent and $\{I_j\}$ is $\mathrm{Ber}(j^{-1})$, we may immediately
write down expressions for the mean and variance of N_n. Thus

$$
\begin{aligned}
E(N_n) &= \sum_{j=1}^{n} \frac{1}{j} \\
&\approx \log n + \gamma
\end{aligned}
\tag{2.5.6}
$$

and

$$
\begin{aligned}
\mathrm{Var}(N_n) &= \sum_{j=1}^{n} \frac{1}{j}\left(1 - \frac{1}{j}\right) \\
&\approx \log n + \gamma - \frac{\pi^2}{6},
\end{aligned}
\tag{2.5.7}
$$

where γ is Euler's constant 0.5772... Records are clearly not common. In a
sequence of 1000 observations we expect to observe only about 7 records.
The precise distribution of N_n can be obtained by using its probability

generating function (pgf). Because of the representation (2.5.5) we can write

$$E(s^{N_n}) = \prod_{j=1}^{n} E(s^{I_j})$$

$$= \prod_{j=1}^{n} \left(1 + \frac{s-1}{j}\right) \qquad (2.5.8)$$

and consequently

$$P(N_n = k) = \text{coeff of } s^k \text{ in } \frac{1}{n!} \prod_{j=1}^{n} (s+j-1)$$

$$= S_n^k / n! \qquad (2.5.9)$$

where S_n^k is a Stirling number of the first kind (see, e.g., Hamming, 1973; p. 166). Since the I_j's have uniformly bounded variances ($\text{Var}(I_j) \leq \frac{1}{4}, \forall j$), the strong law of large numbers holds, i.e.,

$$\lim_{n \to \infty} N_n / \log n = 1 \text{ a.s.} \qquad (2.5.10)$$

The Liapounov condition trivially holds for sums of independent Bernoulli random variables, so N_n obeys the central limit theorem. Specifically

$$(N_n - \log n)/\sqrt{\log n} \xrightarrow{d} N(0, 1). \qquad (2.5.11)$$

Indeed a law of the iterated logarithm holds, i.e.,

$$\text{a.s. } \varlimsup_{n \to \infty} \frac{N_n - \log n}{\sqrt{2 \log n \log \log \log n}} = 1. \qquad (2.5.12)$$

We may make use of the above information regarding the distribution of N_n, in order to discuss the distribution of the kth record time T_k. Note that the events $\{T_k = n\}$ and $\{N_n = k+1, N_{n-1} = k\}$ are equivalent. Consequently

$$P(T_k = n) = P(N_n = k+1, N_{n-1} = k)$$
$$= P(I_n = 1, N_{n-1} = k).$$

But, referring to (2.5.5), the events $\{I_n = 1\}$ and $\{N_{n-1} = k\}$ are independent. Thus, using (2.5.9), we have

$$P(T_k = n) = \frac{1}{n} S_{n-1}^k / (n-1)!$$
$$= S_{n-1}^k / n! , \qquad (2.5.13)$$

in terms of Stirling numbers of the first kind. Note that T_k grows rapidly as k increases. In fact one may verify that

$$E(T_k) = \infty, \quad \forall \, k \geq 1 \tag{2.5.14}$$

and

$$E(\Delta_k) = E(T_k - T_{k-1}) = \infty, \quad \forall \, k \geq 1. \tag{2.5.15}$$

The equivalence of the events $T_k > n$ and $N_n < k+1$, together with the asymptotic normality of N_n [as described in (2.5.11)] yield an asymptotic normal distribution for $\log T_k$. One finds

$$(\log T_k - k)/\sqrt{k} \xrightarrow{d} N(0, 1) \tag{2.5.16}$$

as $k \to \infty$. From (2.5.16) it follows trivially that

$$(\log T_k)/k \xrightarrow{P} 1. \tag{2.5.17}$$

The parallel stronger result involving almost sure convergence was proved by Rényi (1962). He showed first that $(\log T_{k^2})/(k^2) \xrightarrow{a.s.} 1$ using the Borel-Cantelli lemma. The desired result is then easily obtained using the monotonicity of T_k.

Neuts (1967) observed that Δ_k accounts for a disproportionate amount of T_k. Much of the waiting time for the kth record is spent between the $(k-1)$'st and kth record. In fact, he showed by a direct argument that $\log \Delta_k$ and $\log T_k$ have the same asymptotic distribution, i.e., we have

$$(\log \Delta_k - k)/\sqrt{k} \xrightarrow{d} N(0, 1) \tag{2.5.18}$$

and, consequently

$$(\log \Delta_k)/k \xrightarrow{P} 1. \tag{2.5.19}$$

An alternative derivation of (2.5.17) is possible. Consider the ratio Δ_k/T_k. Following Tata (1969) or Glick (1978) we observe that for $x \in (0, 1)$

$$P\left(\frac{\Delta_k}{T_k} > x | T_{k-1} = t\right)$$
$$= P\left(\Delta_k > \left(\frac{x}{1-x}\right) t | T_{k-1} = t\right)$$
$$= P(I_{t+1} = 0, \dots, I_s = 0),$$

where s is the greatest integer $\leq t/(1-x)$ or $[t/(1-x)]$. Thus

$$P\left(\frac{\Delta_k}{T_k} > x | T_{k-1} = t\right) = \frac{t}{[t/(1-x)]} . \tag{2.5.20}$$

It follows that unconditionally (since $T_k \uparrow \infty$ a.s. as $k \uparrow \infty$)

$$\lim_{k \to \infty} P\left(\frac{\Delta_k}{T_k} > x\right) = 1 - x. \qquad (2.5.21)$$

Thus Δ_k/T_k and T_{k-1}/T_k are asymptotically Uniform$(0,1)$. So about half of the waiting time to the kth record is due to the last inter-record wait. We can use (2.5.20) to easily obtain (2.5.18) from (2.5.16) since

$$-\log \Delta_k + \log T_k = -\log(\Delta_k/T_k) \xrightarrow{d} \text{Gamma}(1,1)$$

and so

$$(-\log \Delta_k + \log T_k)/\sqrt{k} \xrightarrow{P} 0.$$

Neuts' direct approach involved computing the exact distribution of Δ_k and then taking limits. By conditioning on the value of R_{k-1}^* (assuming for convenience and with no loss of generality that the X_i's are standard exponential variables), it is clear that, for $j = 1, 2, \ldots$ and $k = 1, 2, \ldots$

$$P(\Delta_k > j) = \int_0^\infty \frac{x^{k-1}}{\Gamma(k)} e^{-x}(1 - e^{-x})^j \, dx. \qquad (2.5.22)$$

The integration in (2.5.22) is readily performed if we expand the term $(1 - e^{-x})^j$. This yields

$$P(\Delta_k > j) = \sum_{m=0}^{j} \binom{j}{m}(-1)^m \frac{1}{(m+1)^k} , \qquad (2.5.23)$$

a result reported in Ahsanullah (1988, p. 11). Summing the expression (2.5.22) over j provides a simple proof that, as observed earlier, $E(\Delta_k) = \infty$ for every k. If we set $j = k + y\sqrt{k}$ and take the limit in (2.5.22), the asymptotic normality of Δ_k is verified.

Using (2.5.22) or (2.5.23) one may directly verify (Ahsanullah, 1988, pp. 9–10)

$$P(\Delta_k = j) = \sum_{m=0}^{j-1} \binom{j-1}{m}(-1)^m \frac{1}{(m+2)^k} \qquad (2.5.24)$$

and

$$E(\Delta_k^{-1}) = k \sum_{m=1}^{\infty} (m+1)^{-(k+1)} \qquad (2.5.25)$$

related to the Riemann zeta function. The pgf can also be easily obtained. It is given by

$$E(s^{\Delta_k}) = \sum_{m=0}^{\infty} \left(\frac{s}{1-s}\right)^{m+1} \frac{(-1)^m}{(m+2)^k} . \qquad (2.5.26)$$

An ingenious alternative view of the Δ_k's was provided by Shorrock (1972a). Assuming a standard exponential parent distribution he showed that

$$\varlimsup_{k \to \infty} \left| \frac{\log \Delta_k - R_{k-1}^*}{\log k} \right| = 1 \text{ a.s.} \qquad (2.5.27)$$

Consequently, the sequence $\log \Delta_k$ shares many of the asymptotic properties of the sequence $\{R_{k-1}^*\}$. Asymptotic normality, a strong law and a law of the iterated logarithm are thus obtained for $\{\log \Delta_k\}$ since they are available for $\{R_k^*\}$. Consequently we may write, in addition to (2.5.18) and (2.5.19),

$$\log \Delta_k / k \xrightarrow{a.s.} 1 \qquad (2.5.28)$$

and

$$\varlimsup_{k \to \infty} \frac{\log \Delta_k - k}{\sqrt{2k \log \log k}} = 1 \text{ a.s.}$$

$$\varliminf_{k \to \infty} \frac{\log \Delta_k - k}{\sqrt{2k \log \log k}} = -1 \text{ a.s.} \qquad (2.5.29)$$

The results (2.5.28) and (2.5.29) remain valid if T_k is put in place of Δ_k in the expressions (these results were first obtained by Rényi, 1962).

2.6   MARKOV CHAINS

There are several Markov chains lurking in the background of any discussion of record values. We catalog some of them in this section. The Markovian character of some sequences can sometimes be exploited to obtain simplified derivations of distributional results related to record values.

First we may observe that the record counting process $\{N_n, n \geq 1\}$ is a non-stationary Markov process with transition probabilities $P(N_n = j \mid N_{n-1} = i)$ given by

$$p_{ij}^{(n-1)} = \begin{cases} \frac{n-1}{n}, & j = i \\ \frac{1}{n}, & j = i + 1. \end{cases} \qquad (2.6.1)$$

Next note that $\{T_n, n \geq 0\}$ forms a stationary Markov chain with initial degenerate distribution $P(T_0 = 1) = 1$ and transition probabilities $P(T_n = k \mid T_{n-1} = j)$ by

$$p_{jk} = \frac{j}{k(k-1)}, \quad k > j. \qquad (2.6.2)$$

Of course $\{R_n\}$ is a Markov chain with $R_0 \sim F$ and with the transitions governed by

$$f_{R_n | R_{n-1}}(r_n | r_{n-1}) = f(r_n)/[1 - F(r_{n-1})], \quad r_n > r_{n-1}. \qquad (2.6.3)$$

Finally, $\{(R_k, \Delta_{k+1}), k \geq 0\}$ is also a Markov chain. The probabilistic structure of $\{R_k, k \geq 0\}$ is described by (2.6.3) and, given the sequence $\{R_k\}$, the Δ_{k+1}'s are conditionally independent geometric random variables with

$$P(\Delta_{k+1} > m | \{R_k, k \geq 0\}) = [F(R_k)]^m \qquad (2.6.4)$$

(an observation due to Strawderman and Holmes, 1970).

2.7 MOMENTS OF RECORD VALUES

Suppose that $\{R_n\}$ is the sequence of record values corresponding to a sequence of observation from a cdf F. It is tempting to assume that existence of moments of F will guarantee existence of moments of the corresponding records. To see that some conditions are needed, Nagaraja (1978) provided the following example. Let

$$F(x) = \begin{cases} -1/(2x^3), & x \leq -1 \\ (1/2) + (x+1)/4(e+1), & -1 < x < e \\ 1 - e(4x(\log x)^k)^{-1}, & x \geq e, k \geq -1. \end{cases} \qquad (2.7.1)$$

Then $E(X)$ exists iff $k > 1$ and $E(R_n)$ exists iff $k > n + 1$. If we take $F(x) = -1/(2x)$ if $x \leq -1$, and the same as above if $x > -1$, it is easy to see that $E(X)$ does not exist and $E(R_n)(n \geq 1)$ exists iff $k > n + 1$. Of course, if the X_i's are bounded then all moments of R_n for any n will exist. A sufficient condition, in the unbounded case, is provided by the following lemma.

Lemma 2.7.1 *If* $E|X|^\gamma$ *exists then* $E(|R_n|^\delta)$ *exists* $\forall\, \delta < \gamma,\ \forall\, n$.

Proof. Let F be the common distribution of the X_i's. Then R_n has the same distribution as $F^{-1}(R_n(U))$, where $R_n(U)$ is the nth record corresponding to i.i.d. Uniform$(0, 1)$ observations. We can then write

$$
\begin{aligned}
E(|R_n|^\delta) &= \int_0^1 |F^{-1}(r)|^\delta \, \frac{[-\log(1-r)]^n}{n!} \, dr \\
&= (n!)^{-1} \int_0^1 \left[|F^{-1}(r)|^\gamma \right]^{\frac{\delta}{\gamma}} \left[[-\log(1-r)]^{\frac{n\gamma}{\gamma-\delta}} \right]^{1-\frac{\delta}{\gamma}} dr \\
&\leq (n!)^{-1} \left[\int_0^1 |F^{-1}(r)|^\gamma dr \right]^{\frac{\delta}{\gamma}} \left[\int_0^1 [-\log(1-r)]^{\frac{n\gamma}{\gamma-\delta}} dr \right]^{1-\frac{\delta}{\gamma}}
\end{aligned}
$$

by Hölder's inequality. The right-hand side is finite since $\int_0^1 |F^{-1}(r)|^\gamma dr = E|X|^\gamma$ which is finite by hypothesis and

$$\int_0^1 [-\log(1-r)]^{\frac{n\gamma}{\gamma-\delta}} dr = \Gamma\left(\frac{n\gamma}{\gamma-\delta} + 1 \right)$$

is finite.   ◯

If the X_i's are unbounded non-negative random variables, then $E(R_n)$ can be obtained by integrating $P(R_n > r)$ and, using an integration by parts argument, Nagaraja (1978) showed that a necessary and sufficient condition for the existence of $E(R_n)$ is that $E(X(\log X)^n) < \infty$.

In a few cases analytic expressions are available for moments of order statistics (including the cases discussed in Section 2.4). The general expression for the δth moment was implicitly provided in the proof of Lemma 2.7.1, namely

$$E(R_n^\delta) = \int_0^1 [F^{-1}(r)]^\delta \, \frac{[-\log(1-r)]^n}{n!} \, dr. \qquad (2.7.2)$$

2.7.1 Weibull Distribution

Here

$$F(x) = \begin{cases} 0, & x < \mu \\ 1 - e^{-(\frac{x-\mu}{\sigma})^c}, & x \geq \mu \end{cases}$$

and

$$F^{-1}(u) = \mu + \sigma[-\log(1-u)]^{\frac{1}{c}}.$$

Moments of the nth record are then given by

$$E(R_n^\delta) = \int_0^1 \left\{ \mu + \sigma[-\log(1-r)]^{\frac{1}{c}} \right\}^\delta \frac{[-\log(1-r)]^n}{n!} \, dr.$$

Making the simplifying assumption that $\mu = 0$, we obtain

$$\begin{aligned} E(R_n^\delta) &= \sigma^\delta \int_0^1 \frac{[-\log(1-r)]^{n+\frac{\delta}{c}}}{n!} \, dr \\ &= \sigma^\delta \Gamma\left(n + 1 + \frac{\delta}{c}\right) \Big/ \Gamma(n+1). \end{aligned} \qquad (2.7.3)$$

In the special case $c = 1$, the exponential case, the mean and variance of R_n are given by

$$E(R_n) = (n+1)\sigma \qquad (2.7.4)$$

$$\text{Var}(R_n) = (n+1)\sigma^2 \qquad (2.7.5)$$

as is obvious from representation (2.4.3) (with $c = 1$, $\mu = 0$).

2.7.2 Power Function Distribution

Here

$$F(x) = \begin{cases} 0, & x \le 0 \\ x^\alpha, & 0 < x < 1 \\ 1, & x \ge 1 \end{cases}$$

and

$$F^{-1}(u) = u^{1/\alpha}, \qquad 0 < u < 1.$$

Consequently the δth moment of the nth record will be given by

$$
\begin{aligned}
E(R_n^\delta) &= \int_0^1 r^{\frac{\delta}{\alpha}} \frac{[-\log(1-r)]^n}{n!} \, dr \\
&= \int_0^\infty (1 - e^{-z})^{\frac{\delta}{\alpha}} \frac{z^n}{n!} e^{-z} \, dz \\
&= \sum_{j=0}^\infty \binom{\delta/\alpha}{j} (-1)^j (1+j)^{-(n+1)}.
\end{aligned}
\qquad (2.7.6)
$$

For example, if $\alpha = 1$, the uniform case, the mean and variance of R_n are given by

$$E(R_n) = 1 - 2^{-(n+1)} \qquad (2.7.7)$$

and

$$\mathrm{Var}(R_n) = 3^{-(n+1)} - 4^{-(n+1)}. \qquad (2.7.8)$$

2.7.3 Pareto Distribution

In this case

$$F(x) = \begin{cases} 0, & x < \sigma \\ 1 - \left(\frac{x}{\sigma}\right)^{-\alpha}, & x \ge \sigma \end{cases}$$

so that

$$F^{-1}(u) = \sigma(1-u)^{-\frac{1}{\alpha}}.$$

Consequently, for $\delta < \alpha$

$$
\begin{aligned}
E(R_n^\delta) &= \sigma^\delta \int_0^1 (1-r)^{-\frac{\delta}{\alpha}} \frac{[-\log(1-r)]^n}{n!} \, dr \\
&= \sigma^\delta \left(1 - \frac{\delta}{\alpha}\right)^{-(n+1)}
\end{aligned}
\qquad (2.7.9)
$$

and the mean and variance are given by

$$E(R_n) = \sigma \left(1 - \frac{1}{\alpha}\right)^{-(n+1)} \qquad (2.7.10)$$

and

$$\text{Var}(R_n) = \sigma^2 \left[\left(1 - \frac{2}{\alpha} \right)^{-(n+1)} - \left(1 - \frac{1}{\alpha} \right)^{-2(n+1)} \right]. \tag{2.7.11}$$

2.7.4 Extreme Value Distribution

Here

$$F(x) = 1 - e^{-e^{(x-\mu)/\sigma}}, \qquad -\infty < x < \infty, \tag{2.7.12}$$

where μ is real and $\sigma > 0$. As noted in (2.4.12), the records for such a distribution admit the representation

$$R_n = \mu + \sigma \log R_n^*, \tag{2.7.13}$$

where, as usual, R_n^* is a Gamma$(n+1, 1)$ random variable. The moment generating function of R_n is readily obtained.

$$\begin{aligned}
M_{R_n}(t) &= e^{t\mu} E(R_n^{*\, t\sigma}) \\
&= e^{t\mu} \Gamma(n+1+t\sigma)/\Gamma(n+1). \tag{2.7.14}
\end{aligned}$$

From this, using known properties of the gamma and digamma functions, we obtain

$$E(R_n) = \mu - \sigma\gamma + \sigma \sum_{j=1}^{n} \frac{1}{j} \tag{2.7.15}$$

and

$$\text{Var}(R_n) = \sigma^2 \left\{ \frac{\pi^2}{6} - \sum_{j=1}^{n} \frac{1}{j^2} \right\}, \tag{2.7.16}$$

where γ is Euler's constant (see Houchens, 1984, pp. 43–44 for details).

2.7.5 Normal Distribution

We say that X is $N(\mu, \sigma^2)$ if its pdf is

$$f(x) = \frac{1}{\sqrt{2\pi}} \exp\left[-\frac{1}{2}(\frac{x-\mu}{\sigma})^2 \right], \qquad -\infty < x < \infty.$$

The standard normal density corresponds to $\mu = 0$, $\sigma = 1$ and, if Z is $N(0, 1)$, then $X = \mu + \sigma Z$ is $N(\mu, \sigma^2)$. Denote the standard normal pdf and cdf by $\varphi(z)$ and $\Phi(z)$, respectively. Since $\Phi^{-1}(z)$ must be approximated numerically, the moments of normal records must be numerically approximated. Table 2.7.1, which gives approximate values for the modes of standard normal records, is abstracted from Houchens (1984). A quick but crude approximation to the mode suggested by the asymptotic normality of R_n in (2.3.27) is $\Phi^{-1}(1 - e^{-n})$. (For $n = 20$, this reduces to 5.87921, slightly less than the value reported in the table.)

The means and variances of the first 10 normal records are presented in Tables 3.6.1 and 3.6.2 in Chapter 3.

Table 2.7.1. Modes of N(0,1) Records

n	Mode	n	Mode
0	0.00000	11	4.27348
1	0.86387	12	4.49131
2	1.45238	13	4.69927
3	1.92095	14	4.89941
4	2.31915	15	5.09229
5	2.67141	16	5.27794
6	2.98919	17	5.45867
7	3.28096	18	5.63393
8	3.55227	19	5.80329
9	3.80639	20	5.96878
10	4.04592		

2.7.6   Covariances and Correlations

The record values R_m and R_n are always positively correlated. However, the correlation coefficient does not exceed $\sqrt{(m+1)/(n+1)}$, where $m < n$. Further, the bound is attained only for the exponential distribution. (See Theorem 4.4.1 for details.) We catalog, without proof, examples in which simple expressions are available for the covariance between the mth and nth record ($m < n$). Further details on single and joint moments will be presented in Chapter 3.

(i) Uniform. $F(x) = x, \; 0 < x < 1$

$$\text{Cov}(R_m, R_n) = 2^{m-n}[3^{-(m+1)} - 4^{-(m+1)}]. \tag{2.7.17}$$

(ii) Exponential. $F(x) = (1 - e^{-x/\sigma}), \; x > 0$

$$\text{Cov}(R_m, R_n) = \sigma^2(m + 1). \tag{2.7.18}$$

(iii) Pareto. $F(x) = 1 - \left(\frac{x}{\sigma}\right)^{-\alpha}, \; x > \sigma$

$$\text{Cov}(R_m, R_n) = \sigma^2 \left(\frac{\alpha}{\alpha - 1}\right)^{n+1} \left[\left(\frac{\alpha - 1}{\alpha - 2}\right)^{m+1} - \left(\frac{\alpha}{\alpha - 1}\right)^{m+1}\right] \tag{2.7.19}$$

provided $\alpha > 2$.

(iv) Extreme value. $F(x) = 1 - \exp[-e^{(x-\mu)/\sigma}]$, $-\infty < x < \infty$

$$\text{Cov}(R_m, R_n) = \sigma^2 \left[\frac{\pi^2}{6} - \sum_{j=1}^{n} \frac{1}{j^2} \right]. \qquad (2.7.20)$$

(v) Normal. No analytic expressions for $\text{Cov}(R_m, R_n)$ are available. A table of approximate covariances for standard normal records may be found in Chapter 3 (Table 3.6.2).

2.8   A DISCRETE INTERLUDE

The absence of ties made possible a relatively straightforward analysis of record sequences corresponding to observations from continuous distributions. If our X_i's have a common discrete distribution, then life is more complicated. It is even possible to entertain alternative definitions of records in such a setting. Up to this point, an observation is considered to be a record if it exceeds in value all previous observations. In the discrete setting one may also consider weak records. An observation is a *weak record* if it is as large as any previous observation (but possibly no larger than some previous observations). Discrete distributions can have any countable subset of the real line as their support but we will focus, as the literature in general has been focused, on the case where the distribution is concentrated on the non-negative integers. Key references for discrete records are Vervaat (1973) and for weak records, Stepanov (1990).

Our basic data will consist of i.i.d. observations $X_1, X_2, \ldots$ with common cdf F concentrated on the non-negative integers and, to avoid terminating record sequences, we assume $F(n) < 1$, $\forall$ n. The cdf $F(x)$ is, of course, defined for every real x but, given our assumptions, is completely determined by its values when $x = 0, 1, 2, \ldots$. It is convenient to introduce new notation to describe aspects of the common distribution of the X_i's. For $k = 0, 1, 2, \ldots$

$$p_k = P(X = k) \qquad (2.8.1)$$
$$q_k = P(X \geq k), \qquad (2.8.2)$$

and

$$r_k = P(X = k)/P(X \geq k). \qquad (2.8.3)$$

In the study of record values from a continuous cdf F a key role was played by the function

$$\psi_F(u) = F^{-1}(1 - e^{-u}) \tag{2.8.4}$$

[introduced in (2.3.14)]. As noted there, the inverse of ψ_F

$$\psi_F^{-1}(x) = -\log(1 - F(x)) \tag{2.8.5}$$

is the hazard function denoted by $H(x)$. The discrete analog of this function, introduced by Vervaat (1973), plays a central role in the discrete setting. It takes the form

$$\psi_d^{-1}(x) = \sum_{j=0}^{[x]} r_j. \tag{2.8.6}$$

A particular sequence of independent Bernoulli random variables plays a key role in the development of the asymptotic theory of records in the discrete case. Recall that in our notation, R_n indicates the nth record value $(n = 0, 1, 2, \ldots)$ occurring at time T_n. (Note that Vervaat's first record value is our zero'th record value.) Now for each $j \geq 0$, define a Bernoulli random variable

$$Z_j = I(R_n = j \text{ for some } n). \tag{2.8.7}$$

It is not difficult to verify that, for each j,

$$P(Z_j = 1) = r_j \tag{2.8.8}$$

and that the Z_j's are independent random variables (see Exercise 2.10). Provided that $\sum_{j=0}^{\infty} r_j(1-r_j) = \infty$, the Z_j's obey the Liapounov condition and so the central limit theorem holds for them. Thus

$$\frac{\left(\sum_{j=0}^{n} Z_j - \sum_{j=0}^{n} r_j\right)}{\sqrt{\sum_{j=0}^{n} r_j(1 - r_j)}} \xrightarrow{d} N(0, 1). \tag{2.8.9}$$

Since the nth record exceeds k iff the number of Z_j's that equal 1 among $Z_0, \ldots, Z_k$ is less than n, we may write the useful observation

$$P(R_n > k) = P(\sum_{j=0}^{k} Z_j < n). \tag{2.8.10}$$

We can use (2.8.10) to obtain the asymptotic distribution of $\psi_d^{-1}(R_n)$ as a first step towards obtaining the asymptotic distribution of R_n. The inverse of ψ_d^{-1} [defined in (2.8.6)] will, naturally, be denoted by ψ_d.

If we assume in addition that

$$\lim_{n \to \infty} \sum_{k=0}^{n} r_k^2 \Big/ \sum_{k=0}^{n} r_k \;=\; p \in [0, 1) \tag{2.8.11}$$

then we may conclude that

$$(\psi_d^{-1}(R_n) - n)/\sqrt{n} \;\xrightarrow{\;d\;}\; N(0, 1-p). \tag{2.8.12}$$

To see that this is true, consider the following relationships which are exactly true or, where indicated, approximately true for large n.

$$
\begin{aligned}
P((\psi_d^{-1}&(R_n) - n)/\sqrt{n} \le x) \\
&= P(R_n \le \psi_d(x\sqrt{n} + n)) \\
&= P(R_n \le \lambda_n) \quad \text{where} \quad \lambda_n = [\psi_d(x\sqrt{n} + n)] \\
&= P\left(\sum_{j=0}^{\lambda_n} Z_j \ge n \right) \\
&= P\left(\frac{\sum_{j=0}^{\lambda_n} Z_j - \sum_{j=0}^{\lambda_n} r_j}{\sqrt{\sum_{j=0}^{\lambda_n}(r_j - r_j^2)}} \ge \frac{n - \sum_{j=0}^{\lambda_n} r_j}{\sqrt{\sum_{j=0}^{\lambda_n}(r_j - r_j^2)}} \right) \\
&\approx 1 - \Phi\left(\frac{n - \psi_d^{-1}\psi_d(x\sqrt{n} + n)}{\sqrt{\psi_d^{-1}(\psi_d(x\sqrt{n} + n))(1 - \sum_{j=0}^{\lambda_n} r_j^2 / \sum_{j=0}^{\lambda_n} r_j)}} \right) \\
&\approx 1 - \Phi\left(\frac{n - x\sqrt{n} - n}{\sqrt{x\sqrt{n} + n\sqrt{1 - p}}} \right) \\
&\approx 1 - \Phi\left(\frac{-x}{\sqrt{1 - p}} \right) = \Phi\left(\frac{x}{\sqrt{1 - p}} \right). \tag{2.8.13}
\end{aligned}
$$

The condition for asymptotic normality of the records themselves is then a close parallel to the corresponding theorem in the continuous case (i.e., Theorem 2.3.1).

Theorem 2.8.1 *If ψ_d is such that*

$$\lim_{s \to \infty} \frac{\psi_d(s + x\sqrt{s}) - \psi_d(s)}{\psi_d(s + \sqrt{s}) - \psi_d(s)} = x, \quad -\infty < x < \infty, \tag{2.8.14}$$

then

$$\frac{R_n - \psi_d(n)}{\psi_d(n + \sqrt{n}) - \psi_d(n)} \;\xrightarrow{\;d\;}\; N(0, 1-p). \tag{2.8.15}$$

Vervaat (1973) provides details together with extensive generalizations including laws of the iterated logarithm and related functional limit theorems. Vervaat does report that condition (2.8.14) is satisfied by the negative binomial distribution. He also points out that if a discrete distribution has a rapidly decreasing tail, then (2.8.11) will not hold and a surfeit of records will be encountered. Such would be the case if the parent distribution was a Poisson distribution. As in the continuous case, it is possible to get non-normal limiting distributions for records. Vervaat (1973) should be consulted for material providing discrete analogs to the limiting behavior exhibited earlier in (2.3.10) and (2.3.13).

What about weak records? Stepanov (1992) presents a discussion which provides an attractive parallel to the above development for ordinary records. The key idea is that whereas a value j was either a record or not; now in terms of weak records, a value j could be attained by more than one weak record. We need to modify our definition of Z_j given by (2.8.7). We will use a tilde to remind us that we are dealing with weak rather than ordinary records. We then define a sequence $\{\tilde{Z}_j, j \geq 0\}$ of random variables by

$$\tilde{Z}_j = m \quad \text{if exactly } m \text{ weak records have the value } j. \qquad (2.8.16)$$

Stepanov provides a neat argument to justify the claim that these $\tilde{Z}_j$'s are independent random variables. Further, the pdf of Z_j is given by

$$P(\tilde{Z}_j = m) = (1 - r_j)r_j^m, \qquad m \geq 0, \qquad (2.8.17)$$

where r_j is defined in (2.8.3). It follows that

$$
\begin{aligned}
E(\tilde{Z}_j) &= r_j/(1 - r_j) & (2.8.18) \\
\text{Var}(\tilde{Z}_j) &= r_j/(1 - r_j)^2 & (2.8.19)
\end{aligned}
$$

and

$$E(\tilde{Z}_j^3) = (r_j^3 + 4r_j^2 - r_j)/(1 - r_j)^3. \qquad (2.8.20)$$

The $\tilde{Z}_j$'s can be shown to satisfy the Liapounov condition and so the central limit theorem will hold for them, i.e.,

$$\frac{\left(\sum_{j=0}^{n} \tilde{Z}_j - \sum_{j=0}^{n} \frac{r_j}{1-r_j}\right)}{\sqrt{\sum_{j=0}^{n} \frac{r_j}{(1-r_j)^2}}} \xrightarrow{d} N(0, 1). \qquad (2.8.21)$$

Denoting the value of the nth weak record by $\tilde{R}_n$, the following parallel to (2.8.10) is clearly valid.

$$P(\tilde{R}_n > k) = P\left(\sum_{j=0}^{k} \tilde{Z}_j < n\right). \tag{2.8.22}$$

In order to discuss the asymptotic distribution of $\tilde{R}_n$ (the nth weak record), it is convenient to introduce the following functions:

$$\tilde{\psi}_d^{-1}(x) = \sum_{j=0}^{[x]} \frac{r_j}{1 - r_j} \tag{2.8.23}$$

$$\tilde{\Lambda}(x) = \sum_{j=0}^{[x]} \frac{r_j}{(1 - r_j)^2}. \tag{2.8.24}$$

If we wish to obtain an asymptotic result for weak records analogous to (2.8.12), certain regularity conditions must be invoked. Specifically we need to assume that

$$\sup_{n \geq 0} r_n = a < 1 \tag{2.8.25}$$

and

$$\lim_{x \to \infty} \tilde{\psi}_d^{-1}(x)/\tilde{\Lambda}(x) = \delta \in [1 - a, 1]. \tag{2.8.26}$$

It then follows that

$$(\tilde{\psi}_d^{-1}(\tilde{R}_n) - n)/\sqrt{n} \xrightarrow{d} N(0, \delta^{-1}). \tag{2.8.27}$$

Stepanov (1992) also describes a related strong law and law of the iterated logarithm for weak records. In addition, he discusses the rate of the convergence described in (2.8.27).

Observe that if it can be verified that $\tilde{\psi}_d$ satisfies condition (2.8.14), then we get asymptotic normality of the weak records themselves, i.e.,

$$\frac{\tilde{R}_n - \tilde{\psi}_d(n)}{\tilde{\psi}_d(n + \sqrt{n}) - \tilde{\psi}_d(n)} \xrightarrow{d} N(0, \delta^{-1}).$$

2.9 GEOMETRIC RESULTS

For most integer valued random variables, the exact distribution of the nth record is, to say the least, cumbersome to describe. There is an exceptional case. The special role played by the exponential distribution when dealing

with records from continuous populations, prepares us for the possibility
that, among discrete distributions, the geometric distribution might be
the friendliest. Such is the case.

Consider independent positive integer valued random variables $\{X_i,$
$i \geq 1\}$ with common pdf given by

$$P(X = j) = pq^{j-1}, \quad j = 1, 2, \ldots , \qquad (2.9.1)$$

where $p \in (0, 1)$ and $q = 1 - p$. If (2.9.1) holds, we say X is geometrically
distributed with parameter p and we write X is Geo(p). Since $P(X > j) =$
q^j, for every j, it is readily verified (using the lack of memory property
of the geometric distribution) that the corresponding sequence of record
values can be represented in the form (for $n = 0, 1, 2, \ldots$)

$$R_n \overset{d}{=} \sum_{j=0}^{n} X_j \qquad (2.9.2)$$

representing the sum of n i.i.d. Geo(p) random variables. The parallel to
the exponential case is striking and not surprising.

The joint pdf of the first $n + 1$ records from a geometric sequence is,
from (2.3.9),

$$f_{R_0, R_1, \ldots, R_n}(r_0, r_1, \ldots, r_n) = \left(\frac{p}{q} \right)^n pq^{r_n - 1},$$
$$1 \leq r_0 < r_1 \cdots < r_n < \infty. \quad (2.9.3)$$

This provides an alternative argument to justify (2.9.2), since we may
sum (2.9.3) over $r_0, r_1, \ldots, r_{n-1}$ to obtain the marginal distribution of the
nth record. At this point, the analogy with the exponential distribution
breaks down. In the continuous case, the transformation (2.3.2) allowed
us to obtain a simple expression for the marginal cdf of the nth record for
a general continuous distribution from the available simple expression for
the marginal cdf of the nth record from an exponential population. The
resulting cdf and pdf were given by (2.3.5) and (2.3.7). In the geometric
case we could obtain the pdf of R_n quite simply, but we relied on the lack
of memory property. In the general discrete case we cannot "transform"
the distribution for R_n to one corresponding to geometric records. We
are, in fact, forced to take the joint discrete pdf (2.3.9) (where the r_i's
are now constrained to be members of the countable set of possible values
of the X_i's) and sum over $r_0, r_1, \ldots, r_{n-1}$. This is easy in the geometric
case [refer to (2.9.3)]. For any other choice of common distribution for

the X_i's, the summation does not yield closed-form expressions. This explains the paucity of papers dealing with the exact distribution of discrete record values, or rather the fact that such papers inevitably focus on the geometric distribution.

For geometric record values [using the representation (2.9.2)] it is clear that (i) J_n and R_{n-1} are independent, (ii) $E(R_n|R_{n-1} = r_{n-1}) = r_{n-1} + c$, and (iii) $J_n \overset{d}{=} R_0$. All of these are essentially characteristic properties of geometric records. Note that (i) and (ii) will hold for general "geometric" distributions whose possible values are of the form $\{a+kb : k = 0, 1, 2, \ldots\}$.

As a consequence of the representation (2.9.2), we see that R_n has a negative binomial distribution with parameters $n + 1$ and p. The pgf of R_n is given by

$$E(s^{R_n}) = \left[\frac{ps}{1 - qs}\right]^{n+1}. \tag{2.9.4}$$

From (2.9.4) we can obtain the factorial moments of R_n. From these or directly from the representation (2.9.2) we get

$$E(R_n) = (n + 1)/p \tag{2.9.5}$$

and

$$\mathrm{Var}(R_n) = (n + 1)q/p^2. \tag{2.9.6}$$

The covariances, most readily obtained using (2.9.2), are of the form

$$\mathrm{Cov}(R_m, R_n) = (m + 1)q/p^2, \quad m < n. \tag{2.9.7}$$

For weak records, the nicest choice of a parent distribution is geometric with possible values $0, 1, 2, \ldots$ (instead of $1, 2, \ldots$). If we let $\tilde{X}_j$'s be such i.i.d. geometric random variables (i.e., $P(\tilde{X}_j = i) = pq^i$, $i = 0, 1, 2, \ldots$). Then it follows that the nth weak record admits the representation

$$\tilde{R}_n \overset{d}{=} \sum_{j=0}^{'n} \tilde{X}_j. \tag{2.9.8}$$

The asymptotic normality of records and weak records for geometric samples is trivially evident from the central limit theorem using representations like (2.9.2) and (2.9.8).

2.10   COUNTING PROCESSES AND k-RECORDS

In Section 2.1 we obtained the joint pdf of the upper record values. There is another approach to the study of their joint behavior that involves *point processes*. This approach, initiated by Dwass (1964), and popularized by the work of Pickands, Shorrock, Resnick, Ignatov, and Goldie, among others, has provided rich dividends in describing the dependence structure of record values and *k-record values* (to be defined later in this section) and record arrival times. We provide a gentle introduction to these results while refraining from providing formal proofs. However, we will provide ample road maps to the interested reader.

2.10.1   A Point Process View

A point process $\boldsymbol{P}$ on a state space $\mathcal{S}$ is a collection of random points and a description of probability assignment to these points. When $\mathcal{S}$ is the real line or a subset thereof, it is convenient to describe $\boldsymbol{P}$ through the properties of the associated counting process $\{N(t), t \in S\}$ where $N(t)$ represents the number of points in $\boldsymbol{P}$ that do not exceed t. When $P\{N(t) - N(t-) = 0 \text{ or } 1 \text{ for all } t\} = 1$, $\boldsymbol{P}$ is called a *simple* point process (Daley and Vere-Jones, 1988, p. 44). Such a process can also be described through the probabilistic structure of interarrival times of the random points of interest.

We have seen that when F is an Exp(1) cdf, the record spacings $J_0, \ldots, J_n$ are i.i.d. Exp(1) random variables. In other words, the associated counting process $\{N(t), t \geq 0\}$ is a homogeneous Poisson process with unit intensity that will be denoted by $N^*(t)$. When F is an arbitrary continuous cdf with $F^{-1}(0) = 0$ and hazard function $H(t)$, since $N(t) = N^*(H(t))$, $N(t)$ is a nonhomogeneous Poisson process with mean function $H(t)$ and rate function $dH(t) = dF(t)/\{1 - F(t)\}$.

We have also seen that when F is a Geo(p) cdf, the record spacings are i.i.d. Geo(p) random variables. An equivalent description that involves the counting process can be given using the indicator random variables Z_j introduced in Section 2.8. Note that $N(t) = \sum_{j \leq t} Z_j$, where Z_j's are i.i.d. Ber(p) random variables, since for the geometric parent, the r_j in (2.8.8) is a constant (namely, p). The associated point process $\boldsymbol{P}$ has support on the set of positive integers and has independent and stationary increments. For an arbitrary discrete F with support on positive integers, $N(t)$ will have independent but possibly nonstationary increments as the r_j's may depend on j.

When F is a cdf with both discrete and continuous components, the associated $\boldsymbol{P}$ can be described as a mixture of the above two scenarios. Formally, we have the following result adapted from Shorrock (1972b). See also Shorrock (1974) and Resnick (1974) for some related discussion. Goldie and Rogers (1984) have named $\{N(t), t \geq 0\}$ satisfying (2.10.1) given below a *Shorrock process* derived from F.

Theorem 2.10.1 (Shorrock, 1972b) *Let F be a cdf having support $\mathcal{S}$ and hazard function $H(x) = -log(1 - F(x))$. Let $\mathcal{D} = \{d_j, j \geq 1\}$ denote the set of the atoms of F where the d_j are arranged in increasing order. For t real, let $N(t) = \#\{j : R_j \leq t\}$. Then,*

$$N(t) = N_c(t) + N_d(t), \qquad (2.10.1)$$

where N_c and N_d are independent point processes on $\mathcal{S}$. Further, $N_c(t)$ is a Poisson process with (continuous) intensity measure Λ_c given by

$$\Lambda_c((-\infty, t]) = H(t) + \sum_{i=1}^{j}\{H(d_i-) - H(d_i)\}, \quad d_j < t \leq d_{j+1}, j \geq 1,$$

with $d_0 = -\infty$. The process $N_d(t)$ is a process with independent increments and can be expressed as

$$N_d(t) = \sum_{d_j \leq t} Z_j,$$

where the Z_j's are mutually independent and Z_j is a Ber(r_j) random variable where $r_j = P(X = d_j)/P(X \geq d_j)$.

2.10.2 k-Record Statistics

There are two subsequences of observations that are called *k-record values* in the literature. To introduce the first, let us define the sequence of *initial ranks* ρ_n given by

$$\rho_n = \#\{j : j \leq n \text{ and } X_n \leq X_j\}, \quad n \geq 1. \qquad (2.10.2)$$

The following result is a generalization of the independence property of the record indicators discussed in Section 2.5.

Lemma 2.10.1 (Rényi, 1962) *When F is continuous, the ρ_n's are independent random variables and ρ_n is uniformly distributed over $\{1, \ldots, n\}$.*

Lemma 2.10.1 turns out to be an extremely valuable tool in the study of k-record processes.

Pickands (1971), Guthrie and Holmes (1975), and Goldie and Rogers (1984) define X_n to be a k-record value if $\rho_n = k$, $n \geq k$. This produces a sequence of k-record values to be denoted by $\{R_n^{(k)}\}$. We will call this a *Type 1 k-record* sequence. Let $\boldsymbol{P}_k$ be the point process that describes this record value sequence. What is interesting is the fact that the processes $\boldsymbol{P}_k, k \geq 1$, are i.i.d.! This result is referred to as the *Ignatov Theorem* in the literature.

Theorem 2.10.2 (Goldie and Rogers, 1984) *For $k \geq 1$, the $\boldsymbol{P}_k$ are i.i.d. point processes. Further, when F is continuous, $\boldsymbol{P}_k$ is a Poisson process with intensity measure Λ where $\Lambda((a, x]) = -log\{1 - F(x)\}$, $x \geq a \equiv F^{-1}(0)$. When F is an arbitrary cdf, $\boldsymbol{P}_k$ is a Shorrock process described in Theorem 2.10.1.*

The above remarkable result implies that $R_n^{(k)} \stackrel{d}{=} R_n$, $n \geq 1$. Ignatov (1978) stated the result for the continuous case. However, before his proof (Ignatov, 1977) appeared in print in 1986, several other independent proofs had arrived (Deheuvels, 1983; Goldie, 1983; Goldie and Rogers, 1984; Stam, 1985). Goldie and Rogers established the result in its most general form. Engelen, Tommasen and Vervaat (1987) provided a simpler proof by treating the discrete case first and then extending the arguments to an arbitrary F. Rogers (1989) suggested further improvements to their proof. Yet another proof was given by Samuels (1992). While at first glance the i.i.d. nature of the $\boldsymbol{P}_k$'s is surprising, sufficient hints are present in Lemma 2.10.1.

We now consider another candidate that is also referred to as a k-record. Let $T_{0(k)} = k$, $R_{0(k)} = X_{k:k}$, and

$$T_{n(k)} = \min\{j : j > T_{(n-1)(k)}, X_j > X_{T_{(n-1)(k)}-k+1:T_{(n-1)(k)}}\}$$

and define $R_{n(k)} = X_{T_{n(k)}-k+1:T_{n(k)}}$ as the nth k-record. In other words, this k-record sequence is the sequence of the kth largest yet seen. We will call this *Type 2 k-record* sequence. Here a k-record is established whenever $\rho_n \geq k$, even though the corresponding X_n may not be a k-record at present unless $k = 1$. However, it will eventually become one. This sequence of k-records was introduced by Dziubdziela and Kopociński (1976) and it has also gained acceptance in the literature. (They actually called them k-th record values.) The sequence $\{R_{n(k)}, n \geq 0\}$ from a (discrete or continuous) cdf F is *identical* in distribution to a first record sequence $\{R_n, n \geq 0\}$ from the cdf $F_{1:k} = 1 - (1 - F)^k$. Consequently, all the distributional properties of the upper record values and record counting statistics do extend to the corresponding k-record sequences (of the second type) in a direct manner.

To visualize the difference between the two definitions of k-records, we consider an example.

Example 2.10.1 Consider the first few 2-records from the following sequence of observed X's

$$1.71, \ 1.23, \ 1.57, \ 1.94, \ 1.84, \ 2.17, \dots \ .$$

The Type 1 2-record sequence $R_n^{(2)}$ (reporting observations with initial rank 2) begins as $1.23, \ 1.57, \ 1.84, \dots$, whereas the Type 2 sequence $R_{n(2)}$ (reporting the second largest observation yet seen) begins as $1.23, \ 1.57, \ 1.71, \ 1.84, \dots \ .$ $\bigcirc$

When $k = 1$, both the sequences of k-records reduce to the sequence of upper records R_n. The k-records, mostly the $R_{n(k)}$'s, surface occasionally in the coming chapters. So please be prepared to recall these brief introductions!

EXERCISES

1. Define a sequence of random variables $\{V_n, n \geq 1\}$ related to the record times $\{T_n, n \geq 1\}$ as follows:

$$V_n = \{T_{n+1}/T_n\},$$

   where $\{x\}$ denotes the smallest integer at least as large as x. Verify that the V_n's are i.i.d. random variables with

$$P(V_n = j) = 1/j(j-1), \quad j = 2, 3, \dots$$

[Galambos and Seneta, 1975]

2. Let $\{E_n, n \geq 1\}$ denote a sequence of i.i.d. standard exponential variables. Verify that the record times $\{T_n, n \geq 1\}$ satisfy

$$T_{n+1} \stackrel{d}{=} [T_n e^{E_n}] + 1.$$

   where $[x]$ denotes the integer part of x. [Williams, 1973]

3. Verify that
$$\lim_{n\to\infty} P(T_n/T_{n+1} > x) = x^{-1}, \quad x > 1.$$

[Tata, 1969]

4. Consider the function g defined as follows:
$$g(0) \;=\; 0$$
$$g(n) \;=\; \sum_{i=1}^{n} \frac{1}{j}, \quad n = 1, 2, \dots .$$

Then consider
$$Y_n = g(T_n) - n.$$

Verify that $\{Y_n\}$ is a martingale with $E(Y_n) = 0, n = 1, 2, \dots$

[Nevzorov, 1990]

5. Let $\{E_n, n \geq 0\}$ be i.i.d. standard exponential random variables and let $\{R_n^*, n \geq 0\}$ denote the record value sequence from a standard exponential distribution. Verify that

$$T_n \overset{d}{=} 1 + \sum_{i=0}^{n-1} \left\{ \left[\frac{E_i}{-\log\{1 - \exp(-R_i^*)\}} + 1 \right] \right\} .$$

Again $[x]$ denotes the integer part of x. [Deheuvels, 1981]

6. We know that $E(\Delta_k) = \infty$ for every k. Verify however that $E(\Delta_k^a) < \infty$ for $0 < a < 1$.

7. Verify that $\{\Delta_k\}$ is a stochastically increasing sequence of random variables; i.e., show that
$$P(\Delta_k > j) < P(\Delta_{k+1} > j), \; \forall \; j.$$

[Tata, 1969]

8. (a) Verify that $\{(R_n, T_n), n \geq 1\}$ is a Markov chain and identify the stationary transition probabilities.

   (b) Verify that $\{(R_n, \Delta_{n-1}), n \geq 1\}$ is a Markov chain and identify the stationary transition probabilities.

9. Since $I_1, I_2, \dots, I_n$ are $\{0, 1\}$ random variables, in order to prove they are independent, it suffices to verify that
$$P(I_1 = 1, I_2 = 1, \dots, I_n = 1) = \prod_{i=1}^{n} P(I_i = 1).$$

Explain why this equality obviously holds.

10. Prove that the Z_j's defined in (2.8.7) are independent random variables.

11. Suppose that F is in the domain of record attraction of the log-normal distribution. Let F_0 be as defined in equation (2.3.16). Use Theorem 2.3.4 to verify that

$$\lim_{x \to \infty} [1 - F(x)] / [1 - F_0(x)] = 1.$$

[Resnick, 1973a]

12. Suppose that ψ_F [defined in (2.3.14)] satisfies

$$\lim_{n \to \infty} \frac{\psi_F^{-1}([\psi_F(n + \sqrt{n}) - \psi_F(n)]x + \psi_F(n)) - n}{x\sqrt{n}} = 1, \ \forall \ x.$$

Verify that F is in the domain of record attraction of the normal distribution.   [Tata, 1969]

13. We may define generalized uniform order statistics as follows. Let n be a natural number, $k \geq 1$, $m_1, \ldots, m_{n-1}$ real, $M_r = \Sigma_{j=r}^{n-1} m_j$, $1 \leq r \leq n - 1$, be parameters such that $\gamma_r = k + n - r + M_r \geq 1$ for all $r \in \{1, \ldots, n - 1\}$, and let $\tilde{m} = (m_1, \ldots, m_{n-1})$, if $n \geq 2$, $\tilde{m}$ real if $n = 1$. If the random variables $U(r, n, \tilde{m}, k)$, $r = 1, \ldots, n$, possess a joint pdf of the form

$$f_{U(1,n,\tilde{m},k),\ldots,U(n,n,\tilde{m},k)}(u_1, \ldots, u_n)$$
$$= k \left(\prod_{j=1}^{n-1} \gamma_j \right) \left(\prod_{i=1}^{n-1} (1 - u_i)^{m_i} \right) (1 - u_n)^{k-1}$$

on the cone $0 \leq u_1 \leq \ldots \leq u_n < 1$, then they are called *uniform generalized order statistics*.

Let F be an arbitrary cdf. The random variables $X(r, n, \tilde{m}, k) = F^{-1}(U(r, n, \tilde{m}, k))$, $r = 1, \ldots, n$, are called *generalized order statistics* based on F. Record values can be subsumed in this scenario by setting $k = 1$ and $m_1 = m_2 = \ldots = m_{n-1} = -1$.   [Kamps, 1995a,b]

14. If the X_i's are i.i.d. geometric random variables, verify that $R_n - R_{n-1}$ and R_{n-1} are independent random variables, for $n = 1, 2, \ldots$. Verify, in addition, that $R_n - R_{n-1} \stackrel{d}{=} R_0$, $\forall \ n$. Show that these properties essentially characterize the geometric distribution (restrict attention to non-negative integer valued X_i's if you wish).

15. Suppose $E(R_n|R_{n-1}) = \alpha R_{n-1} + \beta \ \forall \ n$. What can be said about the common distribution of the X_i's?

16. Verify that, in the discrete case, the record value sequence $\{R_n, n \geq 0\}$ is a Markov chain and identify the corresponding transition probabilities.

17. Assume that the X_i's are non-negative. Show that $E(R_n)$ exists iff $E(X_i(\log X_i)^n)$ exists.   [Nagaraja, 1978]

18. Let R'_n denote the nth *lower* record corresponding to a sequence of Uniform(0,1) random variables. Verify that $R'_n \overset{d}{=} \prod_{i=0}^{n} Y_i$ where the Y_i's are i.i.d. Uniform $(0,1)$ random variables. Use this representation to determine the pdf of R'_n.

19. Confirm that the sequence of expected records $E(R_n), n = 0, 1, 2, \ldots$ corresponding to i.i.d. X_i's determines the common distribution. (Check out Theorem 4.2.1.)

20. (a) Let F be a continuous cdf and g be an arbitrary function such that, $\text{Var}(g(R_i))$ is finite for $i = n, n+1$. Then, show that $\text{Cov}(g(R_n), g(R_{n+1})) \geq 0$.

    (b) Give an example to show that $\text{Cov}(g(R_n), g(R_{n+1}))$ can be negative if F is a discrete cdf.

    (c) For any $m \geq 0, \geq 0, |n - m| \geq 2$ and a continuous cdf F there exists a function g such that $\text{Var}(g(R_i))$ is finite for $i = m, n$ but $\text{Cov}(g(R_n), g(R_m)) < 0$.

                                    [Nagaraja and Nevzorov, 1996]

21. Assume X, the population random variable, is continuous.

    If X has the *increasing failure rate average* (IFRA) property show that R_n also has this property.

    (b) If X has the *new better than used* (NBU) property, show that this property is shared by R_n.

    See Barlow and Proschan (1981) for a description of these reliability properties.   [Gupta and Kirmani, 1988]

22. Suppose R_n has the *increasing failure rate* (IFR) property. Verify that R_{n+1} is also IFR (so if X_1 is IFR so are all R_n's). (See Theorem 4.3.1.)   [Kochar, 1990]

23. Let F_i be the cdf of the random variable Y_i, for $i = 1, 2$. We say that Y_1 is less dispersed than Y_2 if $F_1^{-1}(\beta) - F_1^{-1}(\alpha) \le F_2^{-1}(\beta) - F_2^{-1}(\alpha)$ for all $0 \le \alpha < \beta \le 1$. Now suppose the population cdf F is absolutely continuous and has the *decreasing failure rate* (DFR) property.

  (a) Show that R_m is less dispersed than R_n whenever $m < n$.

  (b) Show that $\mathrm{Var}(R_m) \le \mathrm{Var}(R_n)$. [Kochar, 1996]

24. Suppose that there exists a non-decreasing function $g(x)$ such that

$$\lim_{n \to \infty} \frac{\psi_F^{-1}(a_n + b_n x) - n}{\sqrt{n}} = g(x), \quad \forall\ x.$$

Verify that in this case

$$\lim_{n \to \infty} P((R_n - a_n)/b_n \le x) = \Phi(g(x)).$$

[Tata, 1969]

25. (Rarity of records). How large must n be to ensure that $P(N_n \ge 25) > 0.01$?

26. Verify that the median of T_1 is 2 but its mean is ∞. However $E(T_1 | X_1 = x)$ is finite. Interpret these results in the context of waiting until someone is served slower than you in the supermarket (cf. Feller, 1965, p. 15).

27. Prove Theorem 2.3.2.

28. If the X_i's are i.i.d. $\mathrm{Exp}(\sigma)$ random variables then $R_1 - R_0 \sim \mathrm{Exp}(\sigma)$. Is the converse necessarily true?

29. Suppose $X_1, X_2, \ldots$ are i.i.d. with common cdf

$$F(x) = 1 - \exp(-x/(1 - x)), \quad 0 < x < 1.$$

Determine the limiting distribution and appropriate normalizing constants for R_n. [Resnick, 1973a]

30. The cdf

$$F(x) = 1 - \exp\left\{ -\frac{\alpha^2}{4}(\log x - \sin(\log x))^2 \right\}, \quad x \ge 1$$

is not in the domain of record attraction of the log-normal distribution even though it does satisfy

$$\lim_{x \to \infty} \log(1 - F(x))/\log(1 - F_0(x)) = 1$$

where F_0 is as defined in (2.3.16).   [Resnick, 1973a]

31. Suppose the X_i's are i.i.d. Weibull random variables [with distribu-
    tion as in (2.4.1)]. Prove that if $\mu = 0$, then quotients of successive
    records are independent. In particular set $Q_i = R_i/R_{i-1}$, $i = 1, 2, \ldots$
    and prove that the random variables $\{icQ_i : i = 1, 2, \ldots\}$ are i.i.d.
    standard exponential random variables.

32. Assuming the X_i's are i.i.d. Exp(σ) random variables, prove that

    (a) $R_n - R_{n-1}$ and R_{n-1} are independent;

    (b) $E(R_n|R_{n-1} = r_{n-1}) = r_{n-1} + \sigma$;

    (c) $R_n - R_{n-1} \stackrel{d}{=} R_0$.

33. For the record sequence from the standard uniform distribution
    prove that

    (a) $(1 - R_n)/(1 - R_{n-1})$ and $1 - R_{n-1}$ are independent;

    (b) $E(R_n|R_{n-1} = r_{n-1}) = \frac{1}{2} + \frac{1}{2}r_{n-1}$;

    (c) $(1 - R_n)/(1 - R_{n-1}) \stackrel{d}{=} 1 - R_0 \stackrel{d}{=} X_1$.

34. Suppose the X_i's are i.i.d. Pareto(σ, α) random variables [with cdf
    of the form (2.4.7)]. Prove that

    (a) R_n/R_{n-1} and R_{n-1} are independent;

    (b) $E(R_n|R_{n-1} = r_{n-1}) = \alpha r_{n-1}, \alpha > 1$;

    (c) $R_n/R_{n-1} \stackrel{d}{=} R_0$.

35. Assume the X_i's are i.i.d. extreme value random variables [with cdf
    of the form (2.4.10)]. Prove that, if we define $Y_i = i(R_i - R_{i-1})$,
    $i = 1, 2, \ldots$, then the Y_i's are i.i.d. exponential random variables.

36. The following data (read rowwise) represent the average July tem-
    peratures (in degrees centigrade) of Neuenburg, Switzerland, during
    the period 1864–1993 (from Klüppelberg and Schwere, 1995).

19.0	20.1	18.4	17.4	19.7	21.0	21.4	19.2	19.9	20.4
20.9	17.2	20.2	17.8	18.1	15.6	19.4	21.7	16.2	16.4
19.0	20.6	19.0	20.7	15.8	17.7	16.8	17.1	18.1	18.4
18.7	18.7	18.4	19.2	18.0	18.7	20.7	19.4	19.2	17.4
22.0	21.4	19.3	16.8	18.2	16.2	15.9	22.1	17.5	15.3
16.5	17.4	17.0	18.3	18.3	15.3	18.2	21.5	17.0	21.6
18.2	18.1	17.6	18.2	22.6	19.9	17.1	17.2	17.3	19.4
20.1	20.1	17.0	19.4	17.5	16.8	17.0	19.9	18.2	19.2
18.5	20.8	19.5	21.1	15.8	21.3	21.2	18.8	22.3	18.6
16.8	18.2	17.2	18.4	18.7	21.1	16.3	17.4	18.0	19.5
21.2	16.8	17.4	20.7	18.4	19.8	18.7	20.5	18.3	18.2
18.2	19.2	20.2	18.2	17.4	19.2	16.3	17.4	20.3	23.4
19.2	20.2	19.3	19.0	18.8	20.3	19.7	20.7	19.6	18.1

(a) Obtain the observed values of the following sequences of record statistics (i) T_n (ii) Δ_n (iii) R_n.

(b) Is there evidence of an increasing trend in the data (global warming)? Suggest a test procedure involving suitable record based statistics.

37. Let $X_1, X_2, \ldots$ be i.i.d. Uniform$(0, 1)$ random variables. Verify that the corresponding record sequence $\{R_n\}$ cannot be normalized to obtain a non-trivial limiting distribution. Verify that the same conclusion holds when the common distribution of the X's is (i) a power function distribution, $F(x) = x^\delta$, $0 < x < 1$, $\delta > 0$ and (ii) a Pareto distribution, $F(x) = 1 - x^{-\delta}$, $x > 1$, $\delta > 0$.

38. (a) Show that the Type 2 sequence of records $\{R_{n(k)}, n \geq 0\}$ defined in Section 2.10 is identical in distribution to a first record sequence $\{R_n, n \geq 0\}$ from the cdf $F_{1:k} = 1 - (1 - F)^k$.

(b) Modify Theorem 2.3.4 to describe necessary and sufficient conditions for the domain of attraction results to hold for Type 2 k-records. What are the associated norming constants? Consider both discrete and continuous cases.

39. Use the first 20 values from the data set given in Exercise 2.36 and identify both types of k-record statistics (values, times and inter-record times) for $k = 2, 3, 4, 5$.

CHAPTER 3

MOMENT RELATIONS, BOUNDS AND APPROXIMATIONS

3.1 INTRODUCTION

In Chapter 2, we saw some general properties of record values and also certain distributional results for record values from some specific populations. In this chapter, we derive recurrence relations satisfied by the moments of record values from distributions like the exponential, Weibull, Gumbel, generalized extreme value, Lomax, generalized Pareto, normal and logistic. Next, we discuss some general methods of deriving bounds and approximations for the moments of record values and k-records from an arbitrary distribution. Results presented in this chapter are useful in developing inference procedures, as will be demonstrated in Chapter 5. These developments on record values are quite similar to those present in the order statistics literature; for details on the results on order statistics, interested readers may refer to David (1981), Arnold and Balakrishnan (1989), Balakrishnan and Cohen (1991), and Arnold, Balakrishnan and Nagaraja (1992).

Since the discussion in this chapter will involve both lower and upper record values (depending on the population distribution under consideration), we shall use the following notation for convenience: R_n for the nth upper record value and R'_n for the nth lower record value, for $n = 0, 1, 2, \ldots$.

Let us also denote the moment $E(R_n^k)$ by $\alpha_n^{(k)}$, the product moment $E(R_m^{k_1} R_n^{k_2})$ by $\alpha_{m,n}^{(k_1,k_2)}$, the variance $\mathrm{Var}(R_n)$ by $\sigma_{n,n}$, and the covariance

Cov(R_m, R_n) by $\sigma_{m,n}$. For simplicity, we shall also use the notations α_n for $\alpha_n^{(1)}$, $\alpha_{m,n}$ for $\alpha_{m,n}^{(1,1)}$, and σ_n^2 for $\sigma_{n,n}$. In general, we shall denote the product moment $E\left(\prod_{i=1}^{\ell} R_{n_i}^{k_i}\right)$ by $\alpha_{n_1,n_2,\ldots,n_\ell}^{(k_1,k_2,\ldots,k_\ell)}$ for $0 \leq n_1 < n_2 < \cdots < n_\ell$. We shall also use analogous notations with primes for the single and the product moments of lower record values; for example, we will use $\alpha_n'^{(k)}$ for $E(R_n'^k)$.

3.2   EXPONENTIAL DISTRIBUTION

In Section 2.3, we have already seen that, in the case of the standard exponential distribution, $\{R_n - R_{n-1}, n \geq 1\}$ are i.i.d. Exp(1) random variables and, therefore, R_n is distributed as standard gamma with shape parameter $n + 1$, i.e., Gamma$(n + 1, 1)$, for $n = 0, 1, 2, \ldots$. These distributional properties of exponential records enable us to write down the single and product moments of these record values. For example, we immediately have [see Section 2.3]

$$\alpha_n = n + 1, \ \sigma_n^2 = n + 1 \text{ and } \sigma_{m,n} = \sigma_m^2 = m + 1 \text{ for } m < n. \quad (3.2.1)$$

Hence, it is clear that in this case there is no real need for deriving recurrence relations for single and product moments of record values. However, we proceed to do so in this section, since the approach we take to establish these recurrence relations will then allow us to handle record values from many other populations in an analogous manner.

Theorem 3.2.1 (Balakrishnan and Ahsanullah, 1995) *For $n = 0, 1, 2, \ldots$ and $k = 0, 1, 2, \ldots$,*

$$\alpha_n^{(k+1)} = \alpha_{n-1}^{(k+1)} + (k + 1)\alpha_n^{(k)} \quad (3.2.2)$$

and, consequently, for $m = 0, 1, \ldots, n - 1$,

$$\alpha_n^{(k+1)} = \alpha_m^{(k+1)} + (k + 1)\sum_{i=m+1}^{n} \alpha_i^{(k)} \quad (3.2.3)$$

with $\alpha_n^{(0)} = 1$ and $\alpha_{-1}^{(k)} = 0$.

Proof. From (2.3.7), we have for $n = 0, 1, 2, \ldots$ and $k = 0, 1, 2, \ldots$.

$$\alpha_n^{(k)} = \frac{1}{n!} \int_0^\infty r^k \{-\log(1 - F(r))\}^n f(r) \, dr$$

$$= \frac{1}{n!} \int_0^\infty r^k \{-\log(1 - F(r))\}^n \{1 - F(r)\} \, dr,$$

since $f(r) = 1 - F(r)$ in the case of the standard exponential distribution. Upon integrating by parts treating r^k for integration and the rest of the integrand for differentiation and then simplifying the resulting expression, we derive the relation in (3.2.2). The relation in (3.2.3) is obtained simply by repeated application of (3.2.2). $\quad\bigcirc$

Theorem 3.2.2 (Balakrishnan and Ahsanullah, 1995) *For* $m = 0, 1, 2, \ldots$ *and* $k_1, k_2 = 0, 1, 2, \ldots$,

$$\alpha_{m,m+1}^{(k_1,k_2+1)} = \alpha_m^{(k_1+k_2+1)} + (k_2 + 1)\alpha_{m,m+1}^{(k_1,k_2)}; \tag{3.2.4}$$

for $0 \le m \le n - 2$ *and* $k_1, k_2 = 0, 1, 2, \ldots$,

$$\alpha_{m,n}^{(k_1,k_2+1)} = \alpha_{m,n-1}^{(k_1,k_2+1)} + (k_2 + 1)\alpha_{m,n}^{(k_1,k_2)}; \tag{3.2.5}$$

and, consequently, for $0 \le m \le n - 1$

$$\alpha_{m,n}^{(k_1,k_2+1)} = \alpha_m^{(k_1+k_2+1)} + (k_2 + 1) \sum_{i=m+1}^{n} \alpha_{m,i}^{(k_1,k_2)}. \tag{3.2.6}$$

Proof. From (2.3.11), we have for $0 \le m \le n - 1$ and $k_1, k_2 = 0, 1, 2, \ldots$,

$$\alpha_{m,n}^{(k_1,k_2)} = \frac{1}{m!(n - m - 1)!} \int_0^\infty r_1^{k_1}\{-\log(1 - F(r_1))\}^m \frac{f(r_1)}{1 - F(r_1)} \cdot I(r_1)\, dr_1, \tag{3.2.7}$$

where

$$I(r_1) = \int_{r_1}^\infty r_2^{k_2}\{-\log(1 - F(r_2)) + \log(1 - F(r_1))\}^{n-m-1}\{1 - F(r_2)\}\, dr_2$$

since $f(r_2) = 1 - F(r_2)$. Upon integrating by parts treating $r_2^{k_2}$ for integration and the rest of the integrand for differentiation, we obtain when $n = m + 1$ that

$$I(r_1) = \frac{1}{k_2 + 1}\left[\int_{r_1}^\infty r_2^{k_2+1} f(r_2)\, dr_2 - r_1^{k_2+1}\{1 - F(r_1)\}\right],$$

and when $n \ge m + 2$ that

$$I(r_1) = \frac{1}{k_2 + 1}\bigg[\int_{r_1}^\infty r_2^{k_2+1}\{-\log(1 - F(r_2)) + \log(1 - F(r_1))\}^{n-m-1} \cdot f(r_2)\, dr_2$$

$$- (n - m - 1)\int_{r_1}^\infty r_2^{k_2+1}\{-\log(1 - F(r_2))$$

$$+ \log(1 - F(r_1))\}^{n-m-2} f(r_2)\, dr_2\bigg].$$

The relations in (3.2.4) and (3.2.5) are derived upon substituting the above expressions of $I(r_1)$ into (3.2.7) and simplifying the resulting expressions. The relation in (3.2.6) is obtained simply by repeated application of (3.2.5) and with the help of (3.2.4). ○

Remark 3.2.1 Proceeding along the lines of Theorem 3.2.2, we can easily show that for $0 \leq n_1 < n_2 < \cdots < n_\ell$ and $k_1, k_2, \ldots, k_{\ell+1} = 0, 1, 2, \ldots$

$$\alpha_{n_1,\ldots,n_\ell,n_\ell+1}^{(k_1,\ldots,k_\ell,k_{\ell+1}+1)} = \alpha_{n_1,\ldots,n_{\ell-1},n_\ell}^{(k_1,\ldots,k_{\ell-1},k_\ell+k_{\ell+1}+1)} + (k_{\ell+1}+1)\alpha_{n_1,\ldots,n_\ell,n_\ell+1}^{(k_1,\ldots,k_\ell,k_{\ell+1})},$$
(3.2.8)

and that for $0 \leq n_1 < n_2 < \cdots < n_\ell < n_{\ell+1} - 1$ and $k_1, k_2, \ldots, k_{\ell+1} = 0, 1, 2, \ldots$

$$\alpha_{n_1,\ldots,n_\ell,n_{\ell+1}}^{(k_1,\ldots,k_\ell,k_{\ell+1}+1)} = \alpha_{n_1,\ldots,n_\ell,n_{\ell+1}-1}^{(k_1,\ldots,k_\ell,k_{\ell+1}+1)} + (k_{\ell+1}+1)\alpha_{n_1,\ldots,n_\ell,n_{\ell+1}}^{(k_1,\ldots,k_\ell,k_{\ell+1})}.$$
(3.2.9)

Remark 3.2.2 Balakrishnan and Ahsanullah (1995) have also established more general results of the nature presented in this section by considering the upper record values from the nonidentical exponential model.

3.3   WEIBULL DISTRIBUTION

Let us consider the upper record values $\{R_n, n \geq 0\}$ arising from a sequence of i.i.d. random variables from a Weibull(c) population with pdf

$$f(x) = e^{-x^c}\, cx^{c-1}, \qquad x > 0,$$
(3.3.1)

and cdf

$$F(x) = 1 - e^{-x^c}, \qquad x > 0,$$
(3.3.2)

where $c \in \mathbf{R}^+$ is the shape parameter; see Sections 2.4.1 and 2.7.1.

From (2.3.7), we then have for $n = 0, 1, 2, \ldots$ and $k = 1, 2, \ldots$ [see (2.7.3)]

$$
\begin{aligned}
\alpha_n^{(k)} &= \frac{1}{n!} \int_0^\infty r^k \{-\log(1 - F(r))\}^n\, f(r)\, dr \\
&= \frac{c}{n!} \int_0^\infty e^{-r^c}\, r^{c(n+1)+k-1}\, dr \\
&= \Gamma\left(n + 1 + \frac{k}{c}\right) \Big/ \Gamma(n+1),
\end{aligned}
$$
(3.3.3)

where $\Gamma(\cdot)$ is the complete gamma function.

Further, from (2.3.11), we may similarly derive for $0 \le m < n$

$$\alpha_{m,n} = \frac{\Gamma(m+1+\frac{1}{c})}{\Gamma(m+1)} \frac{\Gamma(n+1+\frac{2}{c})}{\Gamma(n+1+\frac{1}{c})} . \tag{3.3.4}$$

From (3.3.3) and (3.3.4), for example, we immediately have

$$\alpha_n = \frac{\Gamma(n+1+\frac{1}{c})}{\Gamma(n+1)} \quad \text{and}$$

$$\sigma_{m,n} = \frac{\Gamma(m+1+\frac{1}{c})}{\Gamma(m+1)} \left\{ \frac{\Gamma(n+1+\frac{2}{c})}{\Gamma(n+1+\frac{1}{c})} - \frac{\Gamma(n+1+\frac{1}{c})}{\Gamma(n+1)} \right\} . \tag{3.3.5}$$

Hence, it is clear that in this case (as in the case of exponential) there is no need for recurrence relations for single and product moments of record values. However, by using the property that

$$r\,f(r) = c\{1 - F(r)\}\{-\log(1 - F(r))\},$$

the following recurrence relations can be derived by following steps similar to those adopted in the last section.

Theorem 3.3.1 (Balakrishnan and Chan, 1994) *For $n = 0, 1, 2, \ldots$ and $k = 1, 2, \ldots$,*

$$\alpha_{n+1}^{(k)} = \left(1 + \frac{k}{c(n+1)}\right) \alpha_n^{(k)}; \tag{3.3.6}$$

for $m = 0, 1, 2, \ldots$ and $k_1, k_2 = 1, 2, \ldots$,

$$\alpha_{m,m+1}^{(k_1,k_2)} = \frac{(m+1)c}{k_1 + (m+1)c} \alpha_{m+1}^{(k_1+k_2)}; \tag{3.3.7}$$

and for $0 \le m \le n - 2$ and $k_1, k_2 = 1, 2, \ldots$,

$$\alpha_{m,n}^{(k_1,k_2)} = \frac{(m+1)c}{k_1 + (m+1)c} \alpha_{m+1,n}^{(k_1,k_2)}. \tag{3.3.8}$$

Remark 3.3.1 By setting $c = 2$ in the above results, we obtain results for the upper record values from the standard Rayleigh distribution.

3.4   GUMBEL DISTRIBUTION

Let us consider the lower record values $\{R'_n, \ n \geq 1\}$ arising from a sequence of i.i.d. random variables from a Gumbel(0,1) population with pdf

$$f(x) = e^{-e^{-x}} e^{-x}, \qquad -\infty < x < \infty, \tag{3.4.1}$$

and cdf

$$F(x) = e^{-e^{-x}}, \qquad -\infty < x < \infty. \tag{3.4.2}$$

It is easily observed from (3.4.1) and (3.4.2) that the Gumbel distribution satisfies the differential equation

$$f(x) = F(x)\{-\log F(x)\}, \qquad -\infty < x < \infty. \tag{3.4.3}$$

This property can be used to establish some simple recurrence relations for the single and product moments of the lower record values.

Theorem 3.4.1 (Balakrishnan, Ahsanullah and Chan, 1992) *For $n = 0, 1, 2, \ldots$ and $k = 0, 1, 2, \ldots$,*

$$\alpha'^{(k+1)}_{n+1} = \alpha'^{(k+1)}_{n} - \frac{k+1}{n+1}\, \alpha'^{(k)}_{n}; \tag{3.4.4}$$

consequently, for $n = 0, 1, 2, \ldots$ and $k = 0, 1, 2, \ldots$,

$$\alpha'^{(k+1)}_{n+1} = \alpha'^{(k+1)}_{0} - (k+1)\sum_{i=0}^{n} \alpha'^{(k)}_{i} \Big/ (i+1). \tag{3.4.5}$$

Proof. From (2.3.7), we have for $n = 0, 1, 2, \ldots$ and $k = 0, 1, 2, \ldots$

$$
\begin{aligned}
\alpha'^{(k)}_{n} &= \frac{1}{n!} \int_{-\infty}^{\infty} r^k \{-\log F(r)\}^n\, f(r)\, dr \\
&= \frac{1}{n!} \int_{-\infty}^{\infty} r^k \{-\log F(r)\}^{n+1}\, F(r)\, dr
\end{aligned}
$$

upon using (3.4.3). Upon integrating by parts treating r^k for integration and the rest of the integrand for differentiation and then simplifying the resulting expression, we derive the relation in (3.4.4). The relation in (3.4.5) is obtained simply by repeated application of (3.4.4).   ○

Remark 3.4.1 The recurrence relation in (3.4.4) or in (3.4.5) can be used in a simple recursive manner to compute all the single moments of all the lower record values, by starting with the values of $\alpha'^{(k)}_{0} \equiv E(X^k)$.

Remark 3.4.2 By setting $k = 0$ in Theorem 3.4.1, we obtain for $n = 0, 1, 2, \ldots$ that

$$\alpha'_{n+1} = \alpha'_n - \frac{1}{n+1} \tag{3.4.6}$$

and

$$\alpha'_{n+1} = \alpha'_0 - \sum_{i=1}^{n+1} 1/i = 0.57722\ldots - \sum_{i=1}^{n+1} 1/i. \tag{3.4.7}$$

Theorem 3.4.2 (Balakrishnan, Ahsanullah and Chan, 1992) *For $m = 0, 1, 2, \ldots$ and $k_1, k_2 = 0, 1, 2, \ldots$,*

$$\alpha'^{(k_1+1,k_2)}_{m,m+1} = \alpha'^{(k_1+k_2+1)}_{m+1} + \frac{k_1+1}{m+1} \, \alpha'^{(k_1,k_2)}_{m,m+1}; \tag{3.4.8}$$

for $0 \le m \le n-2$ and $k_1, k_2 = 0, 1, 2, \ldots$,

$$\alpha'^{(k_1+1,k_2)}_{m,n} = \alpha'^{(k_1+1,k_2)}_{m+1,n} + \frac{k_1+1}{m+1} \, \alpha'^{(k_1,k_2)}_{m,n}; \tag{3.4.9}$$

and, consequently, for $0 \le m \le n-1$ and $k_1, k_2 = 0, 1, 2, \ldots$,

$$\alpha'^{(k_1+1,k_2)}_{m,n} = \alpha'^{(k_1+k_2+1)}_n + (k_1+1) \sum_{i=m}^{n-1} \alpha'^{(k_1,k_2)}_{i,n} \Big/ (i+1). \tag{3.4.10}$$

Proof. From (2.3.11), we have for $0 \le m \le n-1$ and $k_1, k_2 = 0, 1, 2, \ldots$

$$\alpha'^{(k_1,k_2)}_{m,n} = \frac{1}{m!(n-m-1)!} \int_{-\infty}^{\infty} r_2^{k_2} \, f(r_2) \, I(r_2) \, dr_2, \tag{3.4.11}$$

where

$$
\begin{aligned}
I(r_2) &= \int_{r_2}^{\infty} r_1^{k_1} \{-\log F(r_1)\}^m \{-\log F(r_2) + \log F(r_1)\}^{n-m-1} \\
&\qquad \cdot \frac{f(r_1)}{F(r_1)} \, dr_1 \\
&= \int_{r_2}^{\infty} r_1^{k_1} \{-\log F(r_1)\}^{m+1} \{-\log F(r_2) + \log F(r_1)\}^{n-m-1} \, dr_1
\end{aligned}
$$

upon using (3.4.3). Upon integrating by parts treating $r_1^{k_1}$ for integration and the rest of the integrand for differentiation, we obtain when $n = m+1$ that

$$
\begin{aligned}
I(r_2) &= \frac{1}{k_1+1} \Big[(m+1) \int_{r_2}^{\infty} r_1^{k_1+1} \{-\log F(r_1)\}^m \frac{f(r_1)}{F(r_1)} \, dr_1 \\
&\qquad - r_2^{k_1+1} \{-\log F(r_2)\}^{m+1} \Big],
\end{aligned}
$$

and when $n \geq m + 2$ that

$$
\begin{aligned}
I(r_2) =\ & \frac{1}{k_1 + 1}\Bigg[(m + 1)\int_{r_2}^{\infty} r_1^{k_1+1}\{-\log F(r_1)\}^m \\
& \cdot \{-\log F(r_2) + \log F(r_1)\}^{n-m-1}\, \frac{f(r_1)}{F(r_1)}\, dr_1 \\
& -(n - m - 1)\int_{r_2}^{\infty} r_1^{k+1}\{-\log F(r_1)\}^{m+1} \\
& \cdot \{-\log F(r_2) + \log F(r_1)\}^{n-m-2}\, \frac{f(r_1)}{F(r_1)}\, dr_1\Bigg].
\end{aligned}
$$

The relations in (3.4.8) and (3.4.9) are derived upon substituting the above expressions of $I(r_2)$ into (3.4.11) and simplifying the resulting expressions. The relation in (3.4.10) is obtained simply by repeated application of (3.4.9) and with the use of (3.4.8).   ◯

Remark 3.4.3 The recurrence relations in Theorem 3.4.2 can be applied in a simple recursive manner to compute all the product moments of all the lower record values. By setting $k_1 = 0$ and $k_2 = 1$ in (3.4.10), for example, we get for $0 \leq m \leq n - 1$

$$
\alpha'_{m,n} = \alpha_n'^{(2)} + \sum_{i=m}^{n-1} 1/(i+1)
$$

$$(3.4.12)$$

which, together with the fact that (see Remark 3.4.2)

$$
\alpha'_m = \alpha'_n + \sum_{i=m}^{n-1} 1/(i+1), \tag{3.4.13}
$$

readily yields, the result

$$
\sigma'_{m,n} = \sigma'_{m,m} = \sigma'^2_m. \tag{3.4.14}
$$

This can also be verified easily by direct algebraic calculations.

Remark 3.4.4 Proceeding along the lines of Theorem 3.4.2, we can show that for $0 \leq n_1 < n_2 < \cdots < n_\ell$ and $k_1, k_2, \ldots, k_{\ell+1} = 0, 1, 2, \ldots$

$$
\alpha_{n_1,\ldots,n_{\ell-1},n_\ell,n_\ell+1}^{(k_1,\ldots,k_{\ell-1},k_\ell+1,k_{\ell+1})} = \alpha_{n_1,\ldots,n_{\ell-1},n_\ell+1}^{(k_1,\ldots,k_{\ell-1},k_\ell+k_{\ell+1}+1)} + \frac{k_\ell + 1}{n_\ell + 1}\alpha_{n_1,\ldots,n_\ell,n_\ell+1}^{(k_1,\ldots,k_\ell,k_{\ell+1})},
$$

$$(3.4.15)$$

and for $0 \leq n_1 < n_2 < \cdots < n_\ell < n_{\ell+1} - 1$ and $k_1, k_2, \ldots, k_{\ell+1} = 0, 1, 2, \ldots$

$$\alpha_{n_1,\ldots,n_{\ell-1},n_\ell,n_{\ell+1}}^{(k_1,\ldots,k_\ell-1,k_\ell+1,k_{\ell+1})} = \alpha_{n_1,\ldots,n_{\ell-1},n_\ell+1,n_{\ell+1}}^{(k_1,\ldots,k_\ell-1,k_\ell+1,k_{\ell+1})} + \frac{k_\ell + 1}{n_\ell + 1} \, \alpha_{n_1,\ldots,n_{\ell+1}}^{(k_1,\ldots,k_{\ell+1})}.$$

$$(3.4.16)$$

Remark 3.4.5 It is, of course, possible to derive explicit expressions for the single and the product moments of $\{R'_n, \, n \geq 0\}$ in this case. For example, from (2.3.7), (3.4.1) and (3.4.2) we obtain the moment generating function of R'_n to be

$$M_{R'_n}(t) = E\left(e^{tR'_n}\right) = \frac{1}{n!} \int_{-\infty}^{\infty} e^{-e^{-r}} \, e^{-(n-t)r} \, e^{-r} \, dr$$

$$= \Gamma(n - t + 1)/\Gamma(n + 1);$$

hence, we have the cumulant generating function of R'_n to be

$$K_{R'_n}(t) = \log M_{R'_n}(t) = \log \Gamma(n - t + 1) - \log \Gamma(n + 1), \qquad (3.4.17)$$

which immediately gives the kth cumulant of R'_n to be

$$\kappa_n'^{(k)} = \frac{d^k}{dt^k} \, K_{R'_n}(t)\Big|_{t=0} = (-1)^k \psi^{(k-1)}(n + 1); \qquad (3.4.18)$$

here, $\psi(z) = \frac{d}{dz} \log \Gamma(z) = \Gamma'(z)/\Gamma(z)$ is the psi or digamma function, and $\psi^{(1)}(z), \psi^{(2)}(z), \ldots$ are its successive derivatives known as polygamma functions. These functions have been tabulated quite extensively by Davis (1935) and Abramowitz and Stegun (1965); computational programs for the evaluation of these functions have also been given by Bernardo (1976) and Schneider (1978). Similar expressions can be derived for the product moments as well.

All the results presented above for the Gumbel distribution can be extended to the case of generalized extreme value distribution with pdf

$$f(x) = e^{-(1-\gamma x)^{1/\gamma}} (1 - \gamma x)^{(1/\gamma)-1}, \qquad (3.4.19)$$

having supports $(-\infty, 1/\gamma)$ and $(1/\gamma, \infty)$ according as $\gamma > 0$, and $\gamma < 0$, respectively. Consequently, the cdf is given by

$$F(x) = e^{-(1-\gamma x)^{1/\gamma}}. \qquad (3.4.20)$$

Observe that the limit form of the above distribution as $\gamma \to 0$ is exactly the Gumbel distribution in (3.4.2).

By using the relation [from (3.4.19) and (3.4.20)]

$$(1 - \gamma x)f(x) = F(x)\{-\log F(x)\},$$

the following results can be established for the lower record values from the generalized extreme value distribution.

Theorem 3.4.3 (Balakrishnan, Chan and Ahsanullah, 1993) *For* $n = 0, 1, 2, \ldots$ *and* $k = 0, 1, 2, \ldots$,

$$\alpha_{n+1}^{\prime(k+1)} = \left\{1 + \frac{\gamma(k+1)}{n+1}\right\} \alpha_n^{\prime(k+1)} - \frac{k+1}{n+1}\alpha_n^{\prime(k)}. \tag{3.4.21}$$

Theorem 3.4.4 (Balakrishnan, Chan and Ahsanullah, 1993) *For* $m = 0, 1, 2, \ldots$ *and* $k_1, k_2 = 0, 1, \ldots$,

$$\begin{aligned}
\alpha_{m,m+1}^{\prime(k_1+1,k_2)} &= \frac{1}{m+1+\gamma(k_1+1)} \\
&\quad \cdot \left\{(m+1)\,\alpha_{m+1}^{\prime(k_1+k_2+1)} + (k_1+1)\alpha_{m,m+1}^{\prime(k_1,k_2)}\right\};
\end{aligned} \tag{3.4.22}$$

and for $0 \le m \le n-2$ *and* $k_1, k_2 = 0, 1, 2, \ldots$,

$$\begin{aligned}
\alpha_{m,n}^{\prime(k_1+1,k_2)} &= \frac{1}{m+1+\gamma(k_1+1)} \\
&\quad \cdot \left\{(m+1)\,\alpha_{m+1,n}^{\prime(k_1+1,k_2)} + (k_1+1)\alpha_{m,n}^{\prime(k_1,k_2)}\right\}.
\end{aligned} \tag{3.4.23}$$

Remark 3.4.6 For $m = 0, 1, 2, \ldots$,

$$\sigma_{m,m+1}^{\prime} = \frac{m+1}{m+1+\gamma}\,\sigma_{m+1,m+1}^{\prime}, \tag{3.4.24}$$

and for $0 \le m \le n-2$,

$$\sigma_{m,n}^{\prime} = \frac{n^{(n-m)}}{(n+\gamma)^{(n-m)}}\,\sigma_{n,n}^{\prime}, \tag{3.4.25}$$

where

$$\begin{aligned}
x^{(i)} &= x(x-1)\cdots(x-i+1) \quad \text{for} \quad i = 1, 2, \ldots \\
&= 1 \qquad\qquad\qquad\qquad\ \text{for} \quad i = 0.
\end{aligned} \tag{3.4.26}$$

Remark 3.4.7 From Theorems 3.4.3 and 3.4.4, one may deduce Theorems 3.4.1 and 3.4.2 by letting γ tend to zero.

3.5   LOMAX DISTRIBUTION

Let us consider the upper record values $\{R_n, \ n \geq 1\}$ arising from a sequence of i.i.d. random variables from a Lomax distribution with pdf

$$f(x) = \vartheta(1+x)^{-\vartheta-1}, \qquad x \geq 0, \ \vartheta > 0, \qquad (3.5.1)$$

and cdf

$$F(x) = 1 - (1+x)^{-\vartheta}, \qquad x \geq 0, \ \vartheta > 0. \qquad (3.5.2)$$

Lomax (1954) used this distribution in the analysis of business failure data; this distribution is also known as the Pareto Type II distribution (see Arnold, 1983).

In this case, one may simply derive from Eq. (2.3.7) that for $n = 0, 1, 2, \ldots$ and $k = 1, 2, \ldots$,

$$
\begin{aligned}
\alpha_n^{(k)} &= \frac{1}{n!} \int_0^\infty r^k \left\{ -\log(1 - F(r)) \right\}^n f(r) \, dr \\
&= \frac{\vartheta^{n+1}}{n!} \int_0^\infty u^n \, e^{-\vartheta u} \, (e^u - 1)^k \, du \qquad \text{(with } u = \ln(1+r)) \\
&= \sum_{i=0}^k (-1)^{k-i} \binom{k}{i} \Big/ \left(1 - \frac{i}{\vartheta} \right)^{n+1}, \qquad k < \vartheta. \qquad (3.5.3)
\end{aligned}
$$

One may similarly derive an expression for the product moments of upper record values.

On the other hand, recurrence relations for single and product moments of record values may be derived upon using the relationship [observed from (3.5.1) and (3.5.2)] that

$$(1+x)f(x) = \vartheta\{1 - F(x)\}, \qquad x \geq 0, \ \vartheta > 0. \qquad (3.5.4)$$

These results are presented in the following two theorems.

Theorem 3.5.1 (Balakrishnan and Ahsanullah, 1994b) *For $n = 1, 2, \ldots$ and $k = 0, 1, 2, \ldots$ ($k < \vartheta - 1$),*

$$\alpha_n^{(k+1)} = \frac{k+1}{\vartheta - k - 1} \, \alpha_n^{(k)} + \frac{\vartheta}{\vartheta - k - 1} \, \alpha_{n-1}^{(k+1)}. \qquad (3.5.5)$$

Proof. From (2.3.7), let us consider for $n \geq 1$

$$
\begin{aligned}
\alpha_n^{(k)} + \alpha_n^{(k+1)} &= \int_0^\infty (r^k + r^{k+1}) \, f_n(r) \, dr \\
&= \frac{\vartheta}{n!} \int_0^\infty r^k \left\{ -\log(1 - F(r)) \right\}^n \left\{ 1 - F(r) \right\} dr
\end{aligned}
$$

upon using (3.5.4). Integrating now by parts treating r^k for integration and the rest of the integrand for differentiation, we obtain

$$
\begin{aligned}
\alpha_n^{(k)} + \alpha_n^{(k+1)} &= \frac{\vartheta}{k+1} \left\{ \frac{-1}{(n-1)!} \int_0^\infty r^{k+1} \{-\log(1-F(r))\}^{n-1} \, f(r) \, dr \right. \\
&\quad \left. + \frac{1}{n!} \int_0^\infty r^{k+1} \{-\log(1-F(r))\}^n \, f(r) \, dr \right\} \\
&= \frac{\vartheta}{k+1} \left\{ -\alpha_{n-1}^{(k+1)} + \alpha_n^{(k+1)} \right\}.
\end{aligned}
\tag{3.5.6}
$$

Relation (3.5.5) is derived simply by rewriting (3.5.6). $\bigcirc$

Remark 3.5.1 By repeated application of the relation in (3.5.5), we obtain for $n \geq 1$ and $k = 0, 1, 2, \ldots$,

$$
\alpha_n^{(k+1)} = \vartheta \sum_{i=0}^{k+1} \frac{(k+1)^{(i)}}{(\vartheta - k - 1 + i)^{(i+1)}} \, \alpha_{n-1}^{(k+1-i)},
\tag{3.5.7}
$$

where $x^{(i)}$ is as defined in Eq. (3.4.26).

Proceeding along the lines of Theorem 3.5.1, one may establish the following result.

Theorem 3.5.2 (Balakrishnan and Ahsanullah, 1994b) *For $m = 0, 1, 2, \ldots$ and $k_1, k_2 = 0, 1, \ldots$,*

$$
\alpha_{m,m+1}^{(k_1,k_2+1)} = \frac{k_2 + 1}{\vartheta - k_2 - 1} \, \alpha_{m,m+1}^{(k_1,k_2)} + \frac{\vartheta}{\vartheta - k_2 - 1} \, \alpha_m^{(k_1+k_2+1)};
\tag{3.5.8}
$$

and for $0 \leq m \leq n - 2$ and $k_1, k_2 = 0, 1, 2, \ldots$,

$$
\alpha_{m,n}^{(k_1,k_2+1)} = \frac{k_2 + 1}{\vartheta - k_2 - 1} \, \alpha_{m,n}^{(k_1,k_2)} + \frac{\vartheta}{\vartheta - k_2 - 1} \, \alpha_{m,n-1}^{(k_1,k_2+1)}.
\tag{3.5.9}
$$

Remark 3.5.2 On using the relations in (3.5.5), (3.5.8) and (3.5.9), it can be easily established that

$$
\sigma_{m,m+1} = \frac{\vartheta}{\vartheta - 1} \, \sigma_{m,m}, \qquad m = 0, 1, \ldots,
\tag{3.5.10}
$$

and

$$
\sigma_{m,n} = \frac{\vartheta}{\vartheta - 1} \, \sigma_{m,n-1}, \qquad m \leq n - 2;
\tag{3.5.11}
$$

consequently, for $m = 0, 1, \ldots, n - 1$, we have

$$
\sigma_{m,n} = \left(\frac{\vartheta}{\vartheta - 1} \right)^{n-m} \sigma_{m,m}.
\tag{3.5.12}
$$

3.6   NORMAL DISTRIBUTION

In this section, let us consider the upper record values $\{R_n, \ n \geq 1\}$ arising from a sequence of i.i.d. standard normal variables with pdf

$$f(x) = \frac{1}{\sqrt{2\pi}} e^{-x^2/2}, \qquad -\infty < x < \infty. \qquad (3.6.1)$$

Then, the single and product moments of record values do not have an explicit algebraic form. However, by considering the integral forms

$$\alpha_n^{(k)} = \frac{1}{n!} \int_{-\infty}^{\infty} r^k \left\{-\log(1 - F(r))\right\}^n f(r) \, dr, \quad n \geq 0, \ k \geq 1, \quad (3.6.2)$$

and

$$\alpha_{m,n} = \frac{1}{m!(n-m-1)!} \int_{-\infty}^{\infty} \int_{-\infty}^{r_2} r_1 r_2 \left\{-\log(1 - F(r_1))\right\}^m \frac{f(r_1)}{1 - F(r_1)}$$

$$\cdot \left\{-\log(1 - F(r_2)) + \log(1 - F(r_1))\right\}^{n-m-1} f(r_2) \, dr_1 \, dr_2,$$
$$(3.6.3)$$

Houchens (1984) and Balakrishnan and Chan (1995, 1998) employed numerical quadrature to compute the means, variances and covariances of record values. The values of means, α_n ($n = 0, 1, 2, \ldots, 9$), are presented in Table 3.6.1, while the values of variances and covariances, $\sigma_{m,n}$ ($0 \leq m \leq n \leq 9$), are presented in Table 3.6.2.

In this case, by making use of the property that

$$f'(x) = -x \, f(x), \qquad (3.6.4)$$

some simple recurrence relations can be established for the product moments of record values.

Theorem 3.6.1 (Balakrishnan and Chan, 1995) *For $n = 1, 2, \ldots,$*

$$\alpha_{n-1,n} = \alpha_n^{(2)} - 1 \qquad (3.6.5)$$

and, consequently,

$$\sigma_{n-1,n} = \sigma_{n,n} + \alpha_n \left(\alpha_n - \alpha_{n-1}\right) - 1. \qquad (3.6.6)$$

Proof. For $n \geq 1$, we have from (3.6.2) that

$$\alpha_n^{(2)} = \frac{1}{n!} \int_{-\infty}^{\infty} r^2 \left\{-\log(1 - F(r))\right\}^n f(r) \, dr$$

$$= -\frac{1}{n!} \int_{-\infty}^{\infty} r \left\{-\log(1 - F(r))\right\}^n \, df(r)$$

upon using the relation in (3.6.4). Integrating now by parts, we obtain

$$\alpha_n^{(2)} = 1 + \frac{1}{(n-1)!} \int_{-\infty}^{\infty} r_1 \left\{-\log(1 - F(r_1))\right\}^{n-1} \frac{f(r_1)}{1 - F(r_1)} f(r_1) \, dr_1.$$
$$(3.6.7)$$

Upon using (3.6.4) and noting that

$$f(r_1) = \int_{r_1}^{\infty} -f'(r_2) \, dr_2 = \int_{r_1}^{\infty} r_2 f(r_2) \, dr_2,$$

we may write (3.6.7) as

$$\alpha_n^{(2)} = 1 + \frac{1}{(n-1)!} \int_{-\infty}^{\infty} \int_{r_1}^{\infty} r_1 r_2 \left\{-\log(1 - F(r_1))\right\}^{n-1} \frac{f(r_1)}{1 - F(r_1)}$$
$$\cdot f(r_2) \, dr_2 \, dr_1$$

$$= 1 + \alpha_{n-1,n}.$$

Hence, the relation in (3.6.5). The relation in (3.6.6) then follows very easily. ○

The relations in (3.6.5) and (3.6.6) were used by Balakrishnan and Chan (1995) to assess the accuracy of their numerical procedure employed to evaluate the integrals in (3.6.2) and (3.6.3).

Remark 3.6.1 It is quite interesting to note that the relation in (3.6.5) is very similar to a relation between moments of standard normal order statistics due to Govindarajulu (1963); also see Arnold and Balakrishnan (1989, p. 18).

Table 3.6.1.   Means of the Upper Record Values from Standard Normal

n	α_n	n	α_n
0	0.0000	5	2.7174
1	0.9032	6	3.0339
2	1.4990	7	3.3247
3	1.9687	8	3.5942
4	2.3667	9	3.8471

Table 3.6.2.   Variances and Covariances of the Upper Record Values from Standard Normal

n	0	1	2	3	4	5	6	7	8	9
0	1.0000									
1	0.5956	0.7799								
2	0.4534	0.5953	0.7022							
3	0.3775	0.4964	0.5859	0.6611						
4	0.3292	0.4331	0.5115	0.5773	0.6353					
5	0.2951	0.3885	0.4589	0.5181	0.5702	0.6174				
6	0.2696	0.3550	0.4194	0.4735	0.5212	0.5643	0.6041			
7	0.2495	0.3286	0.3883	0.4385	0.4827	0.5226	0.5595	0.5938		
8	0.2332	0.3073	0.3631	0.4100	0.4514	0.4888	0.5233	0.5554	0.5856	
9	0.2197	0.2895	0.3421	0.3864	0.4253	0.4606	0.4931	0.5234	0.5519	0.5788

Remark 3.6.2 If $\{R'_n,\ n \geq 0\}$ are the lower record values arising from a sequence of i.i.d. standard normal variables, then due to the symmetry of the standard normal distribution we have

$$R'_n \overset{d}{=} -R_n \text{ and } (R'_m, R'_n) \overset{d}{=} (-R_m, -R_n) \qquad (3.6.8)$$

and hence,

$$\alpha'_n = -\alpha_n \text{ and } \sigma'_{m,n} = \sigma_{m,n}. \qquad (3.6.9)$$

As a result, the entries in Tables 3.6.1 and 3.6.2 also yield the means, variances and covariances of the lower record values from the standard normal distribution.

3.7   LOGISTIC DISTRIBUTION

Let $\{R_n,\ n \geq 0\}$ be the upper record values arising from a sequence of i.i.d. variables from a logistic population with pdf

$$f(x) = e^{-x}/(1 + e^{-x})^2, \qquad -\infty < x < \infty, \qquad (3.7.1)$$

and cdf

$$F(x) = 1/(1 + e^{-x}), \qquad -\infty < x < \infty. \tag{3.7.2}$$

For various developments and applications concerning this distribution, interested readers may refer to the recent volume by Balakrishnan (1992).

Theorem 3.7.1 (Balakrishnan, Ahsanullah and Chan, 1995) *With $\zeta(\cdot)$ denoting the Riemann zeta function defined by*

$$\zeta(n) = \sum_{i=1}^{\infty} i^{-n} \tag{3.7.3}$$

(see Abramowitz and Stegun, 1965, for details), we have

$$\alpha_0 = 0 \ \text{and} \ \alpha_{n+1} = \alpha_n + \zeta(n+2), \qquad n \geq 0, \tag{3.7.4}$$

and, consequently,

$$\alpha_n = \sum_{i=2}^{n+1} \zeta(i). \tag{3.7.5}$$

Proof. From (2.3.7), we have for $n = 0, 1, 2, \ldots$

$$
\begin{aligned}
\alpha_n &= \frac{1}{n!} \int_{-\infty}^{\infty} r \left\{ -\log(1 - F(r)) \right\}^n f(r) \, dr \\
&= \int_0^1 \log u \, \frac{1}{n!} \{ -\log(1 - u) \}^n \, du + \int_0^1 \frac{1}{n!} \{ -\log(1 - u) \}^{n+1} \, du
\end{aligned}
$$

since $F^{-1}(u) = \log u - \log(1 - u)$. Setting $v = -\log(1 - u)$, we now get

$$\alpha_n = \int_0^{\infty} \log(1 - e^{-v}) \frac{1}{n!} v^n e^{-v} \, dv + (n+1),$$

which, upon writing $\log(1 - e^{-v}) = -\sum_{i=1}^{\infty} e^{-iv}/i$, immediately gives

$$\alpha_n = (n+1) - S_{n+1}, \tag{3.7.6}$$

where

$$S_n = \sum_{i=1}^{\infty} \frac{1}{i(i+1)^n}, \qquad n = 1, 2, \ldots. \tag{3.7.7}$$

Clearly, we have $S_1 = 1$. For $n \geq 2$, let us write

$$
\begin{aligned}
S_{n+1} &= \sum_{i=1}^{\infty} \frac{1}{i(i+1)^{n+1}} = \sum_{i=1}^{\infty} \frac{1}{(i+1)^n} \left(\frac{1}{i} - \frac{1}{i+1} \right) \\
&= S_n - \sum_{i=1}^{\infty} 1/(i+1)^{n+1} \\
&= S_n + 1 - \zeta(n+1).
\end{aligned}
\tag{3.7.8}
$$

From (3.7.6) and (3.7.8), we have

$$
\alpha_{n+1} - \alpha_n = 1 - (S_{n+2} - S_{n+1}) = \zeta(n+2)
$$

which gives the relationship in (3.7.4). Repeated application of (3.7.4) immediately yields (3.7.5). $\bigcirc$

In a similar manner, the following explicit expressions can be derived for the variances and covariances of the upper record values.

Theorem 3.7.2 (Balakrishnan, Ahsanullah and Chan, 1995) *For* $n = 0, 1, 2, \ldots,$

$$
\sigma_{n,n} = 2(n+1)\zeta(n+2) - (n+1) - S_{n+1}^2 + 2T_{n+1};
\tag{3.7.9}
$$

and for $0 \leq m \leq n - 1$,

$$
\sigma_{m,n} = (m+1)\{\zeta(m+2) + \zeta(n+2) - 1\} - S_{m+1}S_{n+1}
$$

$$
+ \sum_{i=1}^{\infty} \frac{1}{i(i+1)^{n-m}} \sum_{j=1}^{\infty} \frac{1}{j(j+1+i)^{m+1}},
\tag{3.7.10}
$$

where S_n *is as given in (3.7.7) and*

$$
T_n = \sum_{k=2}^{\infty} \frac{1}{k(k+1)^n} \left(1 + \frac{1}{2} + \cdots + \frac{1}{k-1} \right), \qquad n \geq 1.
\tag{3.7.11}
$$

By making use of the expressions of the means, variances and covariances of the upper record values in (3.7.5), (3.7.9) and (3.7.10), respectively, and the tabulated values of the Riemann zeta function (Abramowitz and Stegun, 1965, p. 811), Balakrishnan, Ahsanullah and Chan (1995) have computed these quantities for $n = 0(1)9$. The values of means, α_n $(n = 0, 1, \ldots, 9)$, are presented in Table 3.7.1, while the values of variances and covariances, $\sigma_{m,n}$ $(0 \leq m \leq n \leq 9)$, are presented in Table 3.7.2.

Table 3.7.1. Means of the Upper Record Values from Logistic Distribution

n	α_n	n	α_n
0	0.0000	5	5.9836
1	1.6449	6	6.9919
2	2.8470	7	7.9960
3	3.9293	8	8.9980
4	4.9662	9	9.9990

Table 3.7.2. Variances and Covariances of the Upper Record Values from Logistic Distribution

n	0	1	2	3	4	5	6	7	8	9
0	3.2899									
1	2.4426	2.9882								
2	1.9701	2.6887	3.5414							
3	1.7913	2.5310	3.3885	4.3096						
4	1.7139	2.4636	3.3132	4.2258	5.1779					
5	1.6782	2.4327	3.2788	4.1853	5.1311	6.1016				
6	1.6612	2.4181	3.2625	4.1660	5.1084	6.0754	7.0576			
7	1.6530	2.4110	3.2546	4.1567	5.0974	6.0625	7.0429	8.0323		
8	1.6489	2.4075	3.2508	4.1522	5.0920	6.0563	7.0356	8.0241	9.0180	
9	1.6469	2.4058	3.2489	4.1500	5.0893	6.0532	7.0321	8.0200	9.0134	10.0100

Remark 3.7.1 Due to the fact that $\zeta(\cdot) \approx 1$ for large value of the argument (Abramowitz and Stegun, 1965, p. 811), it is interesting to note from (3.7.4), (3.7.5), (3.7.9) and (3.7.10) that $\alpha_n \approx n+1$, $\sigma_{n,n} \approx n+1$ and $\sigma_{m,n} \approx m+1$ for large m and n. These may also be observed in Tables 3.7.1 and 3.7.2.

Remark 3.7.2 As mentioned earlier in Remark 3.6.2, due to the symmetry of the logistic pdf in (3.7.1), the entries in Tables 3.7.1 and 3.7.2 also yield the means, variances and covariances of the lower record values from the logistic distribution through (3.6.9).

3.8 BOUNDS AND APPROXIMATIONS

Let us consider the nth upper record value, R_n, arising from a sequence of i.i.d. random variables with common cdf $F(x)$ and pdf $f(x)$. By making use of the probability integral transformation $u = F(x)$, we may write from (2.3.7) [recall (2.7.2)]

$$E(R_n) = \alpha_n = \int_0^1 F^{-1}(u) \frac{1}{n!} \{-\log(1-u)\}^n \, du. \tag{3.8.1}$$

Without loss of generality, let us assume that $E(X_i) = 0$ and $\text{Var}(X_i) = E(X_i^2) = 1$. Then, by writing

$$\alpha_n = \int_0^1 F^{-1}(u) \left[\frac{1}{n!} \{-\log(1-u)\}^n - \lambda \right] du \qquad (3.8.2)$$

and applying the Cauchy-Schwarz inequality, we immediately obtain

$$\alpha_n \le \left\{ \binom{2n}{n} - 2\lambda + \lambda^2 \right\}^{1/2}. \qquad (3.8.3)$$

By noting that the right-hand side of (3.8.3) is minimized when $\lambda = 1$, we obtain the universal bound

$$\alpha_n \le \left\{ \binom{2n}{n} - 1 \right\}^{1/2}, \qquad (3.8.4)$$

which was established by Nagaraja (1978). From (3.8.2), he also noted that the bound in (3.8.4) is sharp in the sense that it is achieved for the population with inverse cdf

$$F^{-1}(u) = \frac{1}{\left\{ \binom{2n}{n} - 1 \right\}^{1/2}} \left[\frac{1}{n!} \{-\log(1-u)\}^n - 1 \right], \qquad 0 < u < 1,$$

that is, with cdf

$$F(x) = 1 - \exp\left[-\left\{ n! \left[\left\{ \binom{2n}{n} - 1 \right\}^{1/2} x + 1 \right] \right\}^{1/n} \right],$$

$$-\frac{1}{\left\{ \binom{2n}{n} - 1 \right\}^{1/2}} < x < \infty. \qquad (3.8.5)$$

For $n = 1$, it is of interest to note from (3.8.4) and (3.8.5) that

$$\alpha_1 \le 1$$

and the equality holds iff the population is a translated exponential with cdf

$$F(x) = 1 - e^{-(x+1)}, \qquad x > -1.$$

In general, the bound in (3.8.4) is attained iff the population is a translated Weibull with cdf as in (3.8.5).

If the distribution is symmetric about 0, improvements are possible over the bound in (3.8.4). To see this, we first of all write

$$\alpha_n = \frac{1}{2} \int_0^1 F^{-1}(u) \left[\frac{1}{n!} \{-\log(1-u)\}^n - \frac{1}{n!} \{-\log u\}^n - \lambda \right] du$$

$$(3.8.6)$$

and then apply the Cauchy-Schwarz inequality to derive

$$\alpha_n \leq \frac{1}{2} \left[2\binom{2n}{n} + \lambda^2 - \frac{2}{(n!)^2} \int_0^1 \{\log u \ \log(1-u)\}^n \, du \right]^{1/2}. \quad (3.8.7)$$

By noting that the right-hand side of (3.8.7) is minimized when $\lambda = 0$, we obtain the universal bound

$$\alpha_n \leq \frac{1}{\sqrt{2}} \left\{ \binom{2n}{n} - A_n \right\}^{1/2}, \quad (3.8.8)$$

where

$$A_n = \frac{1}{(n!)^2} \int_0^1 \{\log u \ \log(1-u)\}^n \, du. \quad (3.8.9)$$

This bound was established by Nagaraja (1978), who also noted that the ratio of the bounds in (3.8.8) and (3.8.4) tends to $1/\sqrt{2}$ as $n \to \infty$. (There is a typographical error in his formula of A_n, however.)

Note that, from (3.8.9), we may write

$$A_n = \frac{1}{(n!)^2} \int_0^1 \{-\log u\}^n \left(\sum_{i=1}^{\infty} u^i/i \right)^n du$$

$$= \frac{1}{n!} \sum_{j=n}^{\infty} c_j(n)/(j+1)^{n+1}, \quad (3.8.10)$$

where $c_j(n) = $ coefficient of u^j in $(\sum_{i=1}^{\infty} u^i/i)^n$. These coefficients can be generated easily in a recursive manner as follows. First of all, we have $c_j(1) = 1/j$ for $j = 1, 2, \ldots$. Next, we have

$$c_j(n) = \text{Coefficient of } u^j \text{ in } \left(\sum_{i=1}^{\infty} u^i/i \right)^{n-1} \cdot \left(\sum_{i=1}^{\infty} u^i/i \right)$$

$$= \sum_{k=n-1}^{j-1} \left\{ \text{Coefficient of } u^k \text{ in } \left(\sum_{i=1}^{\infty} u^i/i \right)^{n-1} \right\}$$

$$\cdot \left\{ \text{Coefficient of } u^{j-k} \text{ in } \left(\sum_{i=1}^{\infty} u^i / i \right) \right\}$$

$$= \sum_{k=n-1}^{j-1} c_k (n-1)/(j-k). \tag{3.8.11}$$

Thus, after computing the coefficients $c_j(n)$ from the recurrence relation in (3.8.11), one may compute A_n from (3.8.10) to the desired level of accuracy. Alternatively, from (3.8.11) by writing explicitly

$$c_j(n) =$$

$$\sum_{k_{n-1}=n-1}^{j-1} \sum_{k_{n-2}=n-2}^{k_{n-1}-1} \cdots \sum_{k_2=2}^{k_3-1} \sum_{k_1=1}^{k_2-1} \frac{1}{k_1(k_2 - k_1) \cdots (k_{n-1} - k_{n-2})(j - k_{n-1})},$$

$$j = n, n+1, \ldots,$$

one may compute A_n directly from (3.8.10). For $n = 1(1)6$, A_n has been computed and presented in Table 3.8.1.

Table 3.8.1. Values of A_n in (3.8.9)

n	A_n
1	0.35501708
2	0.03543548
3	0.00165389
4	0.00004469
5	0.00000079
6	0.00000001

For example, the values of the bounds in (3.8.4) and (3.8.8) are reported in Table 3.8.2 for $n = 1(1)9$. Also presented in this table are the exact values for the standard normal distribution taken from Table 3.6.1. It is clear that both bounds become large very quickly even though they are universally "sharp". Hence, some simple improvements over these bounds (that may be distribution-based) will be desirable.

Table 3.8.2. Comparison of Bounds in
(3.8.4) and (3.8.8)

n	Bound in (3.8.4)	Bound in (3.8.8)*	Exact value from standard normal
1	1.0000	0.9069	0.9032
2	2.2361	1.7269	1.4990
3	4.3589	3.1621	1.9687
4	8.3066	5.9161	2.3667
5	15.8430	11.2250	2.7174
6	30.3809	21.4942	3.0339
7	58.5747	41.4246	3.3247
8	113.4116	80.2185	3.5942
9	220.4972	155.9166	3.8471

* While the reported values are computed by using Table 3.8.1, Nagaraja (1978) determined these values by computing A_n by numerical integration using the trapezoidal rule.

One such bound is the "extrapolation-type bound" described below, which is very similar in principle as well as form to the one proposed by Balakrishnan (1990) in the order statistics context. To fix the ideas, let us assume that means of two upper record values are known, say, α_m and α_p $(m < p)$. Then, let us write

$$\alpha_n = \int_0^1 F^{-1}(u) \left[g_n(u) - c\{g_m(u) - \alpha_m F^{-1}(u)\} \right.$$

$$\left. - d\{g_p(u) - \alpha_p F^{-1}(u)\} - \lambda \right] du, \qquad (3.8.12)$$

where c, d and λ will be optimally determined later, and

$$g_\ell(u) = \frac{1}{\ell!} \{-\log(1-u)\}^\ell.$$

Applying the Cauchy-Schwarz inequality to the integral in (4.8.12), we obtain

$$\alpha_n \leq \binom{2n}{n} + c^2 \left\{ \binom{2m}{m} - \alpha_m^2 \right\} + d^2 \left\{ \binom{2p}{p} - \alpha_p^2 \right\} + \lambda^2$$

$$- 2c \left\{ \binom{m+n}{m} - \alpha_m \alpha_n \right\} - 2d \left\{ \binom{p+n}{p} - \alpha_p \alpha_n \right\}$$

$$- 2\lambda(1 - c - d) + 2cd \left\{ \binom{m+p}{m} - \alpha_m \alpha_p \right\}. \qquad (3.8.13)$$

Upon setting $d = -c\alpha_m/\alpha_p$ and using the optimal value of $\lambda = 1 - c(1 - \alpha_m/\alpha_p)$, we get the optimal value of c that minimizes the right-hand side of (4.8.13) to be

$$c_{opt} = \frac{H(m,n) - \frac{\alpha_m}{\alpha_p} H(p,n)}{H(m,m) - 2 \frac{\alpha_m}{\alpha_p} H(m,p) + \frac{\alpha_m^2}{\alpha_p^2} H(p,p)}, \qquad (3.8.14)$$

where

$$H(i,j) = \binom{i+j}{i} - 1. \qquad (3.8.15)$$

The corresponding upper bound is then derived to be

$$\alpha_n \leq \{H(n,n) - \tau\}^{1/2}, \qquad (3.8.16)$$

where

$$\tau = \frac{\left\{H(m,n) - \frac{\alpha_m}{\alpha_p} H(p,n)\right\}^2}{H(m,m) - 2 \frac{\alpha_m}{\alpha_p} H(m,p) + \frac{\alpha_m^2}{\alpha_p^2} H(p,p)} > 0. \qquad (3.8.17)$$

It is quite apparent that the bound in (3.8.16) is an improvement over the bound in (3.8.4) which is not surprising as more information (in the form of α_m and α_p) is utilized in the derivation of the bound in (3.8.16). For example, in the case of the standard normal distribution, by taking $m = 1$ and $p = 2$ and using the values of $\alpha_1 = 0.9032$ and $\alpha_2 = 1.4990$, (3.8.16) yields the upper bound for α_3 and α_4 to be 2.1229 and 4.5194, respectively, as compared to the bounds of 4.3589 and 8.3066 obtained from (3.8.4).

Similar improvements can also be achieved over the bound in (3.8.8) by proceeding in an analogous manner. For example, when the population distribution is symmetric and the values of α_m and α_p are available, we can show that

$$\alpha_n \leq \frac{1}{\sqrt{2}} \{H^*(n,n) - \tau^*\}, \qquad (3.8.18)$$

where

$$H^*(i,j) = \binom{i+j}{i} - \frac{1}{i!j!} \int_0^1 (-\log u)^i (-\log(1-u))^j \, du \qquad (3.8.19)$$

and

$$\tau^* = \frac{\left\{H^*(m,n) - \frac{\alpha_m}{\alpha_p} H^*(p,n)\right\}^2}{H^*(m,m) - 2\frac{\alpha_m}{\alpha_p} H^*(m,p) + \frac{\alpha_m^2}{\alpha_p^2} H^*(p,p)} > 0. \qquad (3.8.20)$$

It is also of interest to mention here that further improvements over the bounds in (3.8.16) and (3.8.18) are possible by minimizing globally with respect to c and d instead of minimizing with respect to c after setting $d = -c\alpha_m/\alpha_p$; refer to Balakrishnan and Bendre (1993) for similar developments in the context of order statistics.

All the preceding discussion has been concerned with the derivation of upper bounds for the mean record values. In the following, we present the orthogonal inverse expansion method of deriving bounds as well as approximations for the mean record values. This method was proposed by Sugiura (1962) in the context of order statistics and also sketched by Nagaraja (1978) for the record value situation. Let $\{\phi_k(u), \ k \geq 0\}$ be a complete orthonormal system of functions in $L^2(0,1)$ (the class of square integrable functions in $(0,1)$), that is,

$$\int_0^1 \phi_i(u)\, du = 0 \text{ and } \int_0^1 \phi_i(u)\phi_j(u)\, du \ = \ 1 \text{ if } i = j$$
$$= \ 0 \text{ if } i \neq j.$$

It is then known that for any function $f(u) \in L^2(0,1)$,

$$f(u) = \lim_{k \to \infty} \sum_{i=0}^{k} a_i \phi_i(u) \text{ and } \int_0^1 f^2(u)\, du = \sum_{i=0}^{\infty} a_i^2,$$

where a_i is the ith Fourier coefficient of $f(u)$ given by

$$a_i = \int_0^1 f(u)\phi_i(u)\, du, \qquad i = 0, 1, 2, \ldots.$$

Further, for any two functions $f(u)$, $g(u) \in L^2(0,1)$ with Fourier coefficients $\{a_i, \ i \geq 0\}$ and $\{b_i, \ i \geq 0\}$, respectively, we may observe that for any $k \geq 0$

$$\left| \int_0^1 f(u)g(u)\, du - \sum_{i=0}^{k} a_i b_i \right|$$

$$= \left| \int_0^1 \left\{ f(u) - \sum_{i=0}^{k} a_i \phi_i(u) \right\} \left\{ g(u) - \sum_{i=0}^{k} b_i \phi_i(u) \right\} du \right|$$

$$\leq \left[\left\{ \int_0^1 f^2(u)\, du - \sum_{i=0}^{k} a_i^2 \right\} \left\{ \int_0^1 g^2(u)\, du - \sum_{i=0}^{k} b_i^2 \right\} \right]^{1/2} \qquad (3.8.21)$$

by the Cauchy-Schwarz inequality, and the equality holds if and only if

$$f(u) - \sum_{i=0}^{k} a_i \phi_i(u) \propto g(u) - \sum_{i=0}^{k} b_i \phi_i(u). \qquad (3.8.22)$$

In (3.8.1), upon setting

$$f(u) = F^{-1}(u) \quad \text{and} \quad g(u) = \frac{1}{n!}\{-\log(1-u)\}^n,$$

we immediately obtain from (3.8.21) that for an arbitrary continuous distribution with mean 0 and variance 1

$$\left| \alpha_n - \sum_{i=0}^{k} a_i b_i \right| \leq \left[\left\{ 1 - \sum_{i=0}^{k} a_i^2 \right\} \left\{ \binom{2n}{n} - \sum_{i=0}^{k} b_i^2 \right\} \right]^{1/2} \qquad (3.8.23)$$

and the equality holds if and only if

$$F^{-1}(u) - \sum_{i=0}^{k} a_i \phi_i(u) \propto \frac{1}{n!}\{-\log(1-u)\}^n - \sum_{i=0}^{k} b_i \phi_i(u) \qquad (3.8.24)$$

with

$$a_i = \int_0^1 F^{-1}(u)\phi_i(u)\,du \text{ and } b_i = \int_0^1 \frac{1}{n!}\{-\log(1-u)\}^n \phi_i(u)du.$$
$$(3.8.25)$$

Now, as a specific example, let us choose the orthonormal system $\{\phi_k(u)\}_{k=0}^\infty$ to be the system of Legendre polynomials given by

$$
\begin{aligned}
\phi_k(u) &= \frac{\sqrt{2k+1}}{k!}\frac{d^k}{du^k}\{u^k(u-1)^k\} \\
&= \sqrt{2k+1}\sum_{i=0}^{k}(-1)^{k-i}\binom{k}{i}\binom{k+i}{i}u^i & (3.8.26) \\
&= \sqrt{2k+1}\sum_{i=0}^{k}(-1)^{i}\binom{k}{i}\binom{k+i}{i}(1-u)^i. & (3.8.27)
\end{aligned}
$$

Making use of the expression in (3.8.26), we can write for $k = 0, 1, 2, \ldots$

$$
\begin{aligned}
a_k &= \sqrt{2k+1}\sum_{i=0}^{k}(-1)^{k-i}\binom{k}{i}\binom{k+i}{i}\int_0^1 F^{-1}(u)u^i\,du \\
&= \sqrt{2k+1}\sum_{i=0}^{k}(-1)^{k-i}\binom{k}{i}\binom{k+i}{i}\alpha_{i+1:i+1}/(i+1), & (3.8.28)
\end{aligned}
$$

where $\alpha_{\ell:\ell}$ denotes the expected value of the largest order statistic in a sample of size ℓ from the distribution $F(\cdot)$. Similarly, by making use of the expression in (3.8.27), we can write for $k = 0, 1, 2, \ldots$

$$b_k = \sqrt{2k+1} \sum_{i=0}^{k} (-1)^i \binom{k}{i} \binom{k+i}{i} \int_0^1 \frac{1}{n!} \{-\log(1-u)\}^n (1-u)^i \, du$$

$$= \sqrt{2k+1} \sum_{i=0}^{k} (-1)^i \binom{k}{i} \binom{k+i}{i} \Big/ (i+1)^{n+1}. \tag{3.8.29}$$

If we take $k = 0$ in (3.8.23), for example, upon using the facts that $a_0 = 0$ and $b_0 = 1$ we simply derive the bound in (3.8.4). Hence, for any $k > 0$ (3.8.23) will yield improvements over the bound in (3.8.4). Further, from (3.8.23) we derive an approximation for α_n as $\sum_{i=0}^{k} a_i b_i$ with bounds being given by $\sum_{i=0}^{k} a_i b_i \pm R$, where R is the R.H.S. of (3.8.23).

For the purpose of illustration, let us consider here the standard normal distribution. In this case, the values of the coefficients a_k, $0 \le k \le 27$, have been computed from (3.8.28) and reported by Arnold and Balakrishnan (1989, p. 95); for example, we have

$$\begin{aligned}
a_0 &= 0, \ a_1 = 0.977205, \ a_2 = 0, \ a_3 = 0.183008, \\
a_4 &= 0, \ a_5 = 0.081699, \ a_6 = 0, \ a_7 = 0.047729, \\
a_8 &= 0, \ a_9 = 0.031880, \ a_{10} = 0, \ a_{11} = 0.023079, \\
a_{12} &= 0, \ a_{13} = 0.017631, \ a_{14} = 0, \ a_{15} = 0.013996, \\
a_{16} &= 0, \ a_{17} = 0.011438, \ a_{18} = 0, \ a_{19} = 0.009560.
\end{aligned}$$

By making use of these values, we computed bounds and approximations for α_n from (3.8.23) for $n = 1(1)9$ with $k = 1(2)19$. These values are presented in Table 3.8.3 along with the exact values taken from Table 3.6.1.

For an arbitrary symmetric distribution with mean 0 and variance 1, improvements can be achieved over the bound in (3.8.23). In this case, since $F^{-1}(1-u) = -F^{-1}(u)$ and $\phi_{2i}(1-u) = \phi_{2i}(u)$ we immediately have

$$a_{2i} = 0, \qquad i = 0, 1, 2, \ldots.$$

Now by applying the Cauchy-Schwarz inequality based on Legendre polynomials of order $i = 1, 3, \ldots, 2k+1, 0, 2, 4, \ldots$ and proceeding as we did before in deriving (3.8.23), we obtain for $k = 0, 1, 2, \ldots$

$$\left| \alpha_n - \sum_{i=0}^{k} a_{2i+1} b_{2i+1} \right|$$

$$\leq \left[\left\{ 1 - \sum_{i=0}^{k} a_{2i+1}^2 \right\} \left\{ \binom{2n}{n} - \sum_{i=0}^{\infty} b_{2i}^2 - \sum_{i=0}^{k} b_{2i+1}^2 \right\} \right]^{1/2}. \quad (3.8.30)$$

Upon using the fact that

$$\sum_{i=0}^{\infty} b_{2i}^2 = \frac{1}{4} \int_0^1 \left[\frac{1}{n!} \{ -\log(1-u) \}^n + \frac{1}{n!} \{ -\log u \}^n \right]^2 du$$

$$= \frac{1}{4} \left[2 \binom{2n}{n} + 2 A_n \right]$$

$$= \frac{1}{2} \left\{ \binom{2n}{n} + A_n \right\}, \quad (3.8.31)$$

where A_n is as defined in (3.8.9), the inequality in (3.8.30) reduces to

$$\left| \alpha_n - \sum_{i=0}^{k} a_{2i+1} b_{2i+1} \right| \leq \left[\left\{ 1 - \sum_{i=0}^{k} a_{2i+1}^2 \right\} \left\{ \frac{\binom{2n}{n} - A_n}{2} - \sum_{i=0}^{k} b_{2i+1}^2 \right\} \right]^{1/2},$$

$$(3.8.32)$$

where a_i and b_i are as given in (3.8.28) and (3.8.29) and A_n as defined in (3.8.9). It is apparent that the bound in (3.8.32) is an improvement over the bound in (3.8.8) for any $k \geq 0$. Furthermore, from (3.8.32) we derive an approximation for α_n as $\sum_{i=0}^{k} a_{2i+1} b_{2i+1}$ with bounds being given by $\sum_{i=0}^{k} a_{2i+1} b_{2i+1} \pm R^*$, where R^* is the R.H.S. of (3.8.32).

Many other developments on bounds and approximations for moments of order statistics can be suitably adapted to develop similar results for moments of record values. Nagaraja (1978) has illustrated this in deriving bounds based on c-comparison and s-comparison following the work of van Zwet (1964). To quote a few results on these lines, let us first define the following.

Table 3.8.3. Approximations and Bounds in (3.8.23) for α_n from the Standard Normal Distribution

n	$k = 1$	$k = 3$	$k = 5$	$k = 7$	$k = 9$	$k = 11$
1	0.8463	0.8866	0.8957	0.8990	0.9005	0.9013
	± 0.1061	± 0.0269	± 0.0117	± 0.0064	± 0.0040	± 0.0027
2	1.2694	1.4140	1.4550	1.4720	1.4807	1.4858
	± 0.3864	± 0.1350	± 0.0679	± 0.0409	± 0.0274	± 0.0197
3	1.4810	1.7435	1.8367	1.8808	1.9054	1.9207
	± 0.8676	± 0.3759	± 0.2132	± 0.1391	± 0.0987	± 0.0742
4	1.5868	1.9397	2.0885	2.1674	2.2150	2.2464
	± 1.7295	± 0.8275	± 0.5061	± 0.3494	± 0.2593	± 0.2020
5	1.6397	2.0506	2.2451	2.3575	2.4302	2.4806
	± 3.3445	± 1.6634	± 1.0569	± 0.7541	± 0.5758	± 0.4597
6	1.6661	2.1110	2.3371	2.4764	2.5710	2.6396
	± 6.4396	± 3.2453	± 2.0950	± 1.5186	± 1.1770	± 0.9528
7	1.6793	2.1429	2.3889	2.5467	2.6581	2.7415
	±12.4299	± 6.2898	± 4.0832	± 2.9786	± 2.3237	± 1.8933
8	1.6860	2.1595	2.4170	2.5866	2.7093	2.8033
	± 24.0806	± 12.1998	± 7.9338	± 5.8000	± 4.5358	± 3.7053
9	1.6893	2.1680	2.4319	2.6084	2.7382	2.8392
	± 46.8096	± 23.7226	± 15.4353	± 11.2916	± 8.8373	± 7.2256

(table continues)

Table 3.8.3 (*continued*)

n	$k = 13$	$k = 15$	$k = 17$	$k = 19$	Exact Value
1	0.9019 ± 0.0020	0.9022 ± 0.0015	0.9024 ± 0.0012	0.9026 ± 0.0009	0.9032
2	1.4890 ± 0.0148	1.4911 ± 0.0116	1.4927 ± 0.0093	1.4938 ± 0.0077	1.4990
3	1.9309 ± 0.0581	1.9380 ± 0.0468	1.9432 ± 0.0387	1.9472 ± 0.0325	1.9687
4	2.2683 ± 0.1629	2.2844 ± 0.1349	2.2966 ± 0.1140	2.3060 ± 0.0979	2.3667
5	2.5174 ± 0.3788	2.5453 ± 0.3196	2.5672 ± 0.2746	2.5846 ± 0.2395	2.7174
6	2.6916 ±0.7952	2.7322 ± 0.6789	2.7648 ± 0.5898	2.7916 ± 0.5196	3.0339
7	2.8064 ± 1.5903	2.8586 ± 1.3662	2.9014 ± 1.1941	2.9372 ± 1.0580	3.3247
8	2.8782 ± 3.1206	2.9394 ± 2.6881	2.9906 ± 2.3558	3.0341 ± 2.0930	3.5942
9	2.9208 ± 6.0913	2.9885 ± 5.2524	3.0459 ± 4.6080	3.0953 ± 4.0983	3.8471

c-ordering. Let $\boldsymbol{F}$ be the class of all distribution functions with positive continuous derivatives on their supports. If F_1 and F_2 are in $\boldsymbol{F}$, we say that F_1 c-precedes F_2 (denoted by $F_1 \underset{c}{\leq} F_2$) if and only if $F_2^{-1} F_1$ is convex on the support of F_1.

s-ordering. Let $\boldsymbol{S}$ be the subclass of $\boldsymbol{F}$ of all symmetric distributions. Let $F(x_0 - x) = 1 - F(x_0 + x)$ for some x_0 and all x if $F \in \boldsymbol{S}$. If F_1 and F_2 are in $\boldsymbol{S}$, we way that F_1 s-precedes F_2 (denoted by $F_1 \underset{s}{\leq} F_2$) if and only if $F_2^{-1} F_1$ is convex for $x > x_0$ with x in the support of F_1.

Then, by using van Zwet's (1964) ideas, the following two results can be established.

Theorem 3.8.1 (Nagaraja, 1978) *If $F_1 \underset{c}{\leq} F_2$, then $F_1(\alpha_n^{[1]}) \leq F_2(\alpha_n^{[2]})$ for all $n \geq 0$ for which $\alpha_n^{[1]}$ and $\alpha_n^{[2]}$, the expected value of the nth upper record from F_1 and F_2, respectively, exist.*

Proof. First of all, for any pair of continuous distributions F_1 and F_2 in **F**, there exists an increasing function $t(\cdot)$ such that, if X is distributed as F_1, then $t(X)$ is distributed as F_2; the function $t(\cdot)$ is uniquely determined by $t(x) = F_2^{-1}[F_1(x)]$. (Prove this!). Since the transformation $t(\cdot)$ is strictly increasing, it will transform the nth record $R_n(F_1)$ from F_1 to the nth record $R_n(F_2)$ from F_2. Through Jensen's inequality, we then obtain

$$h\left(E(R_{n,1})\right) \leq E\left(h(R_{n,1})\right) = E(R_{n,2}) . \qquad (3.8.33)$$

We may rewrite (3.8.33) as

$$F_2^{-1}\left[F_1(\alpha_n^{[1]})\right] \leq \alpha_n^{[2]}$$

yielding

$$F_1\left(\alpha_n^{[1]}\right) \leq F_2\left(\alpha_n^{[2]}\right)$$

for all $n \geq 0$.   ○

Corollary 3.8.1 *By taking $F_2(x) = x$, $0 < x < 1$, and realizing that $\alpha_n^{[2]} = 1 - 2^{-(n+1)}$, Theorem 3.8.1 readily implies that if F_1 is convex on its support, then $F_1(\alpha_n^{[1]}) \leq 1 - 2^{-(n+1)}$. This inequality is simply reversed if F_1 is concave.*

Corollary 3.8.2 *By taking $F_2(x) = 1 - e^{-x}$, $0 \leq x < \infty$, and realizing that $\alpha_n^{[2]} = n + 1$, Theorem 3.8.1 readily implies that if F_1 is IFR, then $F_1(\alpha_n^{[1]}) \leq 1 - e^{-(n+1)}$. This inequality reverses if F_1 is DFR.*

Theorem 3.8.2 *If $F_1 \underset{s}{\leq} F_2$, then $F_1(\alpha_n^{[1]}) \leq F_2(\alpha_n^{[2]})$ for all $n \geq 0$ for which $\alpha_n^{[2]}$ exists.*

Corollary 3.8.3 *By taking $F_2(x) = x$, $0 < x < 1$ (with $x_0 = 1/2$), Theorem 3.8.2 readily yields that if F_1 is the symmetric U-shaped distribution then $F_1(\alpha_n^{[1]}) \leq 1 - 2^{-(n+1)}$.*

Reference may be made to van Zwet (1964), Barlow and Proschan (1981), David (1981), and Arnold and Balakrishnan (1989) in order to find results on bounds and approximations on order statistics which will help develop corresponding results on record values.

3.9 RESULTS FOR k-RECORDS

In Section 2.10, we discussed two types of k-records. In this section, we will develop some results for Type 2 k-records, derive universal bounds for the expected values of k-records for arbitrary as well as symmetric distributions, and improve these bounds by using the principle of greatest convex minorants.

Recall that Type 2 k-record process is defined in terms of the kth largest X yet seen. For a formal definition, let

$$T_{0(k)} = k \qquad \text{with probability 1}$$

and, for $n \geq 1$,

$$T_{n(k)} = \min \left\{ j : j > T_{n-1(k)}, \; X_j > X_{T_{n-1(k)} - k + 1 : T_{n-1(k)}} \right\} , \qquad (3.9.1)$$

where $X_{i:m}$ denotes the ith order statistic in a sample of size m. The k-records are then defined by

$$R_{n(k)} = X_{T_{n(k)} - k + 1 : T_n} \qquad \text{for } n = 0, 1, 2, \ldots . \qquad (3.9.2)$$

As mentioned already in Section 2.10, $\{R_{n(k)}\}$ has precisely the same distribution as the sequence of records (1-records) from the population with distribution $1 - \{1 - F(x)\}^k$. The pdf of $R_{n(k)}$ is given by

$$f_{n(k)}(r) = \frac{k}{n!} \left\{ -\log(1 - F(r))^k \right\}^n (1 - F(r))^{k-1} f(r), \quad -\infty < r < \infty. \qquad (3.9.3)$$

From (3.9.3), we have

$$E(R_{n(k)}) = \frac{k}{n!} \int_{-\infty}^{\infty} r \{ -\log(1 - F(r))^k \}^n (1 - F(r))^{k-1} f(r) \, dr$$

which, upon making the substitution $u = F(r)$, can be rewritten as

$$E(R_{n(k)}) = \int_0^1 F^{-1}(u) g(u) \, du , \qquad (3.9.4)$$

where

$$g(u) = \frac{k^{n+1}}{n!} \left\{ -\log(1 - u) \right\}^n (1 - u)^{k-1} , \quad 0 \leq u \leq 1 . \qquad (3.9.5)$$

Without loss of any generality, let us assume that the original variables X_i's have mean 0 and variance 1; that is,

$$\int_0^1 F^{-1}(u)\, du = 0 \qquad \text{and} \qquad \int_0^1 \{F^{-1}(u)\}^2 du = 1.$$

Then, by applying the Cauchy-Schwarz inequality in (3.9.4), Grudzień and Szynal (1983) established that

$$|E(R_{n(k)})| \leq \left\{ \frac{k^{2n+2}}{(2k-1)^{2n+1}} \binom{2n}{n} - 1 \right\}^{1/2}. \qquad (3.9.6)$$

Note that for the case $k = 1$, the bound in (3.9.6) reduces to the bound in (3.8.4).

Similarly, for symmetric distributions with mean 0 and variance 1, by writing

$$E(R_{n(k)}) = \int_{1/2}^1 F^{-1}(u)\{g(u) - g(1-u)\}\, du, \qquad (3.9.7)$$

where $g(u)$ is as defined in (3.9.5), and by applying the Cauchy-Schwarz inequality, Grudzień and Szynal (1983) showed that

$$|E(R_{n(k)})| \leq \frac{k^{n+1}}{\sqrt{2}} \left\{ \frac{1}{(2k-1)^{2n+1}} \binom{2n}{n} - A_{n(k)} \right\}^{1/2}; \qquad (3.9.8)$$

here,

$$A_{n(k)} = \frac{1}{(n!)^2} \int_0^1 \{\log u \, \log(1-u)\}^n \{u(1-u)\}^{k-1} du. \qquad (3.9.9)$$

Once again, the bound in (3.9.8) reduces to the bound in (3.8.8) when $k = 1$.

As we have already seen in Section 3.8, the bound in (3.9.6) is sharp for the case $k = 1$. However, when $k \neq 1$, the bound in (3.9.6) is not attainable. This is readily observed by noting from (3.9.5) that $g(u)$ is monotonic increasing for $0 \leq u < 1 - e^{-n/(k-1)}$ and monotonic decreasing for $1 - e^{-n/(k-1)} < u \leq 1$. Hence, the principle of greatest convex minorants may be used to improve the bounds in (3.9.6) and (3.9.8); see Raqab (1997). This method has been successfully applied by Moriguti (1953) and Balakrishnan (1993), in order to derive improved bounds for the expected values of order statistics.

To fix the ideas, let $g(u)$ be as defined in (3.9.5) and

$$G(u) = \frac{k^{n+1}}{n!} \int_0^{-\log(1-u)} e^{-ky} \, y^n \, dy, \; 0 \le u \le 1. \tag{3.9.10}$$

Let $G_*(u)$ be the greatest convex minorant of $G(u)$ in the interval $[0,1]$; in other words, $G_*(u)$ is the supremum of all convex functions dominated by $G(u)$ in the interval $[0,1]$. Then, $G_*(u)$ represents the cdf of a random variable which is stochastically larger than the nth k-record $R_{n(k)}$. As a result, we have

$$|E(R_{n(k)})| \le \int_0^1 F^{-1}(u)\{g_*(u) - \lambda\}du, \tag{3.9.11}$$

where

$$g_*(u) = \begin{cases} g(u) & \text{if } 0 \le u \le u_1 \\ g(u_1) & \text{if } u_1 \le u \le 1 \end{cases} \tag{3.9.12}$$

and u_1 satisfies the equation

$$\frac{1 - G(u_1)}{1 - u_1} = g(u_1) = \frac{k^{n+1}}{n!} \{-\log(1 - u_1)\}^n (1 - u_1)^{k-1}. \tag{3.9.13}$$

Now applying the Cauchy-Schwarz inequality in (3.9.11) and then minimizing the R.H.S. with respect to λ, we obtain

$$|E(R_{n(k)})| \le \left\{ \int_0^1 \{g_*(u) - 1\}^2 du \right\}^{1/2}. \tag{3.9.14}$$

Using the expression of $g_*(u)$ in (3.9.12), we can rewrite (3.9.14) as

$$\begin{aligned}
|E(R_{n(k)})| &\le \left[\int_0^{u_1} \{g(u) - 1\}^2 \, du + (1 - u_1)\{g(u_1) - 1\}^2 \right]^{1/2} \\
&= \left[\int_0^{u_1} g^2(u) \, du - 2 \int_0^{u_1} g(u) \, du + u_1 \right. \\
&\qquad \left. + (1 - u_1)\{g^2(u_1) - 2g(u_1) + 1\} \right]^{1/2} \\
&= \left[\frac{k^{2n+2}}{(2k-1)^{2n+1}} \binom{2n}{n} I_{v_1}(2n+1) \right. \\
&\qquad + \frac{k^{2n+2}}{(n!)^2} \{-\log(1 - u_1)\}^{2n} (1 - u_1)^{2k-1} \\
&\qquad \left. - 2 \left\{ I_{v_2}(n+1) + \frac{k^{n+1}}{n!} (-\log(1 - u_1))^n (1 - u_1)^k \right\} + 1 \right]^{1/2} \\
&= B_{n(k)} \quad \text{(say)}, \tag{3.9.15}
\end{aligned}$$

where $I_p(\alpha)$ denotes the incomplete gamma ratio defined by

$$I_p(\alpha) = \frac{1}{\Gamma(\alpha)} \int_0^p e^{-t} t^{\alpha-1}\, dt, \qquad p > 0,\ \alpha > 0,$$

$v_1 = -(2k-1)\log(1-u_1)$, and $v_2 = -k\log(1-u_1)$. The bound in $(3.9.15)$ is sharp and it is attained for a population with $F^{-1}(u) \propto g_*(u) - 1$ for $0 \le u \le 1$. Determining the constant of proportionality from the condition that population variance is equal to 1, we immediately have the population for which the bound in $(3.9.15)$ is attained as

$$F^{-1}(u) = \begin{cases} \frac{1}{\sqrt{B_{n(k)}}}\{g(u) - 1\} & \text{if } 0 \le u \le u_1 \\ \frac{1}{\sqrt{B_{n(k)}}}\{g(u_1) - 1\} & \text{if } u_1 \le u \le 1. \end{cases} \qquad (3.9.16)$$

From $(3.9.10)$, upon writing $1 - G(u_1)$ in a Poisson sum form as

$$1 - G(u_1) = (1 - u_1)^k \sum_{i=0}^n \{-k\log(1 - u_1)\}^i / i!\ ,$$

we can rewrite $(3.9.13)$ as

$$\sum_{i=0}^n \{-k\log(1 - u_1)\}^i / i! = \frac{k^{n+1}}{n!}\ \{-\log(1 - u_1)\}^n$$

or, equivalently, as

$$\sum_{i=0}^{n-1} \{-k\log(1 - u_1)\}^i / i! - \frac{k^n(k-1)}{n!}\ \{-\log(1 - u_1)\}^n = 0. \quad (3.9.17)$$

u_1 needs to be determined numerically from $(3.9.17)$ in order to compute the bound in $(3.9.15)$. However, for the case $n = 1$, $(3.9.17)$ becomes

$$1 - k(k-1)\{-\log(1 - u_1)\} = 0$$

so that u_1 is determined explicitly as

$$u_1 = 1 - e^{-1/\{k(k-1)\}}, \qquad k = 2, 3, \ldots\ . \qquad (3.9.18)$$

Similarly, for the case $n = 2$, $(3.9.17)$ becomes

$$1 - k\log(1 - u_1) - \frac{k^2(k-1)}{2}\ \{\log(1 - u_1)\}^2 = 0$$

from which u_1 is determined explicitly as

$$u_1 = 1 - \exp\left[-\{1 + \sqrt{2k-1}\}/\{k(k-1)\}\right],\ k = 2, 3, \ldots . \qquad (3.9.19)$$

For larger values of n, u_1 can not be determined explicitly and hence has to be found from $(3.9.17)$ be numerical methods.

Remark 3.9.1 For the case when $k = 1$, the function $G(u)$ in (3.9.10) becomes a convex function in $0 \le u \le 1$ and hence the bound in (3.9.15) will not provide any improvement. This is to be expected as the corresponding bound in (3.8.4) is attainable.

For the purpose of illustration, let us consider the case when $k = 2$ and $n = 1$. Eq. (3.9.18) then gives $u_1 = 1 - e^{-1/2} = 0.3935$ which, when used in (3.9.15), yields the bound for $|E(R_{1(2)})|$ to be $B_{1(2)} = 0.3454$. This bound is sharp and is attained for the population with

$$
F^{-1}(u) = \begin{cases} \frac{1}{\sqrt{0.3454}} \, 4\{-\log(1-u)\}(1-u) & \text{if } 0 \le u \le 0.3935 \\[2mm] \frac{0.2131}{\sqrt{0.3454}} & \text{if } 0.3935 \le u \le 1. \end{cases}
$$

$$(3.9.20)$$

Correspondingly, we have the Cauchy-Schwarz bound in (3.9.6) for this case to be $\sqrt{5/27} = 0.4303$.

Table 3.9.1, taken from Raqab (1997), presents values of the bounds computed from (3.9.6) and (3.9.15) for some choices of k and n. As expected, the latter provides an improvement over the former in all cases. From the table, we further observe that the improvement is quite significant whenever n is not much larger than k [when n is much larger than k, u_1 gets closer to 1 so that the bound in (3.9.15) does not yield much of an improvement]; hence, by using the bound in (3.9.15), one may readily obtain an improved bound for the expected value of the difference of two k-records.

Note that the greatest convex minorant principle can be used in a similar manner, in order to improve the bound in (3.9.8) for the case of symmetric distributions. To be specific, let us assume, without loss of any generality, that the population distribution F is symmetric about 0. In this case, we may write

$$
E(R_{n(k)}) = \frac{k^{n+1}}{n!} \int_{1/2}^{1} F^{-1}(u) \left[\{-\log(1-u)\}^n (1-u)^{k-1} \right.
$$
$$
\left. - \{-\log u\}^n u^{k-1} \right] du.
$$

$$(3.9.21)$$

Table 3.9.1. Bounds in (3.9.6) and (3.9.15) for the Mean of the k-records, $|E(R_{n(k)})|$

k	n	u_1	Bound in (3.9.6)	Bound in (3.9.15)
2	2	0.39347	0.43033	0.34507
2	3	0.74488	0.76174	0.73613
2	5	0.96125	1.62534	1.61707
3	3	0.41687	0.63220	0.48066
3	4	0.61896	0.82439	0.73358
3	8	0.94034	1.95985	1.92171
3	10	0.97718	2.80858	2.77564
4	4	0.43299	0.76913	0.56198
4	6	0.68189	1.06685	0.94315
4	10	0.90740	1.92088	1.84058
4	15	0.98124	3.65570	3.58306
5	8	0.71236	1.24237	1.08310
5	12	0.88598	1.93498	1.81150
5	18	0.97273	3.54489	3.42668

As mentioned before, a simple application of the Cauchy-Schwarz inequality in the above integral yields the Grudzień-Szynal bound in (3.9.8). Now, by writing

$$g_1(u) = \frac{k^{n+1}}{n!G(1)} \{-\log(1-u)\}^n (1-u)^{k-1}, \qquad \frac{1}{2} \le u \le 1,$$

$$g_2(u) = \frac{k^{n+1}}{n!G(1)} \{-\log u\}^n u^{k-1}, \qquad \frac{1}{2} \le u \le 1,$$

$$g(u) = \{g_1(u) - g_2(u)\} J_{\{g_1(u) \ge g_2(u)\}}, \tag{3.9.22}$$

where

$$\begin{aligned} J_{\{g_1(u) \ge g_2(u)\}} &= \quad 1 \quad \text{if } g_1(u) \ge g_2(u) \\ &= -1 \quad \text{otherwise,} \end{aligned}$$

and

$$G(u) = \int_{1/2}^{u} g(t)\, dt,$$

it may be easily observed that the bound in (3.9.8) will be achieved iff $F^{-1}(u) \propto g(u)$ for $\frac{1}{2} \le u \le 1$. However, since $g(u)$ is monotonic only when $k = 1$, the Grudzień-Szynal bound in (3.9.8) will be sharp only in the case when $k = 1$. When $k > 1$, we can adopt the greatest convex minorant approach as before and write

$$|E(R_{n(k)})| \le \int_{1/2}^{1} G(1) F^{-1}(u)\{g_*(u) - \lambda\} du, \tag{3.9.23}$$

where

$$g_*(u) = \begin{cases} g(u) & \text{if } \frac{1}{2} \le u \le u_2 \\ g(u_2) & \text{if } u_2 \le u \le 1 \end{cases} \tag{3.9.24}$$

with u_2 satisfying the equation

$$1 - G(u_2) = g(u_2)(1 - u_2). \tag{3.9.25}$$

Applying the Cauchy-Schwarz inequality in (3.9.25) and then minimizing the R.H.S. with respect to λ, we obtain

$$|E(R_{n(k)})| \le \frac{G(1)}{\sqrt{2}} \left\{ \int_{1/2}^{1} \{g_*(u)\}^2 du \right\}^{1/2}$$

$$(= C_{n(k)}, \text{ say}). \tag{3.9.26}$$

Table 3.9.2. Bounds in (3.9.8) and (3.9.28) for the Mean of the k-records, $|E(R_{n(k)})|$

k	n	u_2	Bound in (3.9.8)	Bound in (3.9.28)
2	2	0.72622	0.21236	0.20702
2	3	0.78123	0.74596	0.73473
2	5	0.96125	1.34753	1.34256
3	3	0.63724	0.34841	0.32403
3	4	0.67331	0.76265	0.72328
3	8	0.94034	1.55577	1.53182
3	10	0.97717	2.10809	2.08619
4	4	0.60061	0.45829	0.41896
4	6	0.68807	0.99545	0.92843
4	10	0.90740	1.53129	1.48118
4	15	0.98124	2.67994	2.62867
5	8	0.71299	1.11873	1.03270
5	12	0.88598	1.54015	1.46311
5	18	0.97273	2.60444	2.52411

Upon using the expression of $g_*(u)$ in (3.9.26), we can also express the bound $C_{n(k)}$ in (3.9.28) in terms of incomplete gamma ratio [similar to the one for $B_{n(k)}$ in (3.9.15)]. Clearly, this bound is sharp since it is attained for symmetric distributions with $F^{-1}(u) \propto g_*(u)$ for $\frac{1}{2} \le u \le 1$.

Table 3.9.2, taken from Raqab (1997), presents values of the bounds computed from (3.9.8) and (3.9.28) for some choices of k and n. As expected, the latter provides an improvement over the former in all cases. From the table, we also observe that these two bounds provide an improvement in all cases over the corresponding bounds presented in Table 3.9.1.

Remark 3.9.2 For the class of distributions with

$$\frac{d}{du} F^{-1}(u) = c\, F^{-1}(u)\{-\log(1-u)\}^p(1-u)^{q-1}, \quad 0 < u < 1,$$

$$(3.9.27)$$

where c is the normalizing constant, Kamps (1992) has shown that

$$E(R_{n(k)}) - E(R_{n-1(k)}) = d\, E(R_{n+p(k+q)}), \qquad (3.9.28)$$

where d is some constant involving n, k, p and q. Observe that this general family of distributions includes the exponential, Pareto and Weibull distributions as particular cases for some special choices of p and q. Kamps (1992) has also established some characterizations results based on recurrence relations of this kind.

EXERCISES

1. Prove the recurrence relations in (3.2.8) and (3.2.9).

2. Prove the results presented in Theorem 3.3.1.

3. Prove the recurrence relations in (3.4.15) and (3.4.16).

4. (a) Similar to the explicit expression for the single moments of the lower record values from the Gumbel distribution presented in (3.4.18), can you derive explicit expressions for the product moments and covariances?

   (b) Verify the result in (3.4.14) from these explicit expressions.

5. For the generalized extreme value distribution in (3.4.19), prove the results presented in Theorems 3.4.3 and 3.4.4 and Remark 3.4.6.

6. (a) Similar to the explicit expression for the single moments of the upper record values from the Lomax distribution presented in (3.5.3), can you derive explicit expressions for the product moments and covariances?

   (b) Verify the results in Remark 3.5.2 from these explicit expressions.

7. For the generalized Pareto distribution with pdf

$$f(x) = (1 + \beta x)^{-(1+\beta^{-1})}$$

with supports $x \geq 0$ when $\beta > 0$ and $0 \leq x \leq -\beta^{-1}$ when $\beta < 0$, and corresponding cdf

$$F(x) = 1 - (1 + \beta x)^{-\beta^{-1}},$$

observe the relation

$$(1 + \beta x)f(x) = 1 - F(x).$$

Then, by using this relation, establish the following recurrence relation for the moments of the upper record values:

$$\alpha_{n+1}^{(k+1)} = \frac{1}{1 - (k+1)\beta} \left\{ (k+1)\,\alpha_{n+1}^{(k)} + \alpha_n^{(k+1)} \right\}$$

for $n = 0, 1, 2, \ldots$ and $k = 0, 1, 2, \ldots$, when $(k+1)\beta < 1$.

[Balakrishnan and Ahsanullah, 1994a]

8. For the generalized Pareto distribution considered in the last exercise, prove the following recurrence relations for the product moments and covariances of the upper record values:

   (a) For $m = 0, 1, 2, \ldots$ and $k_1, k_2 = 0, 1, \ldots$,

$$\alpha_{m,m+1}^{(k_1,k_2+1)} = \frac{1}{1 - (k_2+1)\beta} \left\{ (k_2+1)\alpha_{m,m+1}^{(k_1,k_2)} + \alpha_m^{(k_1+k_2+1)} \right\},$$

   (b) for $0 \leq m \leq n - 2$ and $k_1, k_2 = 0, 1, \ldots$,

$$\alpha_{m,n}^{(k_1,k_2+1)} = \frac{1}{1 - (k_2+1)\beta} \left\{ (k_2+1)\alpha_{m,n}^{(k_1,k_2)} + \alpha_{m,n-1}^{(k_1,k_2+1)} \right\}$$

when $(k_2+1)\beta < 1$.

(c) for $m = 0, 1, 2, \ldots,$

$$\sigma_{m,m+1} = \frac{1}{1-\beta} \, \sigma_{m,m};$$

(d) for $0 \le m \le n-2$,

$$\sigma_{m,n} = \frac{1}{(1-\beta)^{n-m}} \, \sigma_{m,m}.$$

(e) Deduce from these relations the results presented in Theorems 3.2.1 and 3.2.2 (for the exponential records) by letting the shape parameter β tend to zero.

[Balakrishnan and Ahsanullah, 1994a)]

9. Verify the explicit algebraic expressions for the variances and co-variances of the upper record values from the logistic distribution presented in Theorem 3.7.2.

10. Prove that the universal bounds in (3.8.4) and (3.8.8) are both sharp (by showing that there exists a distribution for which these bounds are achieved).

11. For symmetric distributions, derive the "extrapolation-type bound" in (3.8.18).

12. Prove Theorem 3.8.2.

13. (a) Derive the bounds in (3.9.6) and (3.9.8) for the expected value of k-record.

    (b) Establish that these bounds are not sharp except for the case when $k = 1$.

14. Prove that the function $g(u)$ in (3.9.24) is monotonic only when $k = 1$, thus establishing that the Grudzień-Szynal bound in (3.9.8) is sharp only for the case when $k = 1$.

15. Derive an expression for the bound $C_{n(k)}$ in (3.9.28) similar to the one for $B_{n(k)}$ in (3.9.15).

CHAPTER 4

CHARACTERIZATIONS

4.1 INTRODUCTION

Well, what is a characterization? In short, it is a condition involving certain properties of a random variable X, which identifies the associated cdf F. The property that uniquely determines F may be based on functions of random variables whose joint distribution is related to that of X. A characterization can be of use in the construction of goodness-of-fit tests and in the examination of the consequences of modeling assumptions made by an applied scientist. For example, the independence of spacings of order statistics of a random sample from a continuous distribution implies the cdf is exponential, and thus can be used to construct a goodness-of-fit test, even with a (Type-II) censored sample. As another example, consider the early characterization result, proved by Cramér in 1937, which states that if X and Y are independent and the sum is assumed to be normal, then each of them must be normal (see, e.g., Cramér, 1970, p. 53). Thus, if a researcher is willing to assume the independence of these random variables and that the sum is normal, it is the same as assuming that they are both normal and are independent. While the former set of conditions may appeal to the researcher, the statistician often finds it more convenient to use the latter set of conditions to develop inference procedures for the data.

By now a substantial number of mathematical statisticians and applied probabilists are convinced about the importance of this area. Consequently, there is a plethora of papers dealing with characterization theorems. Quite a few monographs have appeared as well. We may mention the early work of Kagan, Linnik and Rao (1973), dealing with the normal distribution, and that of Galambos and Kotz (1978), which concentrates on the exponential distribution and related models. Azlarov and Volodin (1986) also deal with exponential characterizations, while the work of Prakasa Rao (1992) puts characterization results in the framework of identifiability of stochastic models.

Within the voluminous characterization literature there are several results involving properties of record values. In the following sections we classify these findings and discuss them. We sprinkle our discussion with formal statements of some of these results in the form of theorems, often with accompanying proofs. Many of them resemble similar characterizations associated with order statistics. In an effort to provide insight into the existence of parallel characterization results, in the concluding section we briefly discuss the similarity between the dependence structures of record values and of order statistics. An elementary introduction to characterizations, based on order statistics, may be found in Arnold, Balakrishnan and Nagaraja (1992, Chapter 6).

For cataloging purposes, we find it convenient to classify the characterization results based on record values into three categories: (i) those that establish one-to-one correspondence between F and certain characteristics of R_n; (ii) those that identify classes of distributions possessing a stated property; and (iii) those that are distribution-specific and are generally motivated by some interesting properties exhibited by the record value sequence. We must point out that neither is this an established way to divide such results, nor does it classify them in a mutually exclusive way.

In Section 4.2 we cover results under (i) and Section 4.3 discusses characterizations that belong to category (ii). Sections 4.4–4.6 consider distribution-specific characterizatons. Section 4.4 is devoted to the numerous exponential characterizations. We discuss other continuous distributions in Section 4.5. Among discrete F, only the geometric parent permits simple characterizations, and we present an overview of these results in Section 4.6. As noted earlier, in the last section we compare the dependence structures of order statistics and of record values from i.i.d. sequences of random variables.

4.2 CHARACTERIZING PROPERTIES OF RECORD VALUES

Our general goal here is to discuss properties of the record value sequence $\{R_n,\ n \geq 0\}$ that will uniquely determine F, the parent cdf. A basic fact is that the cdf of R_n for any one n uniquely identifies F, as can be seen from (2.3.6). Of course, this should be obvious and, hence, is not very exciting! Let us now turn to some not so obvious ones.

4.2.1   The Moment Sequence

The historical moment problem asks the question whether the sequence of moments of a random variable X will uniquely determine its cdf (see, e.g., Shohat and Tamarkin, 1943). In the record value context, one could phrase it as follows: Does the record moment sequence $\{E(R_n), \ n \geq 0\}$ determine F uniquely? In fact, this question has been posed and solved by Kirmani and Beg. Their result can be stated as follows.

Theorem 4.2.1 (Kirmani and Beg, 1984) *Let F be a continuous cdf and assume that δth moment of F exists for some $\delta > 1$. Then F is uniquely determined by the sequence $\{E(R_n), \ n \geq m\}$, where m is a fixed positive integer.*

Proof.   The proof boils down to the completeness property of the sequence of functions $\{(-\log(1-u))^n, \ 0 < u < 1, \ n \geq 0\}$. More precisely, for any $p > 1$, if a function $\eta \in L^p(0, 1)$ satisfies the relation

$$\int_0^1 \eta(u)(-\log(1 - u))^n \, du = 0 \tag{4.2.1}$$

for all $n \geq 0$, then $\eta(u) = 0$ almost everywhere (a.e.) on $(0, 1)$. A simple proof of this basic fact may be found in Lin (1987).

Now suppose, for $i = 1, 2$, $R_n(F_i)$ is the nth upper record value from the continuous cdf F_i. In view of (2.7.2), the assumed condition of the equality of the moment sequences implies that with $\eta(u) = \{F_1^{-1}(u) - F_2^{-1}(u)\}(-\log(1-u))^m$ (4.2.1) holds for all $n \geq 0$.

Recall that we have assumed that δth moment of both F_1 and F_2 exist. Then, from Lemma 2.7.1, it follows that $E\{(R_n(F_1))^p\}$ and $E\{(R_n(F_2))^p\}$ exist for all $p < \delta$. Now take $p = (1 + \delta)/2$ and apply Hölder's inequality to claim that $\eta(u) \in L^p(0, 1)$. Hence, on using the completeness property alluded to above, we can conclude that $F_1^{-1}(u) = F_2^{-1}(u)$ a.e. In other words, $F_1 = F_2$.   ◯

The above proof owes much to the work of Lin (1987), who generalized Theorem 4.2.1 to a characterization based on the jth moment sequence where j is an arbitrary positive integer. We now ask a couple of questions regarding the assumptions made in Theorem 4.2.1.

Firstly, what is the role of the assumption of continuity? Can we relax it to include the class of all cdf's? The answer is 'No', as noted by Lin and Huang (1987). They point out that when F is exponential and G is geometric, if their means coincide, so do the entire record value moment sequences. Second, can a subsequence of the moments of the record values determine F? They provide an example where a subsequence of the moment sequence $\{E(R_n), \ n \geq m\}$ fails to characterize F. This contrasts with a comparable result for the sample maxima, where a similar moment subsequence does indeed characterize the parent cdf.

The above discussion showed that the record mean sequence uniquely determines F. What about record spacings? More precisely, for fixed positive integers m and j, does $\{E(R_{n+j} - R_n), \ n \geq m\}$ identify F? On making use of Theorem 4.2.1, Gupta (1984) shows that F is determined up to a location parameter. It is clear that for a fixed n, the value of $E(R_{n+1} - R_n)$ cannot determine F. But, Gupta notes that the condition $E(R_{n+1} - R_n) = E(X)$ does identify F in the class of *new better than used in expectation* (NBUE) and *new worse than used in expectation* (NWUE) distributions (see, e.g., Barlow and Proschan, 1981, p. 159, for definitions).

4.2.2 Regression of Adjacent Record Values

In Theorem 4.2.1 we needed a countable set of expected records of the form $\{E(R_n), \ n \geq m\}$ to identify F among continuous distributions. But a single regression function (conditional expectation) of the form $E(R_{n+1}|R_n)$ is enough to characterize F! We will see this in the following discussion.

From (2.6.3) we notice that

$$P(R_{n+1} > y \mid R_n = x) = \frac{1 - F(y)}{1 - F(x)} \ , \quad \text{for all } a < x < y < b, \quad (4.2.2)$$

where $a = F^{-1}(0)$ and $b = F^{-1}(1)$. Then, a version of the conditional expectation $E(R_{n+1}|R_n = x)$ is $K(x) = \int_x^b y\,dF(y)/\{1 - F(x)\}$. Note that $K(x)$ is a nondecreasing continuous function and whenever $F(y_1) < F(y_2)$, $K(y_1) < K(y_2)$. On integration by parts, we see that

$$K(x) = x + \int_x^b \{1 - F(y)\}dy/\{1 - F(x)\}. \quad (4.2.3)$$

Further, $K(x)$ uniquely determines F and, in fact, (4.2.3) can be solved to produce F explicitly as

$$1 - F(y) = \frac{K(a)}{K(y)} \exp\left\{ - \int_a^y [K(x) - x]^{-1} dx \right\}. \quad (4.2.4)$$

Of course, there may be nonincreasing versions of $E(R_{n+1}|R_n = x)$ if the support of F has gaps in (a, b). In fact, $E(R_{n+1}|R_n = x)$ will be completely arbitrary in such gaps. As a consequence, we can conclude the above discussion with the following result.

Theorem 4.2.2 (Nagaraja, 1977) *Suppose F is a continuous cdf. If $E(R_{n+1}|R_n = x) = K(x)$ a.s. F, and $K(x)$ is a nondecreasing function on $(F^{-1}(0), F^{-1}(1))$, then F is uniquely determined via (4.2.4).*

This characterization is very similar to the characterization of F based on the conditional mean function $E(X|X > x)$ considered by Kotlarski (1972). Note that the right side of (4.2.2) is nothing but $P(X > y|X > x)$ and thus $K(x) \equiv E(X|X > x)$.

In Theorem 4.2.2 we assumed that $K(x)$ was nondecreasing. If that assumption is removed, it is feasible to have two continuous cdf's for which $E(R_{n+1}|R_n)$ agree with $K(x)$ almost surely with respect to both the distributions. Nagaraja (1977) illustrates this possibility with an example.

From Theorem 2.10.2 it follows that the above result holds for Type 1 k-records as well. In the case of Type 2 k-records, a result generalizing Theorem 4.2.2 is provided by Grudzień and Szynal (1985). The generalization is rather direct, since, as alluded to in Section 2.10 (or from Deheuvels, 1984), we have for all $x < y$ in $(F^{-1}(0), F^{-1}(1))$,

$$ P\left(R_{n+1(k)} > y \mid R_{n(k)} = x\right) = \left\{\frac{1 - F(y)}{1 - F(x)}\right\}^k . \tag{4.2.5}$$

4.3 FAMILIES OF DISTRIBUTIONS

4.3.1 Families Defined by Reliability Properties

In the last section we noted that

$$ P(R_{n+1} > y|R_n = x) = P(X > y|X > x), \qquad x < y, \tag{4.3.1}$$

almost surely. This fact can be exploited to obtain characterizations of certain reliability properties of life distributions in terms of those of record values. Barlow and Proschan (1981) and Ross (1983) serve as useful references in this pursuit.

Note that (4.3.1) can be rephrased as $P(R_{n+1} - R_n > t|R_n = x) = P(X - x > t|X > x)$ for all $t > 0$. In view of Proposition 8.1.3 of Ross (p. 253), we may infer that X is IFR (DFR) iff the conditional distribution of the record spacing $R_{n+1} - R_n$ given $R_n = x$ is stochastically decreasing (increasing) in x. We can also make similar statements regarding NBU and *new worse than used* (NWU) properties in terms of the conditional distribution of the record spacing. Further, since $E(R_{n+1} - R_n | R_n = x) = E(X - x | X > x)$ almost surely, we can use the regression function $E(R_{n+1} | R_n)$ to define reliability properties based on the average remaining life. For example, X has *decreasing mean residual life* (DMRL) distribution iff $E(R_{n+1} - R_n | R_n)$ is decreasing in R_n.

While the above rather simple characterizations are based on the conditional distribution of a record value given the previous one, what about the marginal distribution of the records themselves? We are unaware of any necessary and sufficient conditions; however, we wish to point out the work of Kochar (1990), that explores connections between the reliability properties of X and those of R_n. Let us first establish a basic relationship between their failure rates.

Let $h_n(t)$ be the failure rate function of R_n. Denote by $H_n(t)$ its integrated failure rate function, also referred to as the hazard function; that is, $H_n(t) \equiv \int_{-\infty}^{t} h_n(u)du = -\log(1 - F_{R_n}(t))$. Further, let $h(t)$ $(= h_0(t))$ denote the failure rate of X. Then, from (2.3.5) and (2.3.7) it follows that

$$h_n(t) = h(t) \frac{\{H(t)\}^n/n!}{\sum_{k=0}^{n}\{H(t)\}^k/k!}. \tag{4.3.2}$$

Since the hazard function is uniquely determined by the cdf, (2.3.6) implies that there is a one-to-one relationship between the two failure rates. However, it is not evident from (4.3.2) how reliability properties of $h(t)$ translate into those of $h_n(t)$ and conversely. We will now elaborate on a result that shows that if R_n is DFR for some n, so is X. While this is not a characterization in the purest sense, it indicates how certain assumptions on R_n impose similar constraints on the distribution of X. To facilitate the proof, we present a lemma which is of interest in its own regard.

Lemma 4.3.1 *Let $W(\lambda)$ be a truncated Poisson random variable with parameter λ, being truncated on the right at n, where n is a positive integer. Then $W(\lambda)$ is stochastically increasing in λ.*

Proof. We have to show that, for $j < n$, $P(W(\lambda_1) \leq j) > P(W(\lambda_2) \leq j)$ whenever $\lambda_1 < \lambda_2$. This is equivalent to showing that $g(\lambda_1) < g(\lambda_2)$ where $g(u) = \left\{\sum_{k=0}^{j} u^k/k!\right\}\left\{\sum_{k=0}^{n} u^k/k!\right\}^{-1}$. This can be seen by cross-multiplying and noting the fact that $j < n$ and $\lambda_1 < \lambda_2$. ○

As a consequence of the above lemma, we can conclude that $E(W(\lambda))$ is increasing in λ, a fact that would appear obvious if $W(\lambda)$ were not truncated. It may also be noted that Kochar's Lemma 2.1 presents an algebraic proof of this conclusion. Now we are ready to formally state the result.

Theorem 4.3.1 (Kochar, 1990) *Let F be absolutely continuous. Then, if R_n is DFR for some n, the population random variable X is also DFR.*

Proof. It is sufficient if we can establish that R_{n-1} is DFR whenever R_n is. From (4.3.2) we note that

$$
\begin{aligned}
\frac{h_n(t)}{h_{n-1}(t)} &= \frac{H(t)}{n} \frac{\sum_{k=0}^{n-1}\{H(t)\}^k/k!}{\sum_{k=0}^{n}\{H(t)\}^k/k!} \\
&= \frac{1}{n} \frac{\sum_{k=0}^{n} k\{H(t)\}^k/k!}{\sum_{k=0}^{n}\{H(t)\}^k/k!} \\
&= E(W(H(t)))/n,
\end{aligned}
$$

where W is the truncated Poisson random variable introduced in Lemma 4.3.1. Since $H(t)$ is increasing with t, the ratio of the failure rates is increasing in t. Thus, if h_n is decreasing, so is h_{n-1}. ○

The above argument also implies that if h_{n-1} is increasing, so is, h_n which can be used to prove that if X is IFR, then R_n is also IFR. (See Exercise 1.22.)

4.3.2 Linear Regressions of Adjacent Record Values

In Section 4.2 we have noted that the functional form of the regression function $E(R_{n+1}|R_n)$, if nondecreasing, uniquely identifies F in the class of continuous cdf's with finite $E(R_{n+1})$. We can get more specific about the form of the regression function, say linear. Such an assumption leads to the following result, reported in Nagaraja (1977).

Lemma 4.3.2 *Let F be a continuous cdf. If, for some constants c and d, $E(R_{n+1}|R_n) = cR_n + d$, a.s., then, except for a change of location and scale,*

(i) $F(x) = 1 - (-x)^\theta, \quad -1 < x < 0, \quad$ *if* $0 < c < 1$
(ii) $F(x) = 1 - e^{-x}, \quad x > 0, \quad$ *if* $c = 1$ (4.3.3)
(iii) $F(x) = 1 - x^\theta, \quad x > 1, \quad$ *if* $c > 1$,

where $\theta = c/(1 - c)$. Observe that $c > 0$ always.

Since the assumed form of the regression function is monotonic, the above lemma essentially follows from (4.2.4). Note that the first cdf in (4.3.3) is that of $-X$ where X has a power function cdf, while the last one corresponds to a Pareto distribution. In Section 2.4 we observed the linearity of the regression function for these distributions.

We can now reverse the roles of R_{n+1} and R_n and ask the question: For what distributions is the regression function $E(R_n|R_{n+1})$ linear? One might anticipate families of distributions similar to the ones given by Lemma 4.3.2. To investigate this class, we need to know the conditional distribution of R_n given R_{n+1}. It can be obtained from the conditional pdf of $R_0, \ldots, R_n$ given R_{n+1} described in Section 2.3. For convenience, we recall its implication in the following lemma taken from Nagaraja (1988b).

Lemma 4.3.3 *Assume F is continuous. The conditional joint distribution of $R_0, \ldots, R_n$ given $R_{n+1} = y$, is identical with the joint distribution of the order statistics from a random sample of size $n+1$ from the cdf*

$$F_1(x|y) = \begin{cases} \{H(x)/H(y)\}, & x < y \\ 1, & x \geq y, \end{cases} \qquad (4.3.4)$$

where H is the hazard function of F.

As a consequence of the above result, we can make use of the results of Ferguson (1967), who considered the problem of linear regression in the context of adjacent order statistics. Following that approach, we can conclude that if $E(R_n|R_{n+1}) = cR_{n+1} + d$ a.s., then, except for a change of location and scale,

(i) $F(x) = 1 - \exp(-x^\theta), \ x > 0, \text{ if } 0 < c < 1$

(ii) $F(x) = 1 - \exp(-\exp(x)), \ -\infty < x < \infty, \text{ if } c = 1$ (4.3.5)

(iii) $F(x) = 1 - \exp(-(-x)^\theta), \ x < 0, \text{ if } c > 1,$

where $\theta = c/\{n(1-c)\}$. Here again, c is always positive.

It is interesting to note that the cdf's in (4.3.5) represent the class of possible limit distributions for appropriately normalized sample minimum of a random sample from an arbitrary cdf. (see, e.g., Arnold, Balakrishnan and Nagaraja, 1992, p. 213). Having done all this work, a natural question left to be answered is the following: For what distributions are both $E(R_{n+1}|R_n)$ and $E(R_n|R_{n+1})$ linear in the conditioning random variables? The answer is given by the following result. It represents a characterization of the exponential distribution.

Theorem 4.3.2 (Nagaraja, 1988b) *Let F belong to the class of continuous cdf's for which $E(R_n)$ and $E(R_{n+1})$ are both finite for some $n \geq 0$. If the regressions $E(R_{n+1}|R_n)$ and $E(R_n|R_{n+1})$ are both linear in the conditioning variables, then F is an exponential cdf, possibly with a shift.*

Of course, exponential characterizations indeed are a dime a dozen. We now consider a proper subset of the available results.

4.4   THE EXPONENTIAL DISTRIBUTION

4.4.1   An Incomplete Catalog

"Exponential land" has fertile soil for the cultivation of characterizations. As a consequence, numerous such results have sprouted over the past thirty years or so. In the record value arena, the earliest characterization result was due to Tata (1969), who showed that among absolutely continuous distributions, R_0 and $R_1 - R_0$ are independent iff F is exponential. In this section we attempt to give the flavor of the available results rather than elaborating on each of them.

We begin with a characterization of the exponential parent based on the maximal correlation between record values. This result appears in Nevzorov (1992), as well as in Rohatgi and Székely (1992).

Theorem 4.4.1 *Let F be a continuous cdf and $n > m \geq 0$. Suppose $Var(R_m)$ and $Var(R_n)$ are both finite. Then the correlation of R_m and R_n does not exceed $\sqrt{(m+1)/(n+1)}$. Further, the upper bound is attained iff F is an exponential cdf, possibly with a shift.*

Proof. We follow the work of Rohatgi and Székely, who have used an interesting argument involving a general result by Sarmanov (1958). He showed that for random variables X and Y with finite variances, if $E(Y|X)$ and $E(X|Y)$ are both linear, then the maximum value of $Corr(g(X), s(Y))$ for arbitrary functions g and s is attained when both g and s are linear. Note that $E(R_m^*|R_n^*)$ and $E(R_n^*|R_m^*)$ are both linear. Further, for an arbitrary continuous cdf F, $Corr(R_m, R_n) = Corr(\psi_F(R_m^*), \psi_F(R_n^*))$ where ψ_F is defined by (2.3.14). Consequently the theorem holds. $\bigcirc$

Nevzorov (1992) proved Theorem 4.4.1 for Type 2 k-records. It is easy to use the above approach to prove the general case as $R_{n(k)}$ from the cdf

F behaves like R_n from the cdf $F_{1:k}$ and when F is exponential, so is $F_{1:k}$. Nevzorov's proof is based on expansions involving Laguerre polynomials and it parallels a similar result for order statistics due to Székely and Móri (1985).

A large number of exponential characterizations have been based on independence or identical distribution of certain functions of record values. From the extensive discussion in Section 2.3, we know that if F is an exponential cdf, the following results hold for $j \leq m < n$: (a) R_m and $R_n - R_m$ are independent, (b) R_j and $R_n - R_m$ are independent, (c) $E(R_{n+1} - R_n | R_n)$ does not depend on R_n, (d) $\text{Var}(R_{n+1} - R_n | R_n)$ does not depend on R_n, (e) $R_n - R_m$ and R_{n-m-1} are identically distributed, and (f) $E((R_n - R_m)^s | R_j)$ does not depend on R_j. One can imagine several other manifestations of the i. i. d. nature of the spacings of the upper record values from an exponential distribution. Each of the above properties has been shown to provide a characterization of the exponential distribution in an appropriately chosen class of cdf's.

First we attempt to catalog the available characterization results. Then we discuss a powerful result regarding the integrated Cauchy functional equation due to Lau and Rao (1982). As we see in Section 4.4.3, it can be used to provide several of these characterizations with minimal assumptions on F. We present a result involving first lower record statistics in Section 4.4.4.

In presenting the following list, we note that sometimes several authors proved the same result around the same time, while some results were improved over time with reduced restrictions on F. Some of these papers contain several other characterization results as well. Finally, we remind the reader that by no means can we claim the collection to be exhaustive.

(a) (i) $n = m + 1$ case: Tata (1969), Ahsanullah (1978, 1979), Pfeifer (1982), Lau and Rao (1982), Deheuvels (1984) $(R_{n(k)})$, Witte (1988).

     (ii) $n > m + 1$ case: Srivastava (1981b), Taillie (1981), Nagaraja (1982), Dallas (1981a, 1982), Rao and Shanbhag (1986), Ahsanullah (1987a).

(b) Nayak (1981).

(c) Nagaraja (1977), Ahsanullah (1978), Gupta (1984), Grudzień and Szynal (1985) $(R_{n(k)})$.

(d) Ahsanullah (1981a), Dallas (1981b).

(e)   (i) $n = m + 1$ case : Lau and Rao (1982), Witte (1988).

   (ii) $n > m + 1$ case: Ahsanullah (1981b, 1982, 1987b).

(f) Gupta (1984), Roy (1990).

As we conclude our catalog, we note that some early characterization results for the exponential distribution that are based on record values are discussed in monographs by Galambos and Kotz (1978) and Azlarov and Volodin (1986).

4.4.2   Integrated Cauchy Functional Equation

A beautiful account of the role of functional equations in probability theory in general and in characterization theorems in particular is given by Ramachandran and Lau (1991). They provide a good compilation of the literature on the *integrated Cauchy functional equation* (ICFE). The survey papers of Rao and Shanbhag (1986; 1998) and their monograph (Rao and Shanbhag, 1994) also provide extensive discussions on the ICFE. Our treatment draws heavily upon these sources.

A good starting point for the introduction of ICFE would be to recall what a *Cauchy functional equation* (CFE) is and how handy it is in proving some basic exponential characterizations. On $\mathbf{R}^+$ $(\equiv [0, \infty))$ it is of the form

$$g_0(x + y) = g_0(x)g_0(y), \quad \text{for all } x, y \in \mathbf{R}^+, \qquad (4.4.1)$$

and the characterization of the exponential distribution through the memoryless property is one of its well-known applications.

Now the ICFE can be described in the most general form in the exponential characterization context as follows. We say that g_1 satisfies an ICFE if it is a function defined on $\mathbf{R}^+$ and the following condition holds:

$$g_1(x) = \int_{\mathbf{R}^+} g_1(x + y) d\mu(y), \qquad (4.4.2)$$

for all $x \in \mathbf{R}^+$, where μ is a σ-finite measure on $\mathbf{R}^+$. Note that if g satisfying (4.4.1) is integrable with respect to a σ-finite measure μ_1 on $\mathbf{R}^+$, the integral being a positive quantity c, then it also satisfies (4.4.2) with $\mu = c^{-1}\mu_1$. Lau and Rao (1982) obtained the form of the general solution to the ICFE on $\mathbf{R}^+$. The following refinement of their result is taken from Rao and Shanbhag (1994, p. 29).

Theorem 4.4.2 (Lau-Rao Theorem) *Let g be a non-negative locally integrable function on $\mathbf{R}^+$ that is not a function identically equal to 0 a.e. μ_0, the Lebesgue measure. Suppose g satisfies (4.4.2) a.e. μ_0 for x in $\mathbf{R}^+$, where μ is a σ-finite measure on $\mathbf{R}^+$ satisfying the condition $\mu(\{0\}) < 1$. That is,*

$$g(x) = \int_{\mathbf{R}^+} g(x+y)\, d\mu(y), \qquad a.e. \ \mu_0 \ for \ x \in \mathbf{R}^+. \qquad (4.4.3)$$

Then either μ is arithmetic with some span λ or it is nonarithmetic. In the former case,

$$g(x + n\lambda) = g(x)b^n, n \geq 0, \qquad a.e. \ \mu_0 \ for \ x \in \mathbf{R}^+,$$

where b satisfies the condition $\sum_{n=0}^{\infty} b^n \mu(\{n\lambda\}) = 1$. When μ is nonarithmetic,

$$g(x) \propto e^{\alpha x}, \qquad a.e. \ \mu_0 \ for \ x \in \mathbf{R}^+, \qquad (4.4.4)$$

where α satisfies the condition

$$\int_0^\infty e^{\alpha y}\, d\mu(y) = 1.$$

Several proofs of the above theorem are available. See Chapter 2 of Rao and Shanbhag (1994) or Chapter 2 of Ramachandran and Lau (1991) for details. When $\mathbf{R}^+$ in (4.4.3) is replaced by the real line, the solution of the resulting ICFE can be obtained as a special case of a general result due to Deny (1961).

When the measure μ in Theorem 4.4.2 is arithmetic, we obtain a special case originally due to Shanbhag (1977) often referred to as Shanbhag's Lemma. It serves as an important tool in the characterization of the geometric distribution based on record values (and order statistics). While we will discuss geometric characterizations in Section 4.6, below we give a version of that result convenient for our purpose. For a slightly more general form, see Rao and Shanbhag (1998).

Theorem 4.4.3 (Shanbhag's Lemma) *Let $\{v_n, \ n \geq 0\}$ and $\{w_n, \ n \geq 0\}$ be two sequences of non-negative real numbers such that $v_n \neq 0$ for at least one n, and $w_1 \neq 0$. Then*

$$v_m = \sum_{n=0}^{\infty} v_{m+n} w_n, \qquad m \geq 0, \qquad (4.4.5)$$

iff

$$v_n = v_0 b^n, \ n \geq 1 \ and \ \sum_{n=0}^{\infty} w_n b^n = 1 \qquad for \ some \ b > 0.$$

4.4.3 Characterizations Based on the ICFE and Other Functional Equations

Now we will see how some of the results cataloged in Section 4.4.1 can be proved using Theorem 4.4.2.

Example 4.4.1 (Independence property) As a simple illustration, let us revisit Tata's (1969) pioneering work which claimed that among absolutely continuous distributions with support $\mathbf{R}^+$, the independence of $R_1 - R_0$ and R_0 implies F is exponential. From (4.2.2) it is easy to see that if $P(R_1 - R_0 > y | R_0 = x) = P(R_1 - R_0 > y)$ for all y in $\mathbf{R}^+$ then

$$1 - F(x + y) = \{1 - F(x)\} \; P(R_1 - R_0 > y)$$

for all x and y in $\mathbf{R}^+$. If we assume $F(0) = 0$, this implies $P(R_1 - R_0 > y) = 1 - F(y)$ and consequently the function $g_0(x) = 1 - F(x)$ satisfies the CFE. Thus, we have an exponential characterization. $\bigcirc$

More generally, for a fixed $n \geq 1$, if $R_{n+1} - R_n$ and R_n are independent, then again F must be exponential. Ahsanullah (1978) showed this assuming F is absolutely continuous. One can provide an exponential characterization by assuming F to be merely a continuous cdf. The assumption of independence of $R_{n+1} - R_n$ and R_n can be converted into an ICFE and thus can be handled by Theorem 4.4.2 (see, e.g., Ramachandran and Lau, 1991, pp. 42–43). In fact, on fine-tuning the work of Lau and Rao, Witte (1988) shows that, if $F(F^{-1}(0)) = 0$, without any further assumptions on F, just the independence of $R_{n+1} - R_n$ and R_n implies that F is exponential!

Example 4.4.2 (Constant regression) Now suppose the regression of $R_1 - R_0$ on R_0 is a constant. This is obviously weaker than assuming independence, but here we need to assume that the first moment of $R_1 - R_0$ is finite. From the general result given in Lemma 4.3.2 we can then conclude that F must be exponential. But if we proceed directly, we can see an ICFE almost immediately. Note that

$$E(R_1 - R_0 | R_0 = x) = \int_0^\infty \frac{1 - F(x + y)}{1 - F(x)} \, dy$$

and hence if this is a constant, say, c, we have $cg(x) = \int_0^\infty g(x+y) \, dy$ with $g(x) = 1 - F(x)$. That is, (4.4.3) holds and the associated μ measure is proportional to the Lebesgue measure. Hence (4.4.4) implies that F is an exponential cdf. $\bigcirc$

Next, let us look at a characterization based on identical distributions. Ahsanullah (1979) showed that in the class of continuous distributions having the NWU or NBU property, if $R_{n+1} - R_n$ and X are identically distributed, then F must be exponential. Once again Theorem 4.4.2 comes in handy to dispense with the constraints on the reliability properties of F.

Example 4.4.3 (Identical distribution) Let the population cdf F be continuous and suppose $R_{n+1} - R_n$ also has cdf F. This assumption can be converted into an ICFE by expressing the cdf of $R_{n+1} - R_n$ as an integral by conditioning on R_n and using (4.2.2). In other words, the statement $X \overset{d}{=} R_{n+1} - R_n$ is equivalent to the condition

$$1 - F(x) = \int_0^\infty (1 - F(x+y)) \, \frac{dF_{R_n}(y)}{1 - F(y)} \,, \qquad x \geq 0. \qquad (4.4.6)$$

Equation (4.4.6) above is a special case of (4.4.3) obtained by taking $g(x) = 1 - F(x)$ and $d\mu(y) = dF_{R_n}(y)/(1 - F(y))$. Thus, Theorem 4.4.2 is applicable and we obtain an exponential characterization. Witte (1988), on following this path, notes that, with no other assumptions, the distributional equality itself implies F must be either exponential or of geometric type (with a lattice as its support). ○

Rao and Shanbhag (1994, Section 8.2; 1998) discuss applications of Theorem 4.4.2 that yield several exponential and geometric characterizations based on record values. Using the theorem as a basic tool, they indicate simpler proofs of many known results. We will explore some geometric characterizations in Section 4.6.

Huang and Li (1993) obtain several exponential characterizations based on regressions of functions of the spacings. These generalize some of the earlier results, and involve ICFE's and similar functional equations. For example, in Lemma 4.3.2 we have seen that the statement $E(R_{n+1} - R_n | R_n = x) = c$ for all $x > 0$, implies that F is exponential. What if we impose the condition that $E(g(R_{n+i+1} - R_{n+i}) | R_n = x) = c$, where g is a non-negative nondecreasing function? Assuming F is absolutely continuous, under some additional conditions, Huang and Li obtain an ICFE resulting in an exponential characterization (see Exercise 4.11.) They also consider conditions of the type $E(g(R_{n-i+1} - R_{n-i}) | R_n = x) = E(g(R_{n-i} - R_{n-i-1}) | R_n = x)$, where $1 \leq i < n$. Such a constraint reduces to a functional equation solved by Lau and Prakasa Rao (1990) yielding us an exponential solution. The order statistics property of the conditional distribution of the record values described in Lemma 4.3.3 plays a prominent role in this development.

4.4.4 Lower Record Statistics

While there are plenty of characterizations of the exponential distribution involving properties of upper record values, such is not the case with lower records. There is, however, one such characterization in the literature. It is easy to check that when F is exponential,

$$T_1' R_1' \overset{d}{=} X, \qquad (4.4.7)$$

where R_1' and T_1' denote the first lower record value and record time, respectively. Ahsanullah and Kirmani (1991) proved that if (4.4.7) holds, under certain regularity conditions, F must necessarily be an exponential cdf. The following discussion is a refinement of their work.

First, let us obtain the distribution of $Y \equiv T_1' R_1'$, assuming F is continuous. Let $\bar{F} = 1 - F$ and F_Y denote the cdf of Y. For y real,

$$
\begin{aligned}
\bar{F}_Y(y) &= \sum_{n=2}^{\infty} P\left(T_1' R_1' > y, \ T_1' = n\right) \\
&= \sum_{n=2}^{\infty} P\left(y/n < X_n < X_1; \ X_1 \le X_2, \ldots, X_{n-1}\right) \\
&= \sum_{n=2}^{\infty} \int_{y/n}^{\infty} \{\bar{F}(y/n) - \bar{F}(x)\}\{\bar{F}(x)\}^{n-2}\, dF(x)
\end{aligned}
$$

on conditioning with respect to X_1. Hence we may write

$$\bar{F}_Y(y) = \sum_{n=2}^{\infty} \left\{\bar{F}\left(\frac{y}{n}\right)\right\}^n \frac{1}{n(n-1)}. \qquad (4.4.8)$$

Incidentally, we may recall (2.6.2) to note that

$$P\left(T_1' = n\right) = \frac{1}{n(n-1)}, \qquad n \ge 2. \qquad (4.4.9)$$

We now prove that if (4.4.7) holds, $a \equiv F^{-1}(0)$ is necessarily 0. The proof is by contradiction. If there exists a $y < 0$ such that $\bar{F}(y) < 1$, $\bar{F}(y/n) \le \bar{F}(y) < 1$, and consequently $\{\bar{F}(y/n)\}^n < \bar{F}(y)$ for all $n \ge 2$. This, in view of (4.4.9), contradicts (4.4.7). If $a > 0$, it is easily seen that $\bar{F}(2a) < 1$, while $\bar{F}_Y(2a) = 1$ from (4.4.8). Again, (4.4.7) is contradicted.

Theorem 4.4.4 (Ahsanullah and Kirmani, 1991) *Let F be a continuous cdf satisfying the condition*

$$\lim_{x \to 0+} \frac{F(x) - F(0)}{x} = \lambda, \qquad (4.4.10)$$

where λ is finite and positive. If $T_1' R_1'$ also has cdf F, then F is necessarily exponential (with mean $1/\lambda$).

Proof. We have already shown that if (4.4.7) holds, $F^{-1}(0) = 0$. Define $u(x) = H(x)/x$, for $x > 0$, where H is the hazard function associated with F and take $u(0) = \lim_{x \to 0+} u(x)$. Note that (4.4.10) implies that $u(0) = \lambda$. Also, (4.4.8) can be expressed in terms of $u(x)$ as

$$e^{-xu(x)} = \sum_{n=2}^{\infty} e^{-xu(x/n)} \frac{1}{n(n-1)} , \qquad x \geq 0. \qquad (4.4.11)$$

We will now show that $u(x) = \lambda$ for all $x \geq 0$, where $u(x)$ satisfies (4.4.11). For this purpose, choose a $t > 0$, and let $a_0 = \min\{u(x) : x \in [0, t]\}$ and $a_1 = \max\{u(x) : x \in [0, t]\}$. Define $x_i = \inf\{x : x \in [0, t] \text{ and } u(x) = a_i\}$, $i = 1, 2$. Since $u(x)$ is continuous, the x_i's are well defined and $u(x_i) = a_i$. Note that as t is arbitrary, if we can show that $x_0 = x_1 = 0$, it follows that $u(x) = \lambda$ for all $x \geq 0$.

If $x_0 > 0$, for $0 < x < x_0$, $u(x) > u(x_0)$ and consequently, $x_0 u(x_0) < x_0 u(x_0/n)$. Hence, (4.4.11) fails to hold for $x = x_0$. Therefore, x_0 must be 0. Similarly we can show that x_1 is also 0.   ○

Actually, Ahsanullah and Kirmani assume a priori that $F(0) = 0$ and tacitly use (4.4.10). Recently, Basak (1996) has obtained a generalization of the above theorem that involves k-record statistics. (See Exercise 4.14.)

A result for order statistics comparable to Theorem 4.4.4 states that if $nX_{1:n}$ and X are identically distributed for some $n \geq 2$ and (4.4.10) holds, then F must be exponential (see Galambos and Kotz, 1978, p. 39 and references therein).

4.5   OTHER CONTINUOUS DISTRIBUTIONS

The exponential characterizations given in Section 4.4 can be rephrased as characterizations of other continuous distributions using the identity $-\log(1 - F(R_n)) \equiv H(R_n) \stackrel{d}{=} R_n^*$. For example, if R_{n+1}/R_n and R_n are independent, then F must be a Pareto cdf. In Section 2.4.3 we noted that these two random variables are indeed independent for the Pareto distribution.

Several results pertaining to the power function, Weibull, and Pareto cdf's that parallel the exponential characterizations are possible. In fact, each of the properties of record values from these cdf's we observed in Section 2.4 provides a characterization of the respective cdf. In the case of extreme value parent with cdf given by (2.4.10), $E(R_n - R_{n+1}|R_{n+1})$

does not depend on R_{n+1} and, in fact, from (2.4.12) it can be seen that $R_m - R_n$ and R_n are independent for $m < n$. (Prove it!) The converse result is also true and it yields the following characterization:

Theorem 4.5.1 *Let F be a continuous cdf with support $(-\infty, \infty)$. If $R_m - R_n$ and R_n are independent for some $m < n$, then, except for a change of location and scale,*

$$F(x) = 1 - \exp(-\exp(x)), \qquad -\infty < x < \infty. \qquad (4.5.1)$$

Proof. The idea behind the proof is contained in Lemma 4.3.3. Note that given $R_n = y$, R_m behaves like the mth order statistic from a random sample of size n from the cdf F_1 defined in (4.3.4). On using the incomplete beta form for the cdf of an order statistic (see, e.g., Arnold, Balakrishnan and Nagaraja, 1992, p. 13), we see that, for all $x \leq 0$ and real y,

$$
\begin{aligned}
P(R_m - R_n &\leq x | R_n = y) \\
&= P(R_m \leq x + y | R_n = y) \\
&= \frac{1}{B(n, m-n+1)} \int_0^{H(x+y)/H(y)} u^{m-1}(1-u)^{n-m}\, du.
\end{aligned}
$$

The assumed independence condition implies the expression on the right side is free of y for all $x \leq 0$ and real y. This means $H(x+y)/H(y) = g(x)$, say. On putting $y = 0$, we obtain $g(x) = H(x)/H(0)$ and thus we have

$$g(x + y) = g(x)g(y) \qquad (4.5.2)$$

for all $x \leq 0$ and real y. Since the relationship in (4.5.2) is symmetric in x and y, it holds for all real x, y. Thus, we have a CFE on the real line and hence $g(x) = \exp(\alpha x)$ for some α. Under our assumptions $g(x)$ is increasing and consequently $\alpha > 0$. Now $H(x) = H(0)\exp(\alpha x)$ and, on recalling that $H(x) = -\log(1 - F(x))$, we may conclude that $1 - F(x) = \exp(-c\exp(\alpha x))$, where $c = -\log(1 - F(0))$ is real and positive. Thus $F(\alpha^{-1}(x - \log(c)))$ is the expression given on the right side of (4.5.1). ◯

When X has the cdf F in (4.5.1), $-X$ has Gumbel cdf G_3 given by (2.3.23). While studying the asymptotic properties of upper extremes, Weissman (1978) has proved a result similar to Theorem 4.5.1 that effectively characterizes Gumbel distribution using an independence property of lower record values. Recently, Arnold and Villaseñor (1997) have provided another characterization of the extreme value cdf. They have used the properties of the scaled spacings $R_1 - R_0$ and $2(R_2 - R_1)$ for this purpose.

An interesting characterization of a cdf that is right tail equivalent to a Pareto cdf occurs when we impose the condition $E(X_1|R_1 = x) = E(X_2|R_1 = x)$ for some values of x. In order to compute these conditional expectations one needs to know the corresponding conditional distributions. Since $X_1 \equiv R_0$, Lemma 4.3.3 provides an expression for the conditional cdf of X_1 given R_1. We now present the conditional pdf of X_2 given R_1.

Lemma 4.5.1 (Nagaraja and Nevzorov, 1997) *Let F be an absolutely continuous cdf with pdf f. Then, for any y for which $f(y) > 0$, the conditional cdf of X_2 given $R_1 = y$ has an atom $F(x)/H(y)$ at the point $x = y$ and for $x \neq y$, $f(x|y) = \frac{dF(x|y)}{dx}$ has the form*

$$
f(x|y) = \begin{cases} f(x) \, \frac{H(y)-H(x)}{H(y)}, & \text{if } x < y \\[2mm] 0, \text{ if } x > y, \end{cases}
\tag{4.5.3}
$$

where H is the hazard function of F.

Proof. When $x = y$, the event $\{X_2 \in (y, y+dy), \ R_1 \in (y, y+dy)\}$ is nothing but $\{X_1 < X_2, \ X_2 \in (y, y+dy)\} \cup \{y < X_2 < X_1 < R_1 < y+dy\}$. Hence

$$
P\{X_2 \in (y, y+dy), \ R_1 \in (y, y+dy)\} = F(y)f(y)dy + o(dy).
$$

On recalling $f_{R_1}(y) = H(y)f(y)$, we conclude that the first claim made above holds. For $x + dx < y$, the event

$$
\{X_2 \in (x, x+dx), \ R_1 \in (y, y+dy)\}
$$
$$
\equiv \bigcup_{k=3}^{\infty} \{X_2 \in (x, x+dx); \ X_2 \le X_1 < X_k; \ X_3, \ldots, X_{k-1} \le X_1;
$$
$$
X_k \in (y, y+dy)\},
$$

and consequently we can write

$$
P(X_2 \in (x, x+dx), \ R_1 \in (y, y+dy))
$$
$$
= \sum_{k=3}^{\infty} \int_x^y f(x)f(y)\{F(z)\}^{k-3} f(z) \, dz \, dx \, dy
$$
$$
+ o(dx \, dy).
$$

Hence for $x < y$ in the support of F, the joint pdf of X_2 and R_1 is given by

$$
f_{X_2, R_1}(x, y) = f(x)f(y)\{H(y) - H(x)\}.
$$

Consequently, (4.5.3) also holds.   ○

The above lemma can be used to prove the following result, the proof of which is omitted here.

Theorem 4.5.2 (Nagaraja and Nevzorov, 1997) *For any real x_0, there exists a continuous cdf F such that for all $x \geq x_0$,*

$$E(X_1|R_1 = x) = E(X_2|R_1 = x). \qquad (4.5.4)$$

The cdf F satisfying (4.5.4) has the form given by

$$cF^{-1}(u) + d = \log\left(\frac{u}{1-u}\right) + \frac{1}{1-u}, \quad 0 < F(x_0) \leq u < 1, \quad (4.5.5)$$

for some constants $c > 0$ and d.

Note that, as $u \to 1$ in (4.5.5), $F^{-1}(u) \propto 1/(1-u) = F_0^{-1}(u)$, where $F_0(u)$ is the Pareto$(1,1)$ cdf. As a consequence, the cdf F satisfying (4.5.4) has infinite expectation!

4.6   GEOMETRIC-TAIL DISTRIBUTIONS

We have already noted in Section 2.8 some of the difficulties involved in obtaining closed-form expressions for the marginal as well as the joint distributions of nonconsecutive record values from discrete distributions. However, the stationary Markov property expressed in (2.6.3) still holds. Hence, only the characterizing properties involving adjacent records appear to be tractable. The exception, of course, is the geometric distribution. This explains the existence of numerous characterizations of the geometric distribution, while other distributions are hardly touched.

With $n = 0$ and $m = 1$ the properties (a), (c), (d), (e) and (f) listed in Section 4.4.1 hold for the geometric distribution. To be precise, when X is Geo(p), or when $f(x) = p(1-p)^{x-1}$, $x = 1, 2, \ldots$, the representation in (2.9.2) implies that the following properties hold:

(i) *Independence:* $R_0, R_1 - R_0, R_2 - R_1, \ldots$ are independent random variables.

(ii) *Constant Regression:* $E(R_{n+1} - R_n \mid R_n)$, $E(R_{n+2} - R_{n+1} \mid R_n)$, and $E[(R_1 - R_0)^2 \mid R_0]$ are constants.

(iii) *Identical Distribution:* $R_{n+1} - R_n \stackrel{d}{=} R_0$, $n \geq 0$.

Each of the above properties is shown to be a characteristic property of the geometric or geometric-tail distributions (to be defined below) in the class of distributions having support on positive integers. For details, see Srivastava (1979, 1981a, 1981b), Mohan and Nayak (1982), Ahsanullah and Holland (1984), Deheuvels (1984), Nagaraja, Sen and Srivastava (1989), Stepanov (1990), and Balakrishnan and Balasubramanian (1995). Common techniques used in the proofs of such results are recursion, differencing, and the Shanbhag Lemma (Theorem 4.4.3). As discussed in Section 4.4.3, independence of $R_{n+1} - R_n$ and R_n, as well as the condition $R_{n+1} - R_n \stackrel{d}{=} R_0$, reduces to ICFE for any arbitrary F. Then, from Lau-Rao theorem, the solution turns out to be either exponential or geometric.

We now provide a sample of two results from the literature. The first one involves a simple differencing technique. The second one illustrates an application of Theorem 4.4.3. First, let us develop some notation. We say a cdf F or the associated random variable X has a *geometric-tail distribution*, and write F or X is GeoT(j,p) if for some positive integer j, $f(x) = pP(X \geq x) \equiv p\{1 - F(x-)\}$, $x \geq j$. If X is GeoT(j,p), then $f(x)$ is proportional to $(1 - p)^x$ for all $x \geq j$. Finally, with $j = 1$, a geometric-tail distribution reduces to a geometric distribution.

We now present one of the earliest characterizations of a geometric-tail distribution based on record values.

Theorem 4.6.1 (Srivastava, 1979) *Let F be a cdf with support on the set of positive integers, and have a finite mean. Then $E(R_1 - R_0|R_0 = y) = c$, $y \geq 1$, implies F is a GeoT$(2, 1/c)$ cdf.*

Proof. Since $P(R_1 - R_0 = x \mid R_0 = y) = f(x+y)/\bar{F}(y)$, the constant regression condition can be expressed as

$$\sum_{x=1}^{\infty} xf(x+y) = c\bar{F}(y), \qquad y \geq 1. \qquad (4.6.1)$$

On taking first-order differences on both sides of (4.6.1) we obtain $P(X \geq y) = cf(y)$, $y \geq 2$, where clearly $c > 1$. Further, this means F is GeoT$(2, p)$ with $p = 1/c$.   ○

Huang and Li (1993) consider implications of the assumption that $E(g(R_n - R_m)|R_j)$ is a constant, where $0 \leq j \leq m < n$ are fixed and the monotonic function g satisfies some conditions that are of technical nature. It leads to another characterization of geometric-tail distributions. (See Exercise 4.22.)

Theorem 4.6.2 (Nagaraja, Sen and Srivastava, 1989) *Let F be a cdf with support on the set of non-negative integers. Assume $F^{-1}(1) = \infty$ and $f(x) > 0$, $x = 0, 1, \ldots, n+1$. If the event $\{R_n = n\}$ and the random variable $R_{n+1} - R_n$ are independent, then F is a $Geo\,T(n+1, p)$ cdf for some p, $0 < p < 1$.*

Proof. Since $F^{-1}(1)' = \infty$, the record values involved are well defined. Further, $f(x) > 0$, $x = 0$ to $n+1$, implies $P(R_n = n) > 0$ and $P(R_{n+1} - R_n = 1) > 0$. The independence assumption then leads to the conclusion that $f(x) > 0$ for all $x \geq 0$.

Now, for $x \geq 1$, and $y \geq n$,

$$
\begin{aligned}
P(R_{n+1} - R_n = x,\ R_n = y) &= P(R_{n+1} = x + y \mid R_n = y)P(R_n = y) \\
&= \frac{f(x+y)}{\bar{F}(y)}\,P(R_n = y),
\end{aligned}
$$

is always nonzero. If the event $\{R_n = n\}$ and $R_{n+1} - R_n$ are independent, we then have,

$$
\begin{aligned}
\{f(x+n)/\bar{F}(n)\} &= P(R_{n+1} - R_n = x) \\
&= \sum_{y=n}^{\infty}\{f(x+y)/\bar{F}(y)\}P(R_n = y), \quad x \geq 1.
\end{aligned}
$$

$$(4.6.2)$$

On putting $m = x - 1$, $j = y - n$ and rearranging (4.6.2) we obtain

$$
f(m+n+1) = \sum_{j=0}^{\infty} f(m+n+j+1)\{P(R_n = n+j)\bar{F}(n)/\bar{F}(n+j)\},
$$

$$m \geq 0. \qquad (4.6.3)$$

As n is fixed, denote $f(m+n+1)$ by v_m and $\{P(R_n = n+j)\bar{F}(n)/\bar{F}(n+j)\}$ by w_j. Since $f(y) > 0$ for all $y \geq 0$, $v_m > 0$ for $m > 0$ and $w_1 > 0$. Thus, (4.6.3) takes on the form

$$
v_m = \sum_{j=0}^{\infty} v_{m+j} w_j, \qquad m \geq 0,
$$

and conditions of Theorem 4.4.3 hold. On applying that result we conclude that $f(n+j+1) = f(n+1)\beta^j$, $j \geq 0$ for some $\beta > 0$. Since f has to be a pdf, β is necessarily less than 1. Therefore, the final conclusion is that F is a $GeoT(n+1, 1-\beta)$ cdf. ○

As noted earlier, for the geometric parent, R_j and $R_n - R_m$ are independent whenever $j \le m < n$. Suppose X is geometric and assume that, for fixed $j \le m < n$, $g(R_m, R_n)$ and R_j are independent for all p, $0 < p < 1$, for some function g. What can be said about the form of g? Dallas (1989) posed this question and showed that $g(R_n, R_m)$ depends on R_n and R_m only through the spacing $(R_n - R_m)$. A similar conclusion is drawn when one assumes the constancy of the regression $E(g(R_m, R_n)|R_j)$.

In Section 4.3.2, we examined the consequences of the assumption of the linearity of regression of adjacent record values among continuous cdf's. Korwar (1984) has determined the class of non-negative integer valued random variables for which $E(R_1|R_0 = x)$ has the linear form $c + dx$. This seems to be the only result whose scope goes beyond geometric-tail distributions. As observed in Theorem 4.6.1, $d = 1$ for a GeoT$(2, p)$ distribution. When $d > 1$, one obtains a generalized hypergeometric distribution. For $0 < d < 1$, even though one can solve for F, the resulting distribution would have a finite support. In that case R_1 fails to be well defined!

4.7 DEPENDENCE STRUCTURES OF RECORD VALUES AND ORDER STATISTICS

It is no surprise that there is a close connection between characterizations associated with record values and those related to order statistics. After all, a record value is a sample extreme from a sample of random size. Further, the limit distribution of a lower extreme order statistic, if it exists, coincides with that of an upper record value from one of the cdf's given in (4.3.5) (Nagaraja, 1982). In an attempt to unify characterization results based on order statistics and on record values, Deheuvels (1984) and Gupta (1984) point out an interesting resemblance in their dependence structures. For a continuous F, from (4.2.2) and a similar property of order statistics (see, e.g., Theorem 2.4.1, Arnold, Balakrishnan and Nagaraja, 1992, p. 23), we observe that

$$P(R_{n+1} > y|R_n = x) = P(X_{m:m} > y|X_{m-1:m} = x) \text{ a.s.} \qquad (4.7.1)$$

for arbitrary m and n. Further, both the record value and order statistic sequences possess the Markov property. A similar equivalence holds for $R_{n(k)}$'s, the Type 2 k-record values. To be precise, from (4.2.5) (or from Deheuvels, 1984), we may infer that $P(R_{n+1(k)} > y|R_{n(k)} = x)$

$=P(X_{m-k+1:m} > y|X_{m-k:m} = x)$ a.s. Thus the dependence structures of order statistics and record values are very similar.

We can also examine the dependence structure looking in the reverse direction. In Section 2.3 we noted that, for an exponential parent, the first n record values, conditioned on the $(n+1)$st record value, behave like a random sample from a uniform distribution. More formally, Lemma 4.3.3 indicates that $R_0, \ldots, R_{n-1}$ behave like order statistics from a random sample from the cdf F_1 given by (4.3.4). Similarly, the random variables $X_{1:n+1}, \ldots, X_{n:n+1}$ given $X_{n+1:n+1} = y$ behave like order statistics from a random sample, but from the cdf F_2 given by

$$F_2(x \mid y) = \frac{F(x)}{F(y)}, \qquad x \le y. \qquad (4.7.2)$$

Equation (4.7.1) and the discussion leading to (4.7.2) provide some insight into several interesting parallel characterizations of the uniform and exponential distributions. For example, while F is uniform iff both $E(X_{m+1:n}|X_{m:n})$ and $E(X_{m:n}|X_{m+1:n})$ are linear for some $m < n$, Theorem 4.3.2 indicates that F is exponential iff both $E(R_n|R_{n+1})$ and $E(R_{n+1}|R_n)$ are linear. Similarly, while Székely and Móri (1985) showed that the maximum correlation between $X_{i:n}$ and $X_{j:n}$ is attained when F is uniform, in Theorem 4.4.1 we noted that the correlation between R_m and R_n is the largest whenever F is exponential. The closeness is also evident in the fact that several of the characterizing properties of the exponential distribution based on order statistics and on record values involve ICFEs. Recently, Huang and Su (1994) have examined such parallel characterizations and have made attempts to unify them by viewing the sequences of order statistics and record values as coming from point processes with *order statistics property*. In another attempt at unifying the distribution theory for order statistics and record values, Kamps (1995a,b) has introduced random variables called *generalized order statistics* (see Exercise 2.13). These variables possess the Markov property and the (possibly nonstationary) conditional distribution has a structure that resembles the one given by (4.2.5). The generalization is achieved by replacing k on the right side, there by a non-negative function of k and some auxiliary variables. Consequently, several of the characterizing properties of record values (and order statistics) are shown to hold for these random variables.

We have witnessed further demonstrations of the closeness of the dependence structures of order statistics and record values in Section 4.4.4. As we noted at the end of that section, if (4.4.10) holds, then the properties (i) $T_1'R_1' \overset{d}{=} X$ and (ii) $nX_{1:n} \overset{d}{=} X$ for some fixed n, are equivalent, and are satisfied only by the exponential distribution. It may be worthwhile to point out that R_1' is the sample minimum of a random sample of random size T_1'.

There are also some differences. For example, the conditional distribution of X_i, given the order statistics, is not the same as its distribution, given the record values. In Theorem 4.5.2 we showed that $E(X_1|R_1 = x) = E(X_2|R_1 = x)$ for all $x \geq x_0$ characterized a cdf F satisfying (4.5.5). In contrast, $E(X_1|X_{i:n} = x) \equiv E(X_2|X_{i:n} = x)$ for all i, $1 \leq i \leq n$, $n > 1$, for any continuous cdf! There are subtle differences in the characterization results based on moment sequences of sample maxima and of (upper) record values (see Nagaraja and Nevzorov, 1997). Further, in the discrete case the similarity is less prominent apparently because of the positive probability of ties among the order statistics. Also, in that case, record values are defined only when $F^{-1}(1)$ is not an atom of F. However, a handful of the characterizations of geometric distributions based on records that we encountered in Section 4.7 parallel those involving order statistics.

EXERCISES

Note: In the following assume that records are generated from i.i.d. observations taken from a population represented by the random variable X having cdf F.

1. Give an example to show that Theorem 4.2.2 fails to hold without the assumption that $K(x)$ is nondecreasing.

2. Let n_0 and k be fixed and F be a continuous cdf with finite $(1+\delta)$th moment for some $\delta > 0$. Then show that the sequence of moments $\{E(R_{i+k} - R_i), \ i \geq n_0\}$ determines F up to a location parameter.

   [Gupta, 1984]

3. Let n_0, k, and m be fixed positive integers. Assume F is a continuous cdf with finite $(m + \delta)$th moment for some $\delta > 0$. Then show that the sequence $\{(i + k)!ER_{i+k}^m - i!ER_i^m, \ i \geq n_0\}$ characterizes F.

   [Lin, 1988a]

4. Show that the families of continuous cdf's that satisfy the condition $E(R_n|R_{n+1}) = cR_{n+1} + d$ a.s. are given by (4.3.5).

5. Let F be absolutely continuous. Show that if the (Type 2) k-record $R_{n(k)}$ is IFR, so is $R_{n(j)}$ whenever $j < k$.   [Raqab and Amin, 1997]

6. (a) Prove (4.2.5).

   (b) Generalize Lemma 4.3.2 to (Type 2) k-records by determining the families of continuous distributions for which
   $$E(R_{n+1(k)} \mid R_{n(k)}) = cR_{n(k)} + d \text{ a.s.}$$

   [Grudzień and Szynal, 1985]

7. Give an alternate proof of Theorem 4.4.1.   [Nevzorov, 1992]

8. Assume X is a continuous random variable and $E|X|^{2+\delta}$ is finite for some $\delta > 0$.

   (a) Show that for any $n \geq 1$,
   $$\{E(R_n)\}^2 \leq \frac{n+1}{n} \, E(R_{n-1}^2).$$

   (b) Let c^2 denote the quantity on the right side of the above inequality. Prove that equality holds above iff X is $\text{Exp}(1/c)$.

   [Lin, 1988b]

9. Let X be an absolutely continuous random variable having a monotone failure rate. If $R_n - R_m$ has the same distribution as R_{n-m-1} for some $m < n$, show that X is exponentially distributed.

   [Ahsanullah, 1987b]

10. Let X be an absolutely continuous non-negative random variable with a monotone failure rate. Show that if $R_1 - R_0$ is standard exponential, so is X.   [Arnold and Villaseñor, 1997]

11. Suppose F is absolutely continuous. Let g be a non-decreasing function such that for every $x > 0$, g has a point of increase in $(0, x)$. If for some fixed positive integer j, $E(g(R_j - R_{j-1})|R_j = x) = E(g(R_0)|R_j = x)$ for all $x > 0$, then show that F is an exponential cdf.   [Huang and Li, 1993]

12. Assume F is absolutely continuous with $F^{-1}(0) = 0$. Let g be a non-decreasing function defined for $x \geq 0$ such that $g(0) = 0$ and $E(g(X))$ is finite. Further suppose for positive integers n and i, $E(g(R_{n+i+1} - R_{n+i})|R_n = x) = c$ for all $x > 0$. If there exists a $\xi > 0$ such that $c < \int_0^\infty e^{-\xi x} dg(x) < \infty$, then show that $c = E(g(X))$ and F is exponential.   [Huang and Li, 1993]

13. Let F be a continuous cdf with $F(0) = 0$ and assume that (4.4.10) holds. If $T_1'R_1'$ and the event $\{T_1' = n\}$ are independent for some n, then show that F must be exponential.

[Ahsanullah and Kirmani, 1991]

14. Let F be a continuous cdf satisfying (4.4.10) for some $\lambda > 0$. Let $T_{n(k)}'$ denote the nth lower k-record time and $R_{n(k)}'$ be the corresponding lower (Type 2) k-record. If the random variables $(T_{n(k)}' - k + 1)R_{n(k)}'$ and $X_{1:k}$ are identically distributed, then show that F is an $\text{Exp}(1/\lambda)$ cdf. This generalizes Theorem 4.4.4.   [Basak, 1996]

15. Let n_0, j, k, and m be fixed positive integers such that j and k are, at most, m. Assume F is a continuous cdf with finite $(m + \delta)$th moment for some $\delta > 0$. Suppose

$$ER_i^m = \frac{(i+k)!}{i!} ER_{i+k}^{m-j} \qquad \text{for all } i \geq n_0.$$

Establish that F is a Weibull cdf with $F(x) = 1 - \exp\{-x^{(j/k)}\}$, $x \geq 0$.   [Lin, 1988a]

16. Find the family of continuous distributions for which $E(R_1|X_1 = x) = E(R_1|X_2 = x)$ for all $x \geq x_0 \geq -\infty$.

[Nagaraja and Nevzorov, 1997]

17. Let X be a discrete random variable with support on the set of positive integers. If R_0 and the event $\{R_1 - R_0 = 1\}$ are independent, prove that X has $\text{GeoT}(2, p)$ distribution.   [Srivastava, 1979]

18. Let X be a discrete random variable with finite variance having support on the set of positive integers. If $E((R_1 - R_0)^2|R_0 = x)$ is a constant, show that X must be geometric.

[Balakrishnan and Balasubramanian, 1995]

19. Let F be a discrete cdf. Prove that R_0 and $R_1 - R_0$ are identically distributed iff F has geometric distribution over the lattice set $\{k\gamma, \ k \geq 1\}$ for some $\gamma > 0$.   [Mohan and Nayak, 1982]

20. Let X be a positive integer valued random variable. Establish the following claims:

(a) Let $P(R_{n+1} - R_n = 1) = p > 0$. Then X is $\text{GeoT}(a, p)$ for some integer $a \geq n + 1$, if R_{n-1} and the event $\{R_{n+1} - R_n = 1\}$ are independent.

(b) Assume $f(x) > 0$ for all $x \geq 1$. If $E(R_{n+2} - R_{n+1}|R_n = x) = c$ a.s., then X is $\text{GeoT}(n + 3, 1/c)$.

[Nagaraja, Sen and Srivastava, 1989]

21. Let X be a positive integer valued random variable such that $f(x) > 0$ for all $x \geq 1$.

(a) If $E(R_{n+1(k)} - R_{n(k)} \mid R_{n(k)} = x) = \frac{1}{1-(1-p)^k}$, $x \geq n + 1$, then show that X has a geometric-tail distribution with success parameter p.

(b) If $R_{n+1(k)} - R_{n(k)}$ and $R_{n(k)}$ are independent, show that X has a geometric-tail distribution.   [Stepanov, 1990]

22. Assume X is a discrete random variable with support on positive integers. Let g be a nondecreasing function defined on non-negative integers such that $g(0) = 0$. Suppose for non-negative integers n and i, $E(g(R_{n+i+1} - R_{n+i})|R_n = x) = c$ for all integer $x > n$. Further assume that the constant c satisfies the condition $g(1) < c < \sum_{j=0}^{\infty} e^{-\xi j}(g(n + 1) - g(n)) < \infty$, for some $\xi > 0$. Then show that X is $\text{GeoT}(n + i + 2, p)$ for some p.   [Huang and Li, 1993]

23. Determine the class of non-negative integer valued random variables for which $E(R_1|R_0 = x) = cx + d$, $x = 0, 1, \ldots$.   [Korwar, 1984]

24. Let $\{\tilde{R}_n, n \geq 0\}$ be the sequence of weak (upper) records from a discrete cdf F having pdf $f(x)$. (See Section 2.8.) Assume $f(x) > 0$ for all integer $x \geq 0$ and F has a finite mean. Determine the cdf F's in this family that satisfy the condition $E(\tilde{R}_{n+1}|\tilde{R}_n = x) = \alpha x + \beta$, for all $x \geq 0$.   [Stepanov, 1993]

CHAPTER 5

INFERENCE

5.1 INTRODUCTION

In Chapters 2 and 3, we discussed several properties of record values arising from many specific distributions. A number of these results were later established, in Chapter 4, to be characterizations of those distributions. In this chapter, we will make use of the properties of record values to develop inferential procedures such as point estimation, interval estimation, prediction, and tests of spuriousity. In Section 5.2 we discuss the maximum likelihood estimation of parameters of the one- and two-parameter exponential and uniform distributions, normal, logistic and Gumbel distributions. We also present the likelihood equations for general location-scale distributions. In Section 5.3 we discuss the best linear unbiased estimation of parameters for all these cases, and some other distributions such as Rayleigh and Weibull. In many of these cases, the best linear unbiased estimators turn out to be either the same as the maximum likelihood estimators or simply those adjusted for their bias. In cases such as normal and logistic, where the best linear unbiased estimators cannot be derived explicitly, we present tables to facilitate their computation. In Section 5.4 we discuss the best linear invariant estimation of parameters. The interval estimation of parameters is discussed next in Section 5.5. After describing the unconditional method, we present a conditional inference approach and illustrate it with exponential and extreme value distributions. In Section 5.6, point prediction of future records is presented in which the best linear unbiased prediction, as well as the best linear invariant prediction are detailed. In Section 5.7, the interval prediction of future records is described, in which the conditional, tolerance region and the Bayesian approaches are highlighted. Several examples are then presented in Section

5.8, in order to illustrate the various methods of inference described in the earlier sections. In Section 5.9, we discuss some inferential issues in the case when the available data consist of record values as well as inter-record times. Finally, in Section 5.10, we discuss some distribution-free tests for trends in time-series, using some record statistics.

5.2 MAXIMUM LIKELIHOOD ESTIMATION

Example 5.2.1 (One-parameter Exponential Distribution) Let $R_0, R_1, ...,$ R_n be the upper record values observed from a one-parameter exponential $(\mathrm{Exp}(\sigma))$ distribution with pdf

$$f(x; \sigma) = \frac{1}{\sigma} e^{-x/\sigma}, \qquad x \geq 0, \ \sigma > 0. \qquad (5.2.1)$$

Then, from (2.3.9), we have the likelihood function to be

$$L = \frac{1}{\sigma^{n+1}} e^{-r_n/\sigma}, \qquad 0 \leq r_0 < r_1 < \cdots < r_n < \infty, \qquad (5.2.2)$$

from which we readily get the maximum likelihood estimator (MLE) of σ to be

$$\hat{\sigma} = R_n/(n+1). \qquad (5.2.3)$$

Since R_n/σ is distributed as Gamma$(n+1, 1)$ (see (2.3.1)), we note that

$$E(\hat{\sigma}) = \sigma \qquad \text{and} \qquad \mathrm{Var}(\hat{\sigma}) = \sigma^2/(n+1). \qquad (5.2.4)$$

Further, the likelihood function in (5.2.2) reveals that $\hat{\sigma}$ in (5.2.3) is a complete sufficient statistic and hence is the minimum variance unbiased estimator (MVUE) for σ.   ○

Remark 5.2.1 Samaniego and Whitaker (1986) considered the likelihood estimation of σ by assuming that the available data are of the form $R'_0, K_0,$ $R'_1, K_1, \ldots,$ where $R'_0, R'_1, \ldots$ are successive minima and $K_0, K_1, \ldots$ are the number of trials needed to obtain new records. They treated this estimation problem under both a fixed sample size and an inverse sampling scheme. The nonparametric version of this problem, while estimating certain population quantiles, has also been handled by Samaniego and Whitaker (1988). Data of this nature may be encountered in life-testing and stress-testing experiments. See Section 5.9 for more details on this problem.

Example 5.2.2 (Two-parameter Exponential Distribution) Let $R_0, R_1, \ldots,$ R_n be the upper record values observed from a two-parameter exponential distribution $(\text{Exp}(\mu, \sigma))$ with pdf

$$f(x; \mu, \sigma) = \frac{1}{\sigma} e^{-(x-\mu)/\sigma}, \qquad \mu \le x < \infty, \ \sigma > 0. \tag{5.2.5}$$

From (2.3.9), we have the likelihood function in this case to be

$$L = \frac{1}{\sigma^{n+1}} e^{-(r_n-\mu)/\sigma}, \qquad \mu \le r_0 < r_1 < \cdots < r_n < \infty, \tag{5.2.6}$$

from which we readily obtain the MLE's of μ and σ to be

$$\hat{\mu} = R_0 \qquad \text{and} \qquad \hat{\sigma} = \frac{1}{n+1}(R_n - R_0). \tag{5.2.7}$$

Since $(R_0-\mu)/\sigma, (R_1-R_0)/\sigma, \ldots, (R_n-R_{n-1})/\sigma$ are independent $\text{Exp}(1)$ random variables (see Section 2.3), we have

$$E(\hat{\mu}) = \mu + \sigma \qquad \text{and} \qquad E(\hat{\sigma}) = n\sigma/(n+1); \tag{5.2.8}$$

further,

$$\text{Var}(\hat{\mu}) = \sigma^2, \quad \text{Var}(\hat{\sigma}) = n\sigma^2/(n+1)^2 \quad \text{and} \quad \text{Cov}(\hat{\mu}, \hat{\sigma}) = 0. \tag{5.2.9}$$

$\hat{\mu}$ and $\hat{\sigma}$ are also statistically independent. Note also that $\hat{\mu}$ is a biased estimator for μ, but the bias can be corrected easily, in order to get an unbiased estimator for μ as follows:

$$\begin{aligned} \tilde{\mu} &= \hat{\mu} - \left(\frac{n+1}{n}\right)\hat{\sigma} \\ &= R_0 - \frac{1}{n}(R_n - R_0) \\ &= \frac{n+1}{n}R_0 - \frac{1}{n}R_n. \end{aligned} \tag{5.2.10}$$

Example 5.2.3 (One-parameter Uniform Distribution) Let $R_0, R_1, \ldots, R_n$ be the upper record values available from a one-parameter uniform distribution $(\text{Uniform}(0, \sigma))$ with pdf

$$f(x; \sigma) = \frac{1}{\sigma}, \qquad 0 < x < \sigma. \tag{5.2.11}$$

From (2.3.9), we have the likelihood function to be

$$L = \frac{1}{\sigma \prod_{i=0}^{n-1}(\sigma - r_i)}, \qquad 0 < r_0 < r_1 < \cdots < r_{n-1} < r_n < \sigma$$

which simply yields the MLE of σ to be $\hat{\sigma} = R_n$. From Section 2.4, we also readily have

$$E(\hat{\sigma}) = (1 - 2^{-n-1})\sigma \quad \text{and} \quad \text{Var}(\hat{\sigma}) = (3^{-n-1} - 4^{-n-1})\sigma^2. \quad (5.2.12)$$

An unbiased estimator can be obtained as

$$\tilde{\sigma} = R_n/(1 - 2^{-n-1}) \text{ with } \text{Var}(\tilde{\sigma}) = \sigma^2(3^{-n-1} - 4^{-n-1})/(1 - 2^{-n-1})^2.$$
$$(5.2.13)$$

○

Example 5.2.4 (Two-parameter Uniform Distribution) If we assume that the upper record values $R_0, R_1, \ldots, R_n$ are observed from a two-parameter uniform (Uniform$(\mu, \mu + \sigma)$) distribution with pdf

$$f(x; \mu, \sigma) = \frac{1}{\sigma}, \qquad \mu < x < \mu + \sigma, \qquad (5.2.14)$$

we have the likelihood function to be

$$L = \frac{1}{\sigma \prod_{i=0}^{n-1}(\mu + \sigma - r_i)}, \quad \mu < r_0 < r_1 < \cdots < r_n < \mu + \sigma . \quad (5.2.15)$$

From (5.2.15), we may get the MLE's of μ and σ to be $\hat{\mu} = R_0$ and $\hat{\sigma} = R_n - R_0$; furthermore,

$$E(\hat{\mu}) = \mu + \frac{\sigma}{2}, \quad E(\hat{\sigma}) = (2^{-1} - 2^{-n-1})\sigma,$$

$$\text{Var}(\hat{\mu}) = \sigma^2/12, \quad \text{Var}(\hat{\sigma}) = \sigma^2\left\{3^{-n-1} - 4^{-n-1} + \frac{1}{12}(1 - 2^{-n+1})\right\},$$

$$\text{Cov}(\hat{\mu}, \hat{\sigma}) = -\frac{\sigma^2}{12}(1 - 2^{-n}). \quad (5.2.16)$$

Though the above estimators are biased, they may be easily corrected for bias in order to obtain unbiased estimators of μ and σ as

$$\tilde{\mu} = \hat{\mu} - \hat{\sigma}/2 \quad \text{and} \quad \tilde{\sigma} = \hat{\sigma}/(2^{-1} - 2^{-n-1}). \quad (5.2.17)$$

○

Now, let us assume, in general, that $R_0, R_1, \ldots, R_n$ are the upper record values observed from any location-scale distribution with cdf $F(x; \mu, \sigma) = F\left(\frac{x-\mu}{\sigma}\right)$ and pdf $f(x; \mu, \sigma) = \frac{1}{\sigma} f\left(\frac{x-\mu}{\sigma}\right)$. Then, from (2.3.9), we have the likelihood function to be

$$L = \frac{1}{\sigma^{n+1}} \prod_{i=0}^{n-1} \left\{ \frac{f(r_i^*)}{1 - F(r_i^*)} \right\} f(r_n^*), \quad -\infty < r_0^* < r_1^* < \cdots < r_n^* < \infty, \tag{5.2.18}$$

where $r_i^* = (r_i - \mu)/\sigma$, $i = 0, 1, \ldots, n$, are the upper record values from $F(x)$. The log-likelihood function is obtained from (5.2.18) as

$$\log L = -(n+1) \log \sigma - \sum_{i=0}^{n-1} \log\{1 - F(r_i^*)\} + \sum_{i=0}^{n} \log f(r_i^*),$$

from which we obtain the likelihood equations for μ and σ, respectively, to be

$$\sum_{i=0}^{n-1} \frac{f(r_i^*)}{1 - F(r_i^*)} + \sum_{i=0}^{n} \frac{f'(r_i^*)}{f(r_i^*)} = 0 \tag{5.2.19}$$

and

$$(n+1) + \sum_{i=0}^{n-1} \frac{r_i^* f(r_i^*)}{1 - F(r_i^*)} + \sum_{i=0}^{n} \frac{r_i^* f'(r_i^*)}{f(r_i^*)} = 0. \tag{5.2.20}$$

In the above equations, $f'(x)$ denotes the derivative of $f(x)$. For most location-scale families, (5.2.19) and (5.2.20) need to be solved by numerical methods in order to determine the MLE's $\hat{\mu}$ and $\hat{\sigma}$ based on the given upper record values. However, simplification of these equations may be possible in some cases.

Example 5.2.5 (Normal Distribution) In the case of $N(\mu, \sigma^2)$ distribution, upon denoting $f(\cdot)$ by $\phi(\cdot)$ and $F(\cdot)$ by $\Phi(\cdot)$ and also noting that $\phi'(x) = -x\phi(x)$, we may simplify (5.2.19) and (5.2.20) as, respectively,

$$\sigma \sum_{i=0}^{n-1} \frac{\phi(r_i^*)}{1 - \Phi(r_i^*)} - \sum_{i=0}^{n} r_i + (n+1)\mu = 0, \tag{5.2.21}$$

$$(n+1)\sigma^2 + \sigma \sum_{i=0}^{n-1} \frac{r_i \phi(r_i^*)}{1 - \Phi(r_i^*)} - \sum_{i=0}^{n} r_i^2 + \mu \sum_{i=0}^{n} r_i = 0. \tag{5.2.22}$$

But, these two equations still need to be solved by iterative methods. ○

Example 5.2.6 (Logistic Distribution) In the case of logistic $L(\mu, \sigma^2)$ distribution with standard density function $f(x) = e^{-x}/(1 + e^{-x})^2$ (see Balakrishnan, 1992), we have the relationships $f(x) = F(x)\{1 - F(x)\}$ and $f'(x) = f(x)\{1 - 2F(x)\}$ using which we may simplify (5.2.19) and (5.5.20) as

$$(n + 1) - \sum_{i=0}^{n-1} F(r_i^*) - 2F(r_n^*) = 0 \qquad (5.2.23)$$

and

$$(n + 1)\sigma + \sum_{i=0}^{n} r_i - \sum_{i=0}^{n-1} r_i F(r_i^*) - 2r_n F(r_n^*) = 0. \qquad (5.2.24)$$

Once again, these two equations need to be solved by iterative methods.

◯

Maximum likelihood estimation of the location and scale parameters based on lower record values can be developed exactly along the same lines as presented above for the upper record values. Suppose $R'_0, R'_1, \ldots, R'_n$ are the observed lower record values. Then, similar to (5.2.18), we have the likelihood function in this case to be

$$L = \frac{1}{\sigma^{n+1}} \prod_{i=0}^{n-1} \left\{ \frac{f(r_i'^*)}{F(r_i'^*)} \right\} f(r_n'^*), \quad r_0'^* > r_1'^* > \cdots > r_n'^*, \qquad (5.2.25)$$

where $r_i'^* = (r_i' - \mu)/\sigma$, $i = 0, 1, \ldots, n$, are the lower record values from $F(x)$. In this case, we obtain the likelihood equations for μ and σ to be

$$\sum_{i=0}^{n-1} \frac{f(r_i'^*)}{F(r_i'^*)} - \sum_{i=0}^{n} \frac{f'(r_i'^*)}{f(r_i'^*)} = 0 \qquad (5.2.26)$$

and

$$(n + 1) - \sum_{i=0}^{n-1} \frac{r_i'^* f(r_i'^*)}{F(r_i'^*)} + \sum_{i=0}^{n} \frac{r_i'^* f'(r_i'^*)}{f(r_i'^*)} = 0, \qquad (5.2.27)$$

respectively; in these equations, $f'(x)$ once again denotes the derivative of $f(x)$.

Example 5.2.7 (Gumbel Distribution) In the case of Gumbel(μ, σ) distribution, where $f(x) = e^{-e^{-x}} e^{-x}$, we have the relationships $f(x) = F(x) e^{-x}$ and $f'(x) = f(x)(e^{-x} - 1)$ using which we may simplify (5.2.26) as

$$(n + 1) - e^{-r_n'^*} = 0.$$

This readily yields the MLE of μ to be

$$\hat{\mu} = R'_n + \hat{\sigma} \log(n+1). \qquad (5.2.28)$$

Upon using this expression in (5.2.27) and simplifying, we immediately obtain the MLE of σ to be

$$\hat{\sigma} = \frac{1}{n+1} \sum_{i=0}^{n-1} (R'_i - R'_n). \qquad (5.2.29)$$

From Section 3.4, we readily find that

$$E(\hat{\sigma}) = \frac{n}{n+1} \sigma \quad \text{and} \quad E(\hat{\mu}) = \mu + \sigma \left\{ \gamma - \sum_{i=1}^{n} \frac{1}{i} + \frac{n}{n+1} \log(n+1) \right\},$$
$$(5.2.30)$$

where γ is the Euler constant. Eq. (5.2.30) shows clearly that the estimators $\hat{\mu}$ and $\hat{\sigma}$ are both biased, but that they are asymptotically unbiased. Also, the bias can be corrected, even in case of small samples, in order to get the following unbiased estimators:

$$\tilde{\mu} = R'_n - \tilde{\sigma} \left\{ \gamma - \sum_{i=1}^{n} \frac{1}{i} \right\} \quad \text{and} \quad \tilde{\sigma} = \frac{1}{n} \sum_{i=0}^{n-1} (R'_i - R'_n). \qquad (5.2.31)$$

○

5.3   BEST LINEAR UNBIASED ESTIMATION

Let $R_0, R_1, \ldots, R_n$ be the upper record values observed from a scale-parameter distribution with cdf $F(x/\sigma)$ and pdf $\frac{1}{\sigma} f(x/\sigma)$. Then, the best linear unbiased estimator (BLUE) of σ is (Balakrishnan and Cohen, 1991, p. 74)

$$\sigma^* = \frac{\boldsymbol{\alpha}^T \boldsymbol{\Sigma}^{-1}}{\boldsymbol{\alpha}^T \boldsymbol{\Sigma}^{-1} \boldsymbol{\alpha}} \boldsymbol{R} = \sum_{i=0}^{n} a_i R_i, \qquad (5.3.1)$$

where

$$\begin{aligned} \boldsymbol{R} &= (R_0 \ R_1 \ \ldots \ R_n)^T, \\ \boldsymbol{\alpha} &= (\alpha_0 \ \alpha_1 \ \ldots \ \alpha_n)^T \end{aligned}$$

and

$$\Sigma = \begin{pmatrix} \sigma_{00} & \sigma_{01} & \cdots & \sigma_{0n} \\ \sigma_{01} & \sigma_{11} & \cdots & \sigma_{1n} \\ \cdots & \cdots & \cdots & \cdots \\ \sigma_{0n} & \sigma_{1n} & \cdots & \sigma_{nn} \end{pmatrix}.$$

Further, the variance of σ^* in (5.3.1) is

$$\text{Var}(\sigma^*) = \sigma^2/(\alpha^T \Sigma^{-1} \alpha). \tag{5.3.2}$$

Though the formulas in (5.3.1) and (5.3.2) involve the inverse of the variance-covariance matrix, properties of record values (seen earlier in Chapters 2 and 3) will enable us to derive explicit expressions for the BLUE σ^* and its variance for many specific distributions. For this purpose, we require the following lemma; see, for example, Graybill (1983, p. 198) and Arnold, Balakrishnan and Nagaraja (1992, pp. 174–175).

Lemma 5.3.1 *Let $\Sigma = ((\sigma_{ij}; i,j = 0,1,\ldots,n))$ be a $(n+1) \times (n+1)$ nonsingular symmetric matrix with $\sigma_{ij} = a_i b_j$ for $i \leq j$. Then, Σ^{-1} is a symmetric matrix with its (i,j)-th element (for $i \leq j$) given by*

$$\sigma^{ij} = \begin{cases} \frac{a_1}{a_0(a_1 b_0 - a_0 b_1)} & , \quad i = j = 0 \\[2ex] \frac{a_{i+1}b_{i-1} - a_{i-1}b_{i+1}}{(a_i b_{i-1} - a_{i-1}b_i)(a_{i+1}b_i - a_i b_{i+1})} & , \quad i = j = 1 \text{ to } n-1 \\[2ex] \frac{b_{n-1}}{b_n(a_n b_{n-1} - a_{n-1}b_n)} & , \quad i = j = n \\[2ex] \frac{-1}{a_{i+1}b_i - a_i b_{i+1}} & , \quad j = i+1 \text{ and } i = 0 \text{ to } n-1 \\[2ex] 0 & , \quad j > i+1. \end{cases}$$

Example 5.3.1 (One-parameter Exponential Distribution) Let us assume that the record values are arising from $\text{Exp}(\sigma)$ distribution. In this case, we have (see Section 2.3)

$$\alpha = (1 \; 2 \; \ldots \; n+1)^T$$

and

$$\Sigma = \begin{pmatrix} 1 & 1 & \cdots & \cdots & 1 \\ 1 & 2 & \cdots & \cdots & 2 \\ \vdots & & \ddots & & \vdots \\ 1 & 2 & \cdots & n & n \\ 1 & 2 & \cdots & n & n+1 \end{pmatrix}.$$

With $a_i = i + 1$ (for $i = 0, 1, \ldots, n$) and $b_j = 1$ (for $j \geq i$), Lemma 5.3.1 readily yields

$$\Sigma^{-1} = \begin{pmatrix} 2 & -1 & & & & \\ -1 & 2 & -1 & & 0 & \\ & -1 & & & & \\ & & \ddots & \ddots & \ddots & \\ 0 & & & -1 & 2 & -1 \\ & & & & -1 & 1 \end{pmatrix} \qquad (5.3.3)$$

so that

$$\boldsymbol{\alpha}^T \Sigma^{-1} = (0 \ 0 \ \ldots \ 0 \ 1)$$

and

$$\boldsymbol{\alpha}^T \Sigma^{-1} \boldsymbol{\alpha} = n + 1.$$

Eqs. (5.3.1) and (5.3.2) give the BLUE of σ to be

$$\sigma^* = R_n/(n+1) \qquad (5.3.4)$$

and its variance to be

$$\mathrm{Var}(\sigma^*) = \sigma^2/(n+1). \qquad (5.3.5)$$

We observe that the BLUE σ^* in (5.3.4) is exactly the same as the MLE of σ in (5.2.3); see Example 5.2.1.

This can be derived directly by using the fact that the record spacings $R_0, R_1 - R_0, R_2 - R_1, \ldots, R_n - R_{n-1}$ are once again i.i.d. $\mathrm{Exp}(\sigma)$ random variables (see Section 2.3). Let us take the BLUE of σ to be $\sigma^* = \sum_{i=0}^{n} c_i(R_i - R_{i-1})$, with $R_{-1} \equiv 0$; then, we readily have

$$E(\sigma^*) = \sigma \sum_{i=0}^{n} c_i \quad \text{and} \quad \mathrm{Var}(\sigma^*) = \sigma^2 \sum_{i=0}^{n} c_i^2.$$

Thus, we need to determine c_i's so that $\sum_{i=0}^{n} c_i^2$ is minimized under the restriction $\sum_{i=0}^{n} c_i = 1$; the solution for this problem is easily seen to be $c_i = 1/(n+1)$ for $i = 0, 1, \ldots, n$, which simply yields σ^* to be $R_n/(n+1)$ and its variance to be $\sigma^2/(n+1)$. Similar simplified derivations have been presented recently by Balakrishnan and Rao (1997). $\bigcirc$

Example 5.3.2 (One-parameter Uniform Distribution) Let us assume that the record values are arising from Uniform$(0, \sigma)$ distribution. In this case, we have (see Sections 2.4 and 2.8)

$$\alpha_i = 1 - \frac{1}{2^{i+1}}, \qquad i = 0, 1, \ldots, n,$$

$$\sigma_{ii} = \frac{1}{3^{i+1}} - \frac{1}{4^{i+1}}, \qquad i = 0, 1, \ldots, n,$$

and

$$\sigma_{ij} = 2^{i-j} \left(\frac{1}{3^{i+1}} - \frac{1}{4^{i+1}} \right), \qquad j \geq i+1.$$

Choosing $a_i = 2^i \left(\frac{1}{3^{i+1}} - \frac{1}{4^{i+1}} \right)$ (for $i = 0, 1, \ldots, n$) and $b_j = \frac{1}{2^j}$ (for $j \geq i$), Lemma 5.3.1 yields

$$\sigma^{ii} = 7 \times 3^{i+1}, \qquad i = 0, 1, \ldots, n-1,$$

$$\sigma^{nn} = 4 \times 3^{n+1},$$

$$\sigma^{i,i+1} = -2 \times 3^{i+2}, \qquad i = 0, 1, \ldots, n-1,$$

$$\sigma^{ij} = 0, \qquad j > i+1, \tag{5.3.6}$$

so that

$$\boldsymbol{\alpha}^T \boldsymbol{\Sigma}^{-1} = (-3 \ -3^2 \ \cdots \ -3^n \ \ 2 \times 3^{n+1})$$

and

$$\boldsymbol{\alpha}^T \boldsymbol{\Sigma}^{-1} \boldsymbol{\alpha} = \frac{3}{2} (3^{n+1} - 1).$$

Eqs. (5.3.1) and (5.3.2) give the BLUE of σ to be

$$\sigma^* = \frac{2}{3^{n+1} - 1} \left\{ 2 \times 3^n R_n - 3^{n-1} R_{n-1} - 3^{n-2} R_{n-2} - \cdots - R_0 \right\} \tag{5.3.7}$$

and its variance to be

$$\mathrm{Var}(\sigma^*) = \frac{2\sigma^2}{3(3^{n+1} - 1)}. \tag{5.3.8}$$

Clearly, the BLUE σ^* in (5.3.7) is quite different from the MLE $\hat{\sigma}$ derived earlier; see Example 5.2.3. Furthermore, from (5.3.8) and (5.2.13), we obtain the relative efficiency of σ^* compared to $\tilde{\sigma}$ (MLE adjusted for bias) as

$$RE(\sigma^*) = \frac{\mathrm{Var}(\tilde{\sigma})}{\mathrm{Var}(\sigma^*)} = \frac{3(3^{n+1} - 1) \left(\frac{1}{3^{n+1}} - \frac{1}{4^{n+1}} \right)}{2 \left(1 - \frac{1}{2^{n+1}} \right)^2}. \tag{5.3.9}$$

For example, for $n = 1$, 2 and 3, we get the relative efficiency of the BLUE σ^* to be 103.7%, 109.1% and 115.2%, respectively. It may also be easily observed from (5.3.9) that $RE(\sigma^*) \to 150\%$ as $n \to \infty$, thus revealing that the BLUE σ^* in (5.3.7) is asymptotically 50% more efficient than the estimator $\tilde{\sigma}$ (MLE adjusted for bias) in (5.2.13).   ○

Remark 5.3.1 An interesting application of this result to an auction problem has been presented by Samaniego and Kaiser (1978). They considered an auction in which increasing bids are made in sequence on an object whose value σ is known to each bidder, and assumed that n bids are received with the distribution of each bid being conditionally uniform.

Example 5.3.3 (One-parameter Rayleigh Distribution; Balakrishnan and Chan, 1994) Let us assume that the record values $R_0, R_1, \ldots, R_n$ are arising from a one-parameter Rayleigh distribution with pdf

$$f(x; \sigma) = \frac{x}{\sigma^2} e^{-x^2/(2\sigma^2)}, \qquad x > 0, \ \sigma > 0.$$

In this case, we have (see Section 3.3)

$$\alpha_i = \sqrt{2}\, \Gamma\left(i + \frac{3}{2}\right) \Big/ \Gamma(i+1), \qquad i = 0, 1, \ldots, n,$$

$$\sigma_{ii} = 2\left[i + 1 - \left\{\Gamma\left(i + \frac{3}{2}\right) \Big/ \Gamma(i+1)\right\}^2\right], \qquad i = 0, 1, \ldots, n,$$

and

$$\sigma_{ij} = \frac{2\Gamma(i + \frac{3}{2})}{\Gamma(i+1)} \left\{\frac{\Gamma(j+2)}{\Gamma(j + \frac{3}{2})} - \frac{\Gamma(j + \frac{3}{2})}{\Gamma(j+1)}\right\}, \qquad j \ge i+1.$$

Choosing $a_i = \sqrt{2}\, \Gamma\left(i + \frac{3}{2}\right)/\Gamma(i+1)$ (for $i = 0, 1, \ldots, n$) and $b_j = \sqrt{2}\left\{\frac{\Gamma(j+2)}{\Gamma(j+\frac{3}{2})} - \frac{\Gamma(j+\frac{3}{2})}{\Gamma(j+1)}\right\}$ (for $j \ge i$), Lemma 5.3.1 yields

$$\sigma^{ii} = \frac{8(i+1)^2 + 1}{2(i+1)}, \qquad i = 0, 1, \ldots, n-1,$$

$$\sigma^{nn} = (2n+1)\left\{\frac{\Gamma(n+1)}{\Gamma(n+\frac{1}{2})} - \frac{\Gamma(n+\frac{1}{2})}{\Gamma(n)}\right\} \Big/ \left\{\frac{\Gamma(n+2)}{\Gamma(n+\frac{3}{2})} - \frac{\Gamma(n+\frac{3}{2})}{\Gamma(n+1)}\right\},$$

$$\sigma^{i,i+1} = -(2i+3), \qquad i = 0, 1, \ldots, n-1,$$

$$\sigma^{ij} = 0, \qquad j > i+1, \tag{5.3.10}$$

so that the BLUE of σ is

$$\sigma^* = \frac{\Gamma(n+1)}{\sqrt{2}\, \Gamma(n + \frac{3}{2})} R_n \tag{5.3.11}$$

and its variance is given by

$$\text{Var}(\sigma^*) = \sigma^2 \left[(n+1) \left\{ \frac{\Gamma(n+1)}{\Gamma(n+\frac{3}{2})} \right\}^2 - 1 \right].$$ (5.3.12)

Note that for large n (using Stirling's formula), we may write

$$\sigma^* \simeq R_n / \sqrt{2n}.$$

○

Example 5.3.4 (One-parameter Weibull Distribution; Balakrishnan and Chan, 1994) Let us assume that the record values are observed from the one-parameter Weibull distribution with pdf

$$f(x; \sigma) = \frac{c x^{c-1}}{\sigma^c} e^{-(x/\sigma)^c}, \qquad x > 0, \ \sigma > 0, \ c > 0.$$

In this case, we have (see Section 3.3)

$$\alpha_i = \frac{\Gamma(i+1+\frac{1}{c})}{\Gamma(i+1)}, \qquad i = 0, 1, \ldots, n,$$

$$\sigma_{ii} = \frac{\Gamma(i+1+\frac{2}{c})}{\Gamma(i+1)} - \left\{ \frac{\Gamma(i+1+\frac{1}{c})}{\Gamma(i+1)} \right\}^2, \qquad i = 0, 1, \ldots, n,$$

and

$$\sigma_{ij} = \frac{\Gamma(i+1+\frac{1}{c})}{\Gamma(i+1)} \left\{ \frac{\Gamma(j+1+\frac{2}{c})}{\Gamma(j+1+\frac{1}{c})} - \frac{\Gamma(j+1+\frac{1}{c})}{\Gamma(j+1)} \right\}, \qquad j \geq i.$$

Then, by following exactly the same steps as we did in the case of the Rayleigh distribution, we obtain the BLUE of σ as

$$\sigma^* = \frac{\Gamma(n+1)}{\Gamma(n+1+\frac{1}{c})} R_n$$ (5.3.13)

and its variance as

$$\text{Var}(\sigma^*) = \sigma^2 \left[\frac{\Gamma(n+1)\Gamma(n+1+\frac{2}{c})}{\left\{ \Gamma(n+1+\frac{1}{c}) \right\}^2} - 1 \right].$$ (5.3.14)

Note that if we set $c = 1$, (5.3.13) and (5.3.14) reduce respectively to (5.3.4) and (5.3.5) which are the results for the one-parameter exponential distribution. ○

Next, let us consider the general location-scale family of distributions with cdf $F(x; \mu, \sigma) = F(\frac{x-\mu}{\sigma})$ and pdf $f(x; \mu, \sigma) = \frac{1}{\sigma} f(\frac{x-\mu}{\sigma})$, and assume that the upper record values $R_0, R_1, \ldots, R_n$ are available. Then, the BLUE's of μ and σ are (David, 1981, pp. 128–131; Balakrishnan and Cohen, 1991, pp. 80–82)

$$\mu^* = \frac{\alpha^T \Sigma^{-1} \alpha 1^T \Sigma^{-1} - \alpha^T \Sigma^{-1} 1 \alpha^T \Sigma^{-1}}{(\alpha^T \Sigma^{-1} \alpha)(1^T \Sigma^{-1} 1) - (\alpha^T \Sigma^{-1} 1)^2} R$$

$$= \sum_{i=0}^{n} a_i R_i \qquad (5.3.15)$$

and

$$\sigma^* = \frac{1^T \Sigma^{-1} 1 \alpha^T \Sigma^{-1} - 1^T \Sigma^{-1} \alpha 1^T \Sigma^{-1}}{(\alpha^T \Sigma^{-1} \alpha)(1^T \Sigma^{-1} 1) - (\alpha^T \Sigma^{-1} 1)^2} R$$

$$= \sum_{i=0}^{n} b_i R_i, \qquad (5.3.16)$$

where, as before, R denotes the column vector of the observed upper record values, α denotes the column vector of expected values of the upper record values from the distribution $F(x)$, Σ denotes the variance-covariance matrix of the upper record values from the distribution $F(x)$, and 1 is a column vector (of dimension $n+1$) with all its entries as 1. Furthermore, the variances and covariance of these BLUE's are given by

$$Var(\mu^*) = \frac{\sigma^2 (\alpha^T \Sigma^{-1} \alpha)}{(\alpha^T \Sigma^{-1} \alpha)(1^T \Sigma^{-1} 1) - (\alpha^T \Sigma^{-1} 1)^2}, \qquad (5.3.17)$$

$$Var(\sigma^*) = \frac{\sigma^2 (1^T \Sigma^{-1} 1)}{(\alpha^T \Sigma^{-1} \alpha)(1^T \Sigma^{-1} 1) - (\alpha^T \Sigma^{-1} 1)^2}, \qquad (5.3.18)$$

$$Cov(\mu^*, \sigma^*) = - \frac{\sigma^2 (\alpha^T \Sigma^{-1} 1)}{(\alpha^T \Sigma^{-1} \alpha)(1^T \Sigma^{-1} 1) - (\alpha^T \Sigma^{-1} 1)^2}. \qquad (5.3.19)$$

Though the above formulas involve the inverse of the variance-covariance matrix and appear to be quite complicated, Lemma 5.3.1 and some special properties of record values from many specific distributions will enable explicit derivation of these BLUE's of μ and σ in those cases.

Example 5.3.5 (Two-parameter Exponential Distribution) Let us start
with $\text{Exp}(\mu, \sigma)$ distribution. Then, upon using the expression for $\mathbf{\Sigma}^{-1}$ in
(5.3.3), we obtain

$$\boldsymbol{\alpha}^T \mathbf{\Sigma}^{-1} = (0 \ 0 \ \ldots \ 0 \ 1),$$
$$\mathbf{1}^T \mathbf{\Sigma}^{-1} = (1 \ 0 \ \ldots \ 0 \ 0),$$
$$\boldsymbol{\alpha}^T \mathbf{\Sigma}^{-1} \boldsymbol{\alpha} = n+1, \ \boldsymbol{\alpha}^T \mathbf{\Sigma}^{-1} \mathbf{1} = 1 \text{ and } \mathbf{1}^T \mathbf{\Sigma}^{-1} \mathbf{1} = 1.$$

Eqs. (5.3.15) and (5.3.16) then yield the BLUE's of μ and σ to be

$$\mu^* = \frac{1}{n}\{(n+1)R_0 - R_n\} \tag{5.3.20}$$

and

$$\sigma^* = \frac{1}{n}(R_n - R_0). \tag{5.3.21}$$

From (5.3.17)–(5.3.19), we also obtain the variances and covariance of
these BLUE's to be

$$\text{Var}(\mu^*) = \sigma^2 \left(\frac{n+1}{n}\right), \ \text{Var}(\sigma^*) = \frac{\sigma^2}{n}, \ \text{Cov}(\mu^*, \sigma^*) = -\frac{\sigma^2}{n}. \tag{5.3.22}$$

It should be noted that these BLUE's are exactly the same as the MLE's
adjusted for their bias (see (5.2.7) and (5.2.10)).

In this case also, the BLUE's can be derived by a direct approach as
follows: After noting that the record spacings $R_i - R_{i-1}$ (with $R_{-1} \equiv \mu$),
$i = 0, 1, \ldots, n$, are independently distributed as $\text{Exp}(\sigma)$, we may take the
BLUE's of μ and σ to be

$$\mu^* = c_0 R_0 + \sum_{i=1}^{n} c_i (R_i - R_{i-1}) \text{ and } \sigma^* = \sum_{i=1}^{n} d_i (R_i - R_{i-1}).$$

Then, we readily have

$$E(\mu^*) = c_0(\mu + \sigma) + \sigma \sum_{i=1}^{n} c_i, \ E(\sigma^*) = \sigma \sum_{i=1}^{n} d_i,$$
$$\text{Var}(\mu^*) = \sigma^2 \sum_{i=0}^{n} c_i^2, \text{Var}(\sigma^*) = \sigma^2 \sum_{i=1}^{n} d_i^2 \text{ and Cov}(\mu^*, \sigma^*) = \sigma^2 \sum_{i=1}^{n} c_i d_i.$$

The unbiasedness of μ^* and σ^* place the restrictions $c_0 = 1$, $\sum_{i=1}^{n} c_i = -1$
and $\sum_{i=1}^{n} d_i = 1$. Then, in order to find the *trace-efficient linear unbiased
estimators*, we need to find c_i and d_i so as to minimize

$$\left(1 + \sum_{i=1}^{n} c_i^2\right) + \sum_{i=1}^{n} d_i^2,$$

subject to the conditions $\sum_{i=1}^{n} c_i = -1$ and $\sum_{i=1}^{n} d_i = 1$. The solution for this problem simply turns out to be $c_i = -1/n$ and $d_i = 1/n$ for $i = 1, 2, \ldots, n$, which immediately reveals that the BLUE's in (5.3.20) and (5.3.21) are the trace-efficient linear unbiased estimators. Similarly, in order to find the *determinant-efficient linear unbiased estimators*, we need to find c_i and d_i so as to minimize

$$\left(1 + \sum_{i=1}^{n} c_i^2\right) \sum_{i=1}^{n} d_i^2 - \left(\sum_{i=1}^{n} c_i d_i\right)^2,$$

subject to the conditions $\sum_{i=1}^{n} c_i = -1$ and $\sum_{i=1}^{n} d_i = 1$. The solution for this optimization problem once again turns out to be $c_i = -1/n$ and $d_i = 1/n$ (for $i = 1, 2, \ldots, n$), thus revealing that the BLUE's in (5.3.20) and (5.3.21) are also *determinant-efficient linear unbiased estimators*. As a matter of fact, if we consider more generally the best linear unbiased estimation of the parameter $\theta = \alpha\mu + \beta\sigma$ (for given values of α and β), by taking the BLUE of θ as $\theta^* = d_0 R_0 + \sum_{i=1}^{n} d_i (R_i - R_{i-1})$, we readily have

$$E(\theta^*) = d_0(\mu + \sigma) + \sigma \sum_{i=1}^{n} d_i \quad \text{and} \quad \text{Var}(\theta^*) = \sigma^2 \sum_{i=0}^{n} d_i^2.$$

The unbiasedness of θ^* requires $d_0 = \alpha$ and $\sum_{i=1}^{n} d_i = \beta - \alpha$ and, therefore, we need to determine d_i's $(i = 1, 2, \ldots, n)$ that minimize $\sum_{i=1}^{n} d_i^2$ under the condition $\sum_{i=1}^{n} d_i = \beta - \alpha$. After noting that the solution for this problem is $d_i = (\beta - \alpha)/n$ for $i = 1, 2, \ldots, n$, we observe that $\theta^* = \alpha\mu^* + \beta\sigma^*$; consequently,

$$\text{Var}(\alpha\mu^* + \beta\sigma^*) \leq \text{Var}(\alpha\mu^{**} + \beta\sigma^{**})$$

for any other unbiased estimates (μ^{**}, σ^{**}). Let $C(\mu^*, \sigma^*)$ be the covariance matrix of (μ^*, σ^*). Then, the above inequality can be expressed as

$$(\alpha \ \beta) \left[C(\mu^*, \sigma^*)\right]_{2 \times 2} \begin{pmatrix} \alpha \\ \beta \end{pmatrix} \leq (\alpha \ \beta) \left[C(\mu^{**}, \sigma^{**})\right]_{2 \times 2} \begin{pmatrix} \alpha \\ \beta \end{pmatrix}. \quad (5.3.23)$$

Eq. (5.3.23) implies the complete variance-covariance matrix dominance of the BLUE's, namely,

$$\left[C(\mu^*, \sigma^*)\right]_{2 \times 2} \leq \left[C(\mu^{**}, \sigma^{**})\right]_{2 \times 2}. \quad (5.3.24)$$

As pointed out by Balakrishnan and Rao (1997, 1998), this variance-covariance matrix dominance is a very general result, from which the trace efficiency as well as the determinant efficiency follow as special cases. ○

Example 5.3.6 (Extreme Value Distribution) Let the upper record values $R_0, R_1, \ldots, R_n$ be observed from an extreme value distribution $(EV(\mu, \sigma))$ with pdf

$$f(x; \mu, \sigma) = \frac{1}{\sigma} e^{(x-\mu)/\sigma} e^{-e^{(x-\mu)/\sigma}}, \qquad -\infty < x < \infty.$$

In this case, it is known that the variables $(R_n - R_0)/\sigma, (R_n - R_1)/\sigma, \ldots,$ $(R_n - R_{n-1})/\sigma$ and $(R_n - \mu)/\sigma$ are mutually independent; furthermore, it is known that the variables $(R_n - R_0)/\sigma, (R_n - R_1)/\sigma, \ldots, (R_n - R_{n-1})/\sigma$ are distributed as $Exp(1)$, while $(R_n - \mu)/\sigma$ is distributed as Log-gamma$(n+1)$ (log-gamma with shape parameter $n+1$) ; see Section 2.4.4. By making use of these distributional results and proceeding exactly along the same lines as we just did in the case of the two-parameter exponential distribution, we can derive the BLUE's of μ and σ to be

$$\mu^* = \frac{\alpha_n}{n} \sum_{i=0}^{n-1} R_i + (1 - \alpha_n) R_n = R_n - \frac{\alpha_n}{n} \sum_{i=0}^{n-1} (R_n - R_i)$$

and

$$\sigma^* = R_n - \frac{1}{n} \sum_{i=0}^{n-1} R_i = \frac{1}{n} \sum_{i=0}^{n-1} (R_n - R_i),$$

where $\alpha_n = -\gamma + \sum_{i=1}^{n} 1/i$ with γ being the Euler's constant. Compare these results with the MLE's of μ and σ of the Gumbel distribution based on lower record values (see Example 5.2.7 and, in particular, (5.2.31)). (Do you see some connections?) It is easy to note that $2n\sigma^*/\sigma = \sum_{i=0}^{n-1} 2(R_n - R_i)/\sigma$ has a χ_{2n}^2 distribution (central chi-square distribution with $2n$ degrees of freedom). These exact distributional results have been utilized by Sultan and Balakrishnan (1997c) to develop exact inference (like confidence intervals and tests of hypotheses) for the location parameter μ and the scale parameter σ; see Section 5.7 for an illustrative example. ○

Example 5.3.7 (Two-parameter Uniform Distribution) Let us consider Uniform$(\mu, \mu + \sigma)$ distribution. In this case, upon using the expression of Σ^{-1} in (5.3.6), we obtain

$$\boldsymbol{\alpha}^T \Sigma^{-1} = (-3 \ -3^2 \ \ldots \ -3^n \ 2 \times 3^{n+1}),$$
$$\mathbf{1}^T \Sigma^{-1} = (3 \ -3^2 \ \ldots \ -3^n \ 2 \times 3^{n+1}),$$
$$\boldsymbol{\alpha}^T \Sigma^{-1} \boldsymbol{\alpha} = \frac{3}{2}(3^{n+1} - 1), \ \boldsymbol{\alpha}^T \Sigma^{-1} \mathbf{1} = \frac{3}{2}(3^{n+1} + 1) \text{ and}$$
$$\mathbf{1}^T \Sigma^{-1} \mathbf{1} = \frac{3}{2}(3^{n+1} + 5).$$

Eqs. (5.3.15) and (5.3.16) then yield the BLUE's of μ and σ to be

$$\mu^* = \frac{2}{3^n - 1}\{3^n R_0 + R_1 + 3R_2 + \cdots + 3^{n-2}R_{n-1} - 2 \times 3^{n-1}R_n\}$$

$$(5.3.25)$$

and

$$\sigma^* = \frac{2}{3^n - 1}\{4 \times 3^{n-1}R_n - 2 \times 3^{n-2}R_{n-1} - \cdots - 2 \times 3R_2$$
$$- 2R_1 - (3^n + 1)R_0\}.$$

$$(5.3.26)$$

Further, from (5.3.17)–(5.3.19) we obtain the variances and covariance of these BLUE's to be

$$\mathrm{Var}(\mu^*) = \sigma^2 \frac{(3^{n+1} - 1)}{9(3^n - 1)}, \qquad \mathrm{Var}(\sigma^*) = \sigma^2 \frac{(3^{n+1} + 5)}{9(3^n - 1)}$$

and

$$\mathrm{Cov}(\mu^*, \sigma^*) = -\sigma^2 \frac{(3^{n+1} + 1)}{9(3^n - 1)}.$$

$$(5.3.27)$$

As we did in the case of two-parameter exponential distribution, we can develop direct derivations of trace-efficient linear unbiased estimators and determinant-efficient linear unbiased estimators in this case too. $\bigcirc$

Example 5.3.8 (Two-parameter Rayleigh Distribution) Let us consider the two-parameter Rayleigh distribution with pdf

$$f(x; \mu, \sigma) = \left(\frac{x - \mu}{\sigma^2}\right) e^{-(x-\mu)^2/(2\sigma^2)}, \qquad x \geq \mu, \; \sigma > 0.$$

Upon using the expression of Σ^{-1} in (5.3.10), we get

$$\boldsymbol{\alpha}^T \Sigma^{-1} = \left(0\ 0\ \cdots\ 0\ \frac{1}{\sqrt{2}}\left\{\frac{\Gamma(n + 2)}{\Gamma(n + \frac{3}{2})} - \frac{\Gamma(n + \frac{3}{2})}{\Gamma(n + 1)}\right\}^{-1}\right),$$

$$\mathbf{1}^T \Sigma^{-1} = \left(\frac{3}{2}\ \frac{1}{4}\ \cdots\ \frac{1}{2n}\ (2n + 1)\left\{\frac{\frac{\Gamma(n+1)}{\Gamma(n+\frac{1}{2})} - \frac{\Gamma(n+\frac{1}{2})}{\Gamma(n)}}{\frac{\Gamma(n+2)}{\Gamma(n+\frac{3}{2})} - \frac{\Gamma(n+\frac{3}{2})}{\Gamma(n+1)}} - 1\right\}\right),$$

$$\boldsymbol{\alpha}^T \Sigma^{-1} \boldsymbol{\alpha} = \frac{\Gamma^2(n + \frac{3}{2})}{\Gamma(n + 2)\Gamma(n + 1) - \Gamma^2(n + \frac{3}{2})},$$

$$\boldsymbol{\alpha}^T \Sigma^{-1} \mathbf{1} = \frac{1}{\sqrt{2}}\frac{\Gamma(n + 1)\Gamma(n + \frac{3}{2})}{\Gamma(n + 1)\Gamma(n + 2) - \Gamma^2(n + \frac{3}{2})}$$

and

$$\mathbf{1}^T\Sigma^{-1}\mathbf{1} = 1 + \sum_{i=1}^{n} \frac{1}{2i} + (2n+1)\left\{ \frac{\frac{\Gamma(n+1)}{\Gamma(n+\frac{1}{2})} - \frac{\Gamma(n+\frac{1}{2})}{\Gamma(n)}}{\frac{\Gamma(n+2)}{\Gamma(n+\frac{3}{2})} - \frac{\Gamma(n+\frac{3}{2})}{\Gamma(n+1)}} - 1 \right\}.$$

These expressions, when substituted in (5.3.15)–(5.3.19) and simplified, readily yield the BLUE's of μ and σ and their variances and covariance; see Balakrishnan and Chan (1994). These authors have similarly derived the BLUE's of μ and σ and their variances and covariance in the case of the two-parameter Weibull distribution (of course, with the assumption that the shape parameter c is known). ○

Example 5.3.9 (Normal Distribution; Balakrishnan and Chan, 1995, 1998) For the standard normal distribution, the means, variances and co-variances of the upper record values were presented earlier in Chapter 2. Using these quantities directly in (5.3.15)–(5.3.19), the coefficients a_i and b_i ($i = 0, 1, \ldots, n$) of the BLUE's of μ and σ, respectively, as well as the values of Var(μ^*)/σ^*, Var(σ^*)/σ^2 and Cov(μ^*, σ^*)/σ^2 can be determined numerically for the case when the upper record values $R_0, R_1, \ldots, R_n$ are observed from $N(\mu, \sigma^2)$ distribution. These values, taken from Balakrishnan and Chan (1995, 1998), are all presented in Table 5.3.1 for $n = 1(1)9$. Note that when $n = 1$, the BLUE of μ, turns out to be simply R_0. This is so because R_0 is an unbiased estimator of μ while R_1 is a biased estimator of μ; consequently, an unbiased estimator of μ which is a linear combination of R_0 and R_1 has to be R_0. ○

Example 5.3.10 (Logistic Distribution; Balakrishnan, Ahsanullah and Chan, 1995) Suppose the upper record values $R_0, R_1, \ldots, R_n$ are arising from the logistic $L(\mu, \sigma^2)$ distribution. For the $L(0, 1)$ distribution, the means, variances and covariances of the upper record values were presented earlier in Chapter 3. Upon using these quantities in (5.3.15)–(5.3.19), the coefficients a_i and b_i ($i = 0, 1, \ldots, n$) of the BLUE's of μ and σ, respectively, as well as the values of Var(μ^*)/σ^2, Var(σ^*)/σ^2 and Cov(μ^*, σ^*)/σ^2 can be determined numerically. These values, taken from Balakrishnan, Ahsanullah and Chan (1995), are all presented in Table 5.3.2 for $n = 1(1)9$. Once again, we note that when $n = 1$, the BLUE of μ is simply R_0, for the same reason as mentioned above in Example 5.3.9. ○

Table 5.3.1. Coefficients a_i and b_i of the BLUE's μ^* and σ^*, and the values of $\mathrm{Var}(\mu^*)/\sigma^2$, $\mathrm{Var}(\sigma^*)/\sigma^2$ and $\mathrm{Cov}(\mu^*,\sigma^*)/\sigma^2$ for $N(\mu,\sigma^2)$ distribution

n	a_i	b_i	$\mathrm{Var}(\mu^*)/\sigma^2$	$\mathrm{Var}(\sigma^*)/\sigma^2$	$\mathrm{Cov}(\mu^*,\sigma^*)/\sigma^2$
1	1.0000	−1.1072			
	0.0000	1.1072	1.0000	0.7215	−0.4477
2	0.8639	−0.5860			
	0.3424	−0.2041			
	−0.2063	0.7901	0.9743	0.3449	−0.3493
3	0.7799	−0.4069			
	0.3152	−0.1460			
	0.2073	−0.0926			
	−0.3025	0.6454	0.9475	0.2227	−0.2921
4	0.7211	−0.3150			
	0.2956	−0.1155			
	0.1953	−0.0738			
	0.1455	−0.0537			
	−0.3576	0.5580	0.9230	0.1630	−0.2539
5	0.6768	−0.2588			
	0.2806	−0.0964			
	0.1859	−0.0619			
	0.1389	−0.0453			
	0.1106	−0.0355			
	−0.3927	0.4979	0.9011	0.1279	−0.2261
6	0.6417	−0.2207			
	0.2686	−0.0833			
	0.1784	−0.0537			
	0.1335	−0.0394			
	0.1064	−0.0309			
	0.0883	−0.0253			
	−0.4167	0.4533	0.8816	0.1049	−0.2050

(table continues)

Table 5.3.1 (continued.)

n	a_i	b_i	$\mathrm{Var}(\mu^*)/\sigma^2$	$\mathrm{Var}(\sigma^*)/\sigma^2$	$\mathrm{Cov}(\mu^*,\sigma^*)/\sigma^2$
7	0.6130	−0.1930			
	0.2586	−0.0737			
	0.1721	−0.0477			
	0.1289	−0.0350			
	0.1029	−0.0275			
	0.0854	−0.0226			
	0.0730	−0.0191			
	−0.4339	0.4186	0.8642	0.0886	−0.1881
8	0.5889	−0.1719			
	0.2501	−0.0662			
	0.1668	−0.0430			
	0.1251	−0.0316			
	0.0999	−0.0249			
	0.0830	−0.0205			
	0.0709	−0.0174			
	0.0618	−0.0150			
	−0.4465	0.3906	0.8485	0.0766	−0.1744
9	0.5683	−0.1553			
	0.2428	−0.0603			
	0.1621	−0.0392			
	0.1217	−0.0289			
	0.0973	−0.0228			
	0.0809	−0.0188			
	0.0691	−0.0159			
	0.0603	−0.0138			
	0.0534	−0.0121			
	−0.4560	0.3673	0.8342	0.0674	−0.1629

Table 5.3.2. Coefficients a_i and b_i of the BLUE's μ^* and σ^*, and the values of $\text{Var}(\mu^*)/\sigma^2$, $\text{Var}(\sigma^*)/\sigma^2$ and $\text{Cov}(\mu^*,\sigma^*)/\sigma^2$ for $L(\mu,\sigma^2)$ distribution

n	a_i	b_i	$\text{Var}(\mu^*)/\sigma^2$	$\text{Var}(\sigma^*)/\sigma^2$	$\text{Cov}(\mu^*,\sigma^*)/\sigma^2$
1	1.0000	−0.6079			
	0.0000	0.6079	3.2899	0.5148	−0.5151
2	0.9347	−0.3795			
	0.1546	0.0669			
	−0.0893	0.3126	3.2768	0.3542	−0.4692
3	0.8413	−0.2737			
	0.1734	0.0456			
	0.2101	−0.0265			
	−0.2248	0.2546	3.2026	0.2590	−0.3852
4	0.7688	−0.2136			
	0.1892	0.0325			
	0.2032	−0.0209			
	0.0861	0.0030			
	−0.2473	0.2049	3.1221	0.2037	−0.3185
5	0.7159	−0.1756			
	0.1989	0.0255			
	0.1986	−0.0176			
	0.0855	−0.0025			
	0.0379	0.0003			
	−0.2368	0.1699	3.0530	0.1682	−0.2689
6	0.6767	−0.1495			
	0.2054	0.0212			
	0.1950	−0.0152			
	0.0851	−0.0022			
	0.0380	0.0002			
	0.0169	0.0007			
	−0.2170	0.1447	2.9973	0.1434	−0.2318

(table continues)

Table 5.3.2 (continued.)

n	a_i	b_i	$\text{Var}(\mu^*)/\sigma^2$	$\text{Var}(\sigma^*)/\sigma^2$	$\text{Cov}(\mu^*, \sigma^*)/\sigma^2$
7	0.6468	-0.1303			
	0.2099	0.0183			
	0.1921	-0.0133			
	0.0847	-0.0020			
	0.0381	0.0002			
	0.0171	0.0006			
	0.0077	0.0009			
	-0.1964	0.1259	2.9530	0.1252	-0.2033
8	0.6234	-0.1156			
	0.2134	0.0161			
	0.1897	-0.0118			
	0.0843	-0.0018			
	0.0381	0.0001			
	0.0172	0.0006			
	0.0078	0.0005			
	0.0036	0.0004			
	-0.1776	0.1116	2.9173	0.1111	-0.1810
9	0.6047	-0.1040			
	0.2160	0.0145			
	0.1878	-0.0106			
	0.0841	-0.0016			
	0.0381	0.0001			
	0.0173	0.0005			
	0.0079	0.0004			
	0.0037	0.0003			
	0.0017	0.0002			
	-0.1614	0.1002	2.8884	0.0999	-0.1630

5.4   BEST LINEAR INVARIANT ESTIMATION

In the last section, we discussed the best linear unbiased estimation of the parameters of a distribution based on record values. In this section, we will describe the best linear invariant estimation which deals with optimal linear estimation through minimizing the mean square errors (in the class of linear estimators).

Example 5.4.1 (One-parameter Exponential Distribution) For the $\text{Exp}(\sigma)$ distribution, the BLUE of σ was derived in Example 5.3.1 to be $\sigma^* = R_n/(n+1)$ and its variance as $\text{Var}(\sigma^*) = \sigma^2/(n+1)$. In this case, instead of minimizing the variance among all linear unbiased estimators, we may minimize the mean square error among all linear estimators. This readily yields the estimator $\tilde{\sigma}^* = R_n/(n+2)$ with its mean square error being $\sigma^2/(n+2)$. The efficiency of this estimator, relative to the BLUE σ^*, is then $(n+2)/(n+1)$.   ◯

   With a similar motivation, for a general location-scale family of distributions, Mann (1969) derived the best linear invariant estimators (BLIE's), having minimum mean square error and invariance with respect to location parameter μ, of μ and σ as

$$\tilde{\mu}^* = \mu^* - \left(\frac{V_3}{1+V_2}\right)\sigma^* \quad \text{and} \quad \tilde{\sigma}^* = \sigma^*/(1+V_2), \qquad (5.4.1)$$

where μ^* and σ^* denote the BLUE's of μ and σ as before, and $V_1 = \text{Var}(\mu^*)/\sigma^2$, $V_2 = \text{Var}(\sigma^*)/\sigma^2$ and $V_3 = \text{Cov}(\mu^*, \sigma^*)/\sigma^2$. Furthermore, the mean square errors of these BLIE's are given by

$$MSE(\tilde{\mu}^*) = \sigma^2\left(V_1 - \frac{V_3^2}{1+V_2}\right) \quad \text{and} \quad MSE(\tilde{\sigma}^*) = \sigma^2 V_2/(1+V_2).$$

$$(5.4.2)$$

Example 5.4.2 (Two-parameter Exponential Distribution) For $\text{Exp}(\mu, \sigma)$ distribution, upon noting from (5.3.22) that $V_1 = (n+1)/n$, $V_2 = 1/n$ and $V_3 = -1/n$, (5.4.1) yields

$$\tilde{\mu}^* = \mu^* + \sigma^*/(n+1) \qquad \text{and} \qquad \tilde{\sigma}^* = n\sigma^*/(n+1).$$

Substituting now the expressions of μ^* and σ^* from (5.3.20) and (5.3.21) and simplifying, we obtain the BLIE's of μ and σ as

$$\tilde{\mu}^* = \frac{1}{n+1}\{(n+2)R_0 - R_n\} \quad \text{and} \quad \tilde{\sigma}^* = \frac{1}{n+1}(R_n - R_0). \quad (5.4.3)$$

In addition, we get

$$MSE(\tilde{\mu}^*) = \sigma^2(n+2)/(n+1) \quad \text{and} \quad MSE(\tilde{\sigma}^*) = \frac{\sigma^2}{n+1}. \qquad (5.4.4)$$

○

Example 5.4.3 (Two-parameter Uniform Distribution) For Uniform(μ, $\mu + \sigma$) distribution, upon observing from (5.3.27) that

$$V_1 = \frac{3^{n+1} - 1}{9(3^n - 1)}, \quad V_2 = \frac{3^{n+1} + 5}{9(3^n - 1)} \quad \text{and} \quad V_3 = -\frac{3^{n+1} + 1}{9(3^n - 1)},$$

(5.4.1) readily gives

$$\tilde{\mu}^* = \mu^* + \frac{3^{n+1} + 1}{4(3^{n+1} - 1)} \sigma^* \quad \text{and} \quad \tilde{\sigma}^* = \frac{9(3^n - 1)}{4(3^{n+1} - 1)} \sigma^*.$$

Substituting the expressions of μ^* and σ^* from (5.3.25) and (5.3.26) and simplifying, we obtain the BLIE's of μ and σ as

$$\tilde{\mu}^* = \frac{1}{3^{n+1} - 1} \left\{ \frac{1}{2} (3^{n+2} + 1) R_0 + \sum_{i=1}^{n-1} 3^i R_i - 2 \times 3^n R_n \right\} \qquad (5.4.5)$$

and

$$\tilde{\sigma}^* = \frac{9}{3^{n+1} - 1} \left\{ -\frac{1}{2} (3^n + 1) R_0 - \sum_{i=1}^{n-1} 3^{i-1} R_i + 2 \times 3^{n-1} R_n \right\}. \qquad (5.4.6)$$

In addition, we get

$$MSE(\tilde{\mu}^*) = \sigma^2 \left\{ \frac{3^{n+2} - 1}{12(3^{n+1} - 1)} \right\} \quad \text{and} \quad MSE(\tilde{\sigma}^*) = \sigma^2 \left\{ \frac{3^{n+1} + 5}{4(3^{n+1} - 1)} \right\}.$$

$$(5.4.7)$$

○

For all other cases as well, we can similarly derive the BLIE's of μ and σ from the corresponding BLUE's. For example, from Tables 5.3.1 and 5.3.2 we can easily determine the coefficients $\tilde{a}_i$ and $\tilde{b}_i$ of the BLIE's $\tilde{\mu}^*$ and $\tilde{\sigma}^*$ (using (5.4.1)) and the values of $MSE(\tilde{\mu}^*)/\sigma^2$ and $MSE(\tilde{\sigma}^*)\sigma^2$ (using (5.4.2)) for $N(\mu, \sigma^2)$ and $L(\mu, \sigma^2)$ distributions, respectively.

5.5 INTERVAL ESTIMATION AND TESTS OF HYPOTHESES

So far, we have discussed different methods for the point estimation of location and scale parameters based on record values. Though the exact sampling distributions of these estimators are intractable, in general, it is possible to derive them in a few cases.

Example 5.5.1 (Two-parameter Exponential Distribution) Let us assume that $R_0, R_1, \ldots, R_n$ are upper record values observed from $\mathrm{Exp}(\mu, \sigma)$ distribution. Then, in Section 5.2 (see Example 5.2.2) we derived the MLE's of μ and σ to be

$$\hat{\mu} = R_0 \quad \text{and} \quad \hat{\sigma} = \frac{1}{n+1}(R_n - R_0).$$

We have also seen in this case (in Section 2.3) that $(R_0 - \mu)/\sigma$, $(R_1 - R_0)/\sigma, \ldots, (R_n - R_{n-1})/\sigma$ are i.i.d. $\mathrm{Exp}(1)$ random variables. Consequently, the pivotal quantity

$$Z_1 = \frac{\hat{\mu} - \mu}{\sigma} = \frac{R_0 - \mu}{\sigma} \tag{5.5.1}$$

is distributed as $\mathrm{Exp}(1)$. Using this distributional result, confidence intervals for μ can be constructed and also tests of hypotheses concerning μ (when σ is known) can be carried out easily. Since $(n+1)\hat{\sigma} = R_n - R_0 = \sum_{i=1}^{n}(R_i - R_{i-1})$, we observe that the pivotal quantity

$$Z_2 = \frac{2(n+1)\hat{\sigma}}{\sigma} = \sum_{i=1}^{n} \frac{2(R_i - R_{i-1})}{\sigma} \tag{5.5.2}$$

is distributed as χ_{2n}^2. Using this distributional result, confidence intervals for σ can be constructed and also tests of hypotheses concerning σ can be carried out. Similarly, in the case when σ is unknown, interval estimation and tests of hypotheses for μ can be based on the pivotal quantity

$$Z_3 = \frac{\hat{\mu} - \mu}{\left(\frac{n+1}{n}\right)\hat{\sigma}} = \frac{R_0 - \mu}{\frac{1}{n}\sum_{i=1}^{n}(R_i - R_{i-1})}. \tag{5.5.3}$$

It is easy to verify that the pivotal quantity Z_3 in (5.5.3) has a $F_{2,2n}$ distribution (central F distribution with $(2, 2n)$ degrees of freedom). Note here that if we had used the BLUE's of μ and σ derived in Section 5.3 (see Example 5.3.5) instead of the MLE's, we would have based the confidence intervals and tests of hypotheses exactly on the same pivotal quantities. It is so because of the close relationship that exists in this case between the two sets of estimators.   ○

Even though these exact distributional results are simple and elegant, these type of exact results are possible only very few cases (Try Gumbel!). In all other cases, the required percentage points of the pivotal quantities Z_1, Z_2 and Z_3 may be determined through Monte Carlo simulations. These percentage points may then be used to construct confidence intervals and also carry out tests of hypotheses for the location and scale parameters. For this reason, Sultan and Balakrishnan (1997a,b,c) studied the higher-order moments of record values (moments and product moments of order up to four) and used them to determine the mean, variance and the coefficients of skewness and kurtosis of the pivotal quantities based on BLUE's. They then used these quantities to develop Edgeworth series approximations (see Johnson, Kotz and Balakrishnan, 1994, pp. 25–30) for the distributions of the pivotal quantities and then evaluated their accuracy in terms of probability coverage and width of the confidence intervals based on these Edgeworth series approximations. They developed this method of inference for the Rayleigh, Weibull and Gumbel populations.

However, if the conditional method of inference for location and scale parameters, as introduced by Fisher (1934) and explained in detail by Lawless (1982), is adopted here, then the conditional marginal distributions of the pivotal quantities can be used to construct exact conditional confidence intervals for the parameters of interest for any location-scale distribution; see Chan (1998). To be specific, let $R_0, R_1, \ldots, R_n$ be the upper record values observed from a distribution belonging to the location-scale family with pdf $f(x; \mu, \sigma)$ and cdf $F(x; \mu, \sigma)$ (introduced earlier in Section 5.2) given by

$$f(x; \mu, \sigma) = \frac{1}{\sigma} f\left(\frac{x-\mu}{\sigma}\right)$$

and

$$F(x; \mu, \sigma) = F\left(\frac{x-\mu}{\sigma}\right),$$

where $f(x)$ is the pdf of the standard member of the family ($\mu = 0$, $\sigma = 1$) and $F(x)$ is its corresponding cdf. Then, the joint pdf of $R_0, R_1, \ldots, R_n$ is (see (2.3.9))

$$f(r_0, r_1, \ldots, r_n; \mu, \sigma) = \frac{1}{\sigma^{n+1}} f\left(\frac{r_n - \mu}{\sigma}\right) \prod_{i=0}^{n-1} \left\{ \frac{f\left(\frac{r_i - \mu}{\sigma}\right)}{1 - F\left(\frac{r_i - \mu}{\sigma}\right)} \right\}. \quad (5.5.4)$$

From (5.5.4), it is evident that the joint pdf preserves the location-scale
structure and so the standardized variables $\frac{R_0-\mu}{\sigma}, \frac{R_1-\mu}{\sigma}, \ldots, \frac{R_n-\mu}{\sigma}$ have
a joint distribution functionally independent of μ and σ. This, in fact,
enables the development of conditional inference here. Let $(\tilde{\mu}, \tilde{\sigma})$ be any
pair of equivariant estimators; for example, they may be the MLE's or the
BLUE's or the BLIE's of μ and σ. Then,

$$Z_1 = \frac{\tilde{\mu} - \mu}{\tilde{\sigma}} \quad \text{and} \quad Z_2 = \frac{\tilde{\sigma}}{\sigma} \qquad (5.5.5)$$

are pivotal quantities, and the vector $\boldsymbol{A} = (A_0, A_1, \ldots, A_n)$ where $A_i = \frac{R_i-\tilde{\mu}}{\tilde{\sigma}}$ (for $i = 0, 1, \ldots, n$) is an ancillary statistic (of dimension only $n - 1$). Hence, inference for parameters μ and σ can be based on the joint
distribution of (Z_1, Z_2) conditional on the observed value $\boldsymbol{a}$ of $\boldsymbol{A}$. The
following theorem presents the joint pdf of Z_1 and Z_2 given $\boldsymbol{A} = \boldsymbol{a}$.

Theorem 5.5.1 (Chan, 1998) *Let $\tilde{\mu}$ and $\tilde{\sigma}$ be any pair of equivariant
estimators of μ and σ based on the observed values of $R_0, R_1, \ldots, R_n$. Let
$Z_1 = (\tilde{\mu} - \mu)/\tilde{\sigma}$, $Z_2 = \tilde{\sigma}/\sigma$ and $A_i = (r_i - \tilde{\mu})/\tilde{\sigma}$ for $i = 0, 1, \ldots, n$. Then
the joint pdf of Z_1 and Z_2, given $\boldsymbol{a} = (a_0, a_1, \ldots, a_n)$, is*

$$f(z_1, z_2 \mid \boldsymbol{a}) = C(\boldsymbol{a}, n) z_2^n \, f(a_n z_2 + z_1 z_2) \prod_{i=0}^{n-1} \left\{ \frac{f(a_i z_2 + z_1 z_2)}{1 - F(a_i z_2 + z_1 z_2)} \right\},$$

$$(5.5.6)$$

where $C(\boldsymbol{a}, n)$ is the normalizing constant.

From the joint density function in (5.5.6), the marginal conditional
pdf's of Z_1 and Z_2, given the ancillary statistic $\boldsymbol{a} = (a_0, a_1, \ldots, a_n)$, can
be obtained as

$$f(z_1 \mid \boldsymbol{a}) = \int_0^\infty f(z_1, z_2 \mid \boldsymbol{a}) \, dz_2 \qquad (5.5.7)$$

and

$$f(z_2 \mid \boldsymbol{a}) = \int_{-\infty}^\infty f(z_1, z_2 \mid \boldsymbol{a}) \, dz_1, \qquad (5.5.8)$$

respectively. These marginal conditional pdf's can be used to develop
conditional inference for the parameters μ and σ.

Note that, under this approach, conditional inference can also be developed for any quantile of the underlying distribution. Realizing that the

p-th quantile of the distribution (x_p) is given by $r_p = \mu + \sigma\, w_p$, where w_p is such that $F(w_p) = 1 - p$, we can use the pivotal quantity

$$Z_p = \frac{(\tilde{\mu} - \mu) - \sigma\, w_p}{\tilde{\sigma}} = \frac{\tilde{\mu} - r_p}{\tilde{\sigma}}, \qquad (5.5.9)$$

in order to develop conditional inference for the p-th quantile r_p for any $0 < p < 1$.

Example 5.5.2 (Extreme Value Distribution) Let us take $EV(\mu, \sigma)$ distribution considered earlier in Example 5.3.6. Here, we have

$$f(x) = e^x\, e^{-e^x}, \qquad -\infty < x < \infty, \qquad (5.5.10)$$

and

$$F(x) = 1 - e^{-e^x}, \qquad -\infty < x < \infty, \qquad (5.5.11)$$

respectively. In this case, as we have already seen in Example 5.3.6, the BLUE's of μ and σ are given by

$$\mu^* = \frac{\alpha_n}{n} \sum_{i=0}^{n-1} R_i + (1 - \alpha_n) R_n \qquad (5.5.12)$$

and

$$\sigma^* = R_n - \frac{1}{n} \sum_{i=0}^{n-1} R_i, \qquad (5.5.13)$$

where $\alpha_n = -\gamma + \sum_{i=1}^{n} \frac{1}{i}$ and $\gamma = 0.5772157\ldots$ is Euler's constant. It is easy to establish that these BLUE's are equivariant estimators. Now, upon using the expressions of $f(x)$ and $F(x)$ in (5.5.11) and (5.5.12), we obtain from (5.5.6) the conditional joint pdf of Z_1 and Z_2, given $\boldsymbol{a} = (a_0, a_1, \ldots, a_n)$, to be

$$f(z_1, z_2 \mid \boldsymbol{a}) = C(\boldsymbol{a}, n) z_2^n \exp\left\{ z_2 \sum_{i=0}^{n} a_i + (n+1) z_1 z_2 - e^{a_n z_2 + z_1 z_2} \right\},$$
$$-\infty < z_1 < \infty, \ 0 < z_2 < \infty. \qquad (5.5.14)$$

From (5.5.15), the conditional marginal distributions of the pivotal quantities Z_1, Z_2 and Z_p, conditioned on $\boldsymbol{a}$, can all be derived.

For example, the conditional distribution of the pivotal quantity Z_2, given $\boldsymbol{a} = (a_0, a_1, \ldots, a_n)$, can be shown to be distributed as Gamma$(n, \frac{1}{n})$; that is, the conditional pdf of Z_2, given $\boldsymbol{a}$, is

$$h_2(z \mid \boldsymbol{a}) = \frac{1}{\Gamma(n)} n^n z^{n-1} e^{-nz}, \qquad z > 0; \qquad (5.5.15)$$

for a proof, see Chan (1998). Similarly, the conditional cdf of Z_1, given $\boldsymbol{a}$, is given by

$$P(Z_1 \leq t \mid \boldsymbol{a}) = \int_0^\infty h_2(z \mid \boldsymbol{a}) \, I_{e^{e^{a_n z + tz}}} \, (n+1) \, dz, \qquad (5.5.16)$$

where $I_p(\alpha)$ denotes the incomplete gamma ratio defined as

$$I_p(\alpha) = \frac{1}{\Gamma(\alpha)} \int_0^p e^{-u} u^{\alpha-1} \, du.$$

Proceeding along the same lines, the conditional cdf of Z_p, given $\boldsymbol{a}$, can be shown to be

$$P(Z_p \leq t \mid \boldsymbol{a}) = \int_0^\infty h_2(z \mid \boldsymbol{a}) \, I_{e^{e^{a_n z + tz + w_p}}} \, (n+1) \, dz, \qquad (5.5.17)$$

where $w_p = \log\{-\log(1-p)\}$.

From the conditional distributions of the pivotal quantities Z_2, Z_1 and Z_p in (5.5.16), (5.5.17) and (5.5.18), respectively, conditional confidence intervals and conditional tests of hypotheses for the parameters σ, μ and r_p can all be developed. For example, from (5.5.16) the $100(1-\alpha)\%$ conditional confidence interval for σ can be obtained as

$$\left[\frac{2n\sigma^*}{\chi^2_{2n, 1-\frac{\alpha}{2}}} \, , \, \frac{2n\sigma^*}{\chi^2_{2n, \frac{\alpha}{2}}} \right], \qquad (5.5.18)$$

where $\chi^2_{m,\beta}$ denotes the β-th percentile of the χ^2_m distribution. However, the percentage points of Z_1 and Z_p, which are needed for the development of conditional inference for the parameters μ and r_p, need to be determined from (5.5.17) and (5.5.18) through numerical integration. Chan (1998) has recommended the usage of Simpson's composite algorithm (see Algorithm 4.1 in Burden and Faires, 1985) for this purpose. Yet, it is useful here to note that the index parameter of the incomplete gamma ratios in (5.5.17) and (5.5.18) is an integer; therefore, the incomplete gamma ratio in (5.5.18) can be written as

$$I_{e^{e^{a_n z + tz + w_p}}} \, (n+1) = 1 - e^{e^{a_n z + tz + w_p}} \sum_{i=0}^n \frac{e^{i(a_n z + tz + w_p)}}{i!} \, .$$

Consequently, only single integration is needed for the evaluation of the conditional cdf of Z_1 or Z_p. $\bigcirc$

Remark 5.5.1 Though the illustration of the conditional inference here has been made with the BLUE's of μ and σ, all the results will be exactly the same if we use any other set of equivalent estimators (like MLE's or BLIE's) in place of the BLUE's.

5.6   POINT PREDICTION

Prediction of future records becomes a problem of great interest. For example, while studying the record rainfalls or snowfalls, having observed the record values until the present time, we will be naturally interested in predicting the amount of rainfall or snowfall that is to be expected when the present record is broken for the first time in future.

Specifically, let $R_0, R_1, \ldots, R_n$ be the first $n + 1$ upper record values observed from a specific distribution. Then, we may be interested in predicting (either point prediction or interval prediction) the value of the next record (R_{n+1}), or, more generally, the value of the mth record (R_m) for some $m > n$.

5.6.1   Best Linear Unbiased Prediction

For any location-scale distribution (with μ as the location parameter and σ as the scale parameter), the best linear unbiased predictor (BLUP) of R_m can be obtained from the results on the general linear model due to Goldberger (1962). The BLUP of R_m is then given by

$$R_m^* = \mu^* + \alpha_m \sigma^* + \boldsymbol{\omega}^T \boldsymbol{\Sigma}^{-1}(\boldsymbol{R} - \mu^* \boldsymbol{1} - \sigma^* \boldsymbol{\alpha}), \qquad (5.6.1)$$

where μ^*, σ^* are the BLUE's of μ and σ discussed in Section 5.3, $\boldsymbol{R}$ is the vector of observed record values, $\boldsymbol{1}$ is a vector of 1's, $\boldsymbol{\alpha}$ is the vector of means of record values from the standard distribution, $\boldsymbol{\Sigma} = ((\sigma_{ij}))$ is the variance-covariance matrix of the standard record values, and $\boldsymbol{\omega}^T = (\sigma_{0,m}, \sigma_{1,m}, \ldots, \sigma_{n,m})$.

Example 5.6.1 (Two-parameter Exponential Distribution; Ahsanullah, 1980) For $\text{Exp}(\mu, \sigma)$ distribution, we have the expressions of $\boldsymbol{\alpha}$, $\boldsymbol{\Sigma}^{-1}$, μ^* and σ^* as in Section 5.3 (see Example 5.3.5), and $\alpha_m = m + 1$ and $\boldsymbol{\omega}^T = (1, 2, \ldots, n + 1)$ so that $\boldsymbol{\omega}^T \boldsymbol{\Sigma}^{-1} = (0\ 0\ \cdots\ 0\ 1)$. Upon substitution into (5.6.1), we obtain the BLUP of R_m to be

$$
\begin{aligned}
R_m^* &= \mu^* + (m+1)\sigma^* + \{R_n - \mu^* - (n+1)\sigma^*\} \\
&= R_n + (m-n)\sigma^* \\
&= R_n + \frac{m-n}{n}(R_n - R_0) \\
&= \frac{1}{n}\{mR_n - (m-n)R_0\}.
\end{aligned}
\qquad (5.6.2)
$$

From (5.6.2), we may also readily derive the variance of the BLUP of R_m as

$$\text{Var}(R_m^*) = \frac{\sigma^2}{n^2} \{m^2(n+1) + (m-n)^2 - 2m(m-n)\}$$

$$= \frac{\sigma^2}{n}(m^2 + n). \tag{5.6.3}$$

From (5.6.3), we observe that the precision of the BLUP of R_m decreases as m increases which is to be expected.

An alternate and simpler way of deriving the BLUP has been described recently by Doganaksoy and Balakrishnan (1997). For the BLUP of R_{n+1}, for example, all that this method requires is to equate the BLUE of μ (or σ) based on $(R_0, R_1, \ldots, R_n)$ and $(R_0, R_1, \ldots, R_n, R_{n+1})$ and then solve for R_{n+1}. In the case of the two-parameter exponential distribution, for example, we have from Example 5.3.5 the BLUE of σ based on $(R_0, R_1, \ldots, R_n)$ to be

$$\sigma_n^* = \frac{1}{n}(R_n - R_0). \tag{5.6.4}$$

Similarly, the BLUE of σ based on $(R_0, \ldots, R_n, R_{n+1})$ is

$$\sigma_{n+1}^* = \frac{1}{n+1}(R_{n+1} - R_0). \tag{5.6.5}$$

Upon equating (5.6.5) with (5.6.4) and solving for R_{n+1}, we obtain the BLUP of R_{n+1} as

$$R_{n+1}^* = R_0 + \frac{n+1}{n}(R_n - R_0) = \frac{1}{n}\{(n+1)R_n - R_0\}, \tag{5.6.6}$$

which agrees with the expression in (5.6.2). Since the BLUE remains unchanged if the BLUP of a future record is taken as an observed value of that record (see Doganaksoy and Balakrishnan, 1997), by repeated application of the expressions in (5.6.5) and (5.6.6), we readily obtain the BLUP of R_m as

$$R_m^* = R_0 + \frac{m}{n}(R_n - R_0) = \frac{1}{n}\{mR_n - (m-n)R_0\},$$

which is exactly the same as the expression in (5.6.2). $\bigcirc$

Remark 5.6.1 One can proceed exactly along the same lines and make use of the results presented earlier in Section 5.3, in order to derive the BLUP's for the case of two-parameter uniform, extreme value, Rayleigh, Weibull, logistic and normal distributions.

Example 5.6.2 (Two-parameter Uniform Distribution) For Uniform$(\mu, \mu + \sigma)$ distribution considered earlier in Example 5.3.7, we have from (5.3.25) the BLUE of μ based on $(R_0, R_1, \ldots, R_n)$ to be

$$\mu_n^* = \frac{2}{3^n - 1} \{3^n R_0 + R_1 + 3R_2 + \cdots + 3^{n-2} R_{n-1} - 2 \times 3^{n-1} R_n\}.$$

$$(5.6.7)$$

Similarly, the BLUE of μ based on $(R_0, R_1, \ldots, R_n, R_{n+1})$ is

$$\mu_{n+1}^* = \frac{2}{3^{n+1} - 1} \{3^{n+1} R_0 + R_1 + 3R_2 + \cdots + 3^{n-1} R_n - 2 \times 3^n R_{n+1}\}.$$

$$(5.6.8)$$

Upon equating (5.6.8) with (5.6.7) and solving for R_{n+1}, we obtain the BLUP of R_{n+1} as

$$
\begin{aligned}
R_{n+1}^* &= \frac{1}{2 \times 3^n} \left[3^{n+1} R_0 + R_1 + 3R_2 + \cdots + 3^{n-1} R_n \right. \\
&\quad \left. - \frac{3^{n+1} - 1}{3^n - 1} \left\{ 3^n R_0 + R_1 + 3R_2 + \cdots + 3^{n-2} R_{n-1} - 2 \times 3^{n-1} R_n \right\} \right] \\
&= -R_0 - R_1 - 3R_2 - \cdots - 3^{n-2} R_{n-1} + \left(3^n + \frac{3^{n-1} - 1}{2} \right) R_n \\
&= 3^n R_n - R_0 + \sum_{i=1}^{n-1} 3^{i-1} (R_n - R_i).
\end{aligned}
$$

$$(5.6.9)$$

By repeated application, we can once again determine the BLUP of R_m. $\bigcirc$

Example 5.6.3 (Extreme Value Distribution) For $EV(\mu, \sigma)$ distribution considered earlier in Example 5.3.6, we have the BLUE of σ based on $(R_0, R_1, \ldots, R_n)$ as

$$\sigma_n^* = \frac{1}{n} \sum_{i=0}^{n-1} (R_n - R_i).$$

$$(5.6.10)$$

Similarly, the BLUE of σ based on $(R_0, R_1, \ldots, R_n, R_{n+1})$ is given by

$$\sigma_{n+1}^* = \frac{1}{n+1} \sum_{i=0}^{n} (R_{n+1} - R_i).$$

$$(5.6.11)$$

Upon equating (5.6.11) with (5.6.10) and solving for R_{n+1}, we obtain the BLUP of R_{n+1} as

$$R_{n+1}^* = R_n - \frac{1}{n} \sum_{i=0}^{n-1} R_i + \frac{1}{n+1} \sum_{i=0}^{n} R_i$$

$$= \frac{n+2}{n+1} R_n - \frac{1}{n(n+1)} \sum_{i=0}^{n-1} R_i$$

$$= R_n + \frac{1}{n(n+1)} \sum_{i=0}^{n-1} (R_n - R_i). \qquad (5.6.12)$$

Remark 5.6.2 In the case of normal and logistic distributions, by adopting this approach the BLUP's of future records can be easily determined from the BLUE's tabulated in Tables 5.3.1 and 5.3.2.

5.6.2 Best Linear Invariant Prediction

From the results of Mann (1969), it is known that the best linear invariant predictor (BLIP) of R_m is given by

$$\tilde{R}^*_m = R^*_m - \left(\frac{V_4}{1+V_2}\right)\sigma^*, \qquad (5.6.13)$$

where $\mathrm{Var}(\sigma^*) = \sigma^* V_2$ and

$$\sigma^2 V_4 = \mathrm{Cov}\left(\sigma^*, \ (1-\boldsymbol{\omega}^T\boldsymbol{\Sigma}^{-1}\mathbf{1})\mu^* + (\alpha_m - \boldsymbol{\omega}^T\boldsymbol{\Sigma}^{-1}\boldsymbol{\alpha})\sigma^*\right). \quad (5.6.14)$$

Example 5.6.4 (Two-parameter Exponential Distribution; Ahsanullah, 1980) From Example 5.3.5, we have

$$
\begin{aligned}
\mu^* &= \{(n+1)R_0 - R_n\}/n, \\
\sigma^* &= (R_n - R_0)/n, \\
\boldsymbol{\Sigma}^{-1}\mathbf{1} &= (1\ 0\ \cdots\ 0\ 0)^T, \\
\boldsymbol{\Sigma}^{-1}\boldsymbol{\alpha} &= (0\ 0\ \cdots\ 0\ 1)^T, \\
\boldsymbol{\omega}^T &= (1\ 2\ \cdots\ n\ n+1),
\end{aligned}
$$

$\alpha_m = m+1$, and $\mathrm{Var}(\sigma^*) = \sigma^2 V_2 = \sigma^2/n$. Upon substituting these expressions in (5.6.14), we obtain

$$\sigma^2 V_4 = \mathrm{Cov}(\sigma^*, (m-n)\sigma^*) = (m \div n)\mathrm{Var}(\sigma^*) = \sigma^2 (m-n)/n.$$

With $V_2 = 1/n$ and $V_4 = (m-n)/n$, we then obtain from (5.6.13)

$$\tilde{R}^*_m = \{(m+1)R_n - (m-n)R_0\}/(n+1). \qquad (5.6.15)$$

Proceeding similarly, we can derive the BLIP's for some other distributions such as two-parameter uniform, extreme value, two-parameter Rayleigh and Weibull distributions. Also, in the case of normal and logistic distributions, the best linear invariant predictors can be computed from (5.6.13), by using the corresponding tables of BLUE's in Tables 5.3.1 and 5.3.2 and the tables of means, variances and covariances of record values presented earlier in Chapters 2 and 3.

5.6.3 Asymptotic Linear Prediction

While discussing the asymptotic linear prediction of extreme order statistics, Nagaraja (1984) observed that the asymptotic prediction of the order statistics $X_{m:N}$ based on $X_{1:N}, X_{2:N}, \ldots, X_{n+1:N}$ reduces to that of predicting the m-th upper record value R_m based on $R_0, R_1, \ldots, R_n$. Specifically, let there exist constants a_N^* and $b_N^* > 0$ such that

$$P\left(\frac{X_{1:N} - a_N^*}{b_N^*} \le x\right) = 1 - \{1 - F(a_N^* + b_N^* x)\}^N \to G^*(x) \quad (5.6.16)$$

as $N \to \infty$ for all x in the support of G^*. Then, it is known that G^* can be one of the following three cdf's (up to location and scale change):

$$
\begin{aligned}
G_1^*(x) &= 1 - e^{-(-x)^{-c}} &, \quad x < 0, \ c > 0 \ (Frech\acute{e}t),\\
G_2^*(x) &= 1 - e^{-x^c} &, \quad x > 0, \ c > 0 \ (Weibull),\\
G_3^*(x) &= 1 - e^{-e^x} &, \quad -\infty < x < \infty \ (Extreme \ value);
\end{aligned}
$$

see, for example, Section 8.3 of Arnold, Balakrishnan and Nagaraja (1992). If the condition in (5.6.16) holds, then the limiting distribution of $\frac{X_{k+1:N} - a_N^*}{b_N^*}$ is that of R_k, the k-th upper record value from the distribution G^*. Hence, $X_{1:N}, X_{2:N}, \ldots, X_{n+1:N}$ behave approximately as the first $(n+1)$ upper record values, $R_0, R_1, \ldots, R_n$, from the distribution $G^*(\mu + \sigma x)$ with $\mu = a_N^*$ and $\sigma = b_N^*$, if N is large.

In addition to discussing the BLUP and the BLIP of R_m for the three distributions above, Nagaraja (1984) has shown that, in the case of the Weibull distribution, when c is unknown, no linear unbiased estimators exist for the location μ or the scale σ or the shape parameter c. However, when c is known, BLUE's and BLIE's can be derived for μ and σ (see Sections 5.3 and 5.4). Nagaraja (1984) then discussed the BLUP and the BLIP of R_m based on the record values $R_0, R_1, \ldots, R_n$. The coefficients of record values in BLUE's, BLUP's and BLIP's for the Weibull distribution, taken from Nagaraja (1984), are presented in Tables 5.6.1 and 5.6.2.

Table 5.6.1. Coefficients of Record Values in the BLUP's, BLIP's and BLUE's for the Weibull Distribution when $n = 4$

$$c = 2.0$$

	0	1	2	3	4	MSE/σ^2
			i			
R_5^*	-0.1440	-0.0240	-0.0160	-0.0120	1.1960	0.0548
$\tilde{R}_5^*$	-0.1194	-0.0199	-0.0133	-0.0010	1.1626	0.0532
R_6^*	-0.2760	-0.0460	-0.0307	-0.0230	1.3757	0.1180
$\tilde{R}_6^*$	-0.2297	-0.0383	-0.0255	-0.0191	1.3126	0.1123
R_9^*	-0.6220	-0.1037	-0.0691	-0.0518	1.8466	0.3420
$\tilde{R}_9^*$	-0.5212	-0.0869	-0.0579	-0.0434	1.7094	0.3151
μ^*	1.4400	0.2400	0.1600	0.1200	-0.9600	0.4800
σ^*	-0.6603	-0.1100	-0.0734	-0.0550	0.8987	0.1521

$$c = 2/3$$

	0	1	2	3	4	MSE/σ^2
			i			
R_5^*	-0.3571	0.0179	0.0071	0.0036	1.3286	19.2857
$\tilde{R}_5^*$	-0.2692	0.0135	0.0054	0.0027	1.2477	18.1172
R_6^*	-0.7440	0.0372	0.0149	0.0074	1.6845	52.2054
$\tilde{R}_6^*$	-0.5564	0.0278	0.0111	0.0056	1.5119	46.8819
R_9^*	-2.0640	0.1032	0.0413	0.0206	2.8988	262.2638
$\tilde{R}_9^*$	-1.5159	0.0758	0.0303	0.0152	2.3946	216.8445
μ^*	1.1905	-0.0595	-0.0238	-0.0119	-0.0952	4.2857
σ^*	-0.0992	0.0050	0.0020	0.0010	0.0913	0.4893

Table 5.6.2. Coefficients of Record Values in the BLUP's, BLIP's and BLUE's for the Weibull Distribution when $n = 9$

$c = 2.0$

i	R_{10}^*	$\tilde{R}_{10}^*$	R_{11}^*	$\tilde{R}_{11}^*$	R_{14}^*	$\tilde{R}_{14}^*$	μ^*	σ^*
0	-0.0530	-0.0488	-0.1036	-0.0954	-0.2438	-0.2250	1.0605	-0.3396
1	-0.0088	-0.0081	-0.0173	-0.0159	-0.0406	-0.0375	0.1767	-0.0566
2	-0.0059	-0.0054	-0.0115	-0.0106	-0.0271	-0.0250	0.1178	-0.0377
3	-0.0044	-0.0041	-0.0086	-0.0080	-0.0203	-0.0187	0.0884	-0.0283
4	-0.0035	-0.0033	-0.0069	-0.0064	-0.0163	-0.0150	0.0707	-0.0226
5	-0.0029	-0.0027	-0.0058	-0.0053	-0.0135	-0.0125	0.0589	-0.0189
6	-0.0025	-0.0023	-0.0049	-0.0045	-0.0116	-0.0107	0.0505	-0.0162
7	-0.0022	-0.0020	-0.0043	-0.0040	-0.0102	-0.0094	0.0442	-0.0141
8	-0.0020	-0.0018	-0.0038	-0.0035	-0.0090	-0.0083	0.0393	-0.0126
9	1.0853	1.0785	1.1668	1.1536	1.3924	1.3621	-0.7070	0.5466
MSE/σ^2	0.0259	0.0257	0.0534	0.0528	0.1441	0.1408	0.3535	0.0615

$c = 2/3$

i	R_{10}^*	$\tilde{R}_{10}^*$	R_{11}^*	$\tilde{R}_{11}^*$	R_{14}^*	$\tilde{R}_{14}^*$	μ^*	σ^*
0	-0.1698	-0.1472	-0.3472	-0.3002	-0.9229	-0.7926	1.1317	-0.0345
1	0.0085	0.0074	0.0174	0.0150	0.0461	0.0396	-0.0566	0.0017
2	0.0034	0.0029	0.0069	0.0060	0.0185	0.0159	-0.0226	0.0007
3	0.0017	0.0015	0.0035	0.0030	0.0092	0.0079	-0.0113	0.0003
4	0.0010	0.0008	0.0020	0.0017	0.0053	0.0045	-0.0065	0.0002
5	0.0006	0.0005	0.0012	0.0011	0.0033	0.0028	-0.0040	0.0001
6	0.0004	0.0004	0.0008	0.0007	0.0022	0.0019	-0.0027	0.0001
7	0.0003	0.0002	0.0006	0.0005	0.0015	0.0013	-0.0019	0.0001
8	0.0002	0.0002	0.0004	0.0004	0.0011	0.0010	-0.0014	0.0000
9	1.1537	1.1332	1.3144	1.2718	1.8356	1.7176	-0.0247	0.0312
MSE/σ^2	29.7917	29.2638	70.6449	68.3590	273.4775	255.9214	4.0741	0.2314

5.7   INTERVAL PREDICTION

While all the details presented in the last section are concerning the point prediction of a future record, prediction intervals can also be constructed by using the distributions of appropriate pivotal quantities. In some cases, exact distribution can be derived for the pivotal quantity, while in other cases, the required percentage points may have to be either approximated or determined through Monte Carlo simulations. We shall present in this Section examples of both kinds.

5.7.1   Prediction Intervals Based on BLUE's

Example 5.7.1 (Two-parameter Exponential Distribution) For $\text{Exp}(\mu, \sigma)$ distribution, in order to construct a prediction interval for the record value R_m, let us consider the BLUP of R_m (see Example 5.6.1) given by

$$R_m^* = R_n + \frac{m - n}{n} (R_n - R_0) = R_n + (m - n)\sigma^*.$$

It is clear from the above formula that, if we were to use the pivotal quantity $(R_m - R_m^*)/\sigma^*$ for constructing a prediction interval for R_m, we may as well consider the pivotal quantity as

$$P = \frac{R_m - R_n}{\sigma^*(m-n)}.$$ (5.7.1)

Since

$$\frac{2(R_m - R_n)}{\sigma} = \sum_{i=n+1}^{m} \frac{2(R_i - R_{i-1})}{\sigma} \overset{d}{=} \chi^2_{2(m-n)}$$

and

$$\frac{2n\sigma^*}{\sigma} = \frac{2(R_n - R_0)}{\sigma} = \sum_{i=1}^{n} \frac{2(R_i - R_{i-1})}{\sigma} \overset{d}{=} \chi^2_{2n},$$

and that these two variables are statistically independent, we have from (5.7.1) that

$$P = \left\{ \frac{2(R_m - R_n)}{2(m-n)\sigma} \right\} \bigg/ \left\{ \frac{2n\sigma^*}{2n\sigma} \right\} \overset{d}{=} F_{2(m-n),2n}.$$ (5.7.2)

Hence, the necessary percentage points of the pivotal quantity P in (5.7.1) can be taken from the tables of percentage points of $F_{2m-2n,2n}$ distribution, using which prediction intervals for R_m can be constructed. $\bigcirc$

Example 5.7.2 (Normal and Logistic Distributions) For $N(\mu, \sigma^2)$ and $L(\mu, \sigma^2)$ distributions, the BLUE's of the location and scale parameters (μ and σ) are presented in Tables 5.3.1 and 5.3.2. By making use of these estimators, one may consider two pivotal quantities

$$P_1 = \frac{R_{n+1} - R_n}{\sigma^*} \quad \text{and} \quad P_2 = \frac{R_{n+1} - \mu^*}{\sigma^*}$$ (5.7.3)

for the purpose of constructing prediction intervals for R_{n+1}. Through Monte Carlo simulations (based on 5,001 runs), Balakrishnan and Chan (1995, 1998) and Balakrishnan, Ahsanullah and Chan (1995) have determined selected percentage points of the pivotal quantities P_1 and P_2 in (5.7.3) for the normal and logistic distributions, respectively. These values are presented in Tables 5.7.1–5.7.4. It should be mentioned here that these authors have also proposed the pivotal quantity P_2 for testing the spuriosity of the current record value R_{n+1} at a desired level of significance, with large values of P_2 supporting the spuriosity of R_{n+1}; see Section 5.8. $\bigcirc$

Table 5.7.1. Simulated Percentage Points of the Pivotal Quantity P_1 for $N(\mu, \sigma^2)$ Distribution

p	n				
	2	3	4	5	6
0.005	0.0030	0.0028	0.0019	0.0015	0.0014
0.01	0.0072	0.0060	0.0047	0.0032	0.0032
0.025	0.0145	0.0141	0.0098	0.0080	0.0078
0.05	0.0310	0.0234	0.0211	0.0171	0.0166
0.1	0.0618	0.0487	0.0434	0.0373	0.0333
0.9	4.5331	1.9162	1.3255	1.0222	0.9066
0.95	10.0115	3.1695	1.9651	1.4585	1.2656
0.975	19.4769	4.8885	2.8291	2.0058	1.6566
0.99	50.2796	8.3427	3.9870	2.6667	2.3044
0.995	103.4032	12.4844	4.9967	3.1599	2.8145

Table 5.7.2. Simulated Percentage Points of the Pivotal Quantity P_2 for $N(\mu, \sigma^2)$ Distribution

p	n				
	2	3	4	5	6
0.90	5.4363	3.4200	3.2937	3.3863	3.6309
0.95	10.9146	4.6883	3.9400	3.8241	3.9846
0.975	20.3801	6.4064	4.8063	4.4057	4.3646
0.99	51.1828	9.8381	5.9341	5.0637	4.9973
0.995	104.3064	13.9672	6.9805	5.5404	5.5519

Table 5.7.3. Simulated Percentage Points of the Pivotal Quantity P_1 for $L(\mu, \sigma^2)$ Distribution

p	n				
	2	3	4	5	6
0.01	0.0098	0.0108	0.0125	0.0106	0.0102
0.025	0.0292	0.0291	0.0277	0.0229	0.0256
0.05	0.0581	0.0591	0.0531	0.0505	0.0503
0.1	0.1268	0.1240	0.1142	0.1033	0.1104
0.9	10.0706	4.1691	3.4384	3.0907	2.9258
0.95	22.4724	6.3768	4.9800	4.3303	3.9349
0.975	46.6467	9.1746	7.0918	5.9316	5.1046
0.99	115.7234	14.9839	10.8573	8.5048	7.1207

Table 5.7.4. Simulated Percentage Points of the
Pivotal Quantity P_2 for $L(\mu, \sigma^2)$ Distribution

	n				
p	2	3	4	5	6
0.9	11.7156	7.1213	7.5213	8.1117	9.0774
0.95	24.1174	9.3364	9.0945	9.4247	10.0833
0.975	48.2917	11.9950	11.2842	11.1417	11.3239
0.99	117.3684	17.8432	14.6475	13.7101	13.2169
0.995	160.4331	24.4229	18.2532	15.8817	14.7741

5.7.2   Conditional Prediction Intervals

In Section 5.5, we discussed the conditional inference for the location and
scale parameters of any distribution. We may use the same conditional
approach to develop conditional prediction intervals for future records.
For this purpose, we start with the joint pdf of $R_0, R_1, \ldots, R_{n+1}$ given by

$$f(r_0, r_1, \ldots, r_{n+1}; \mu, \sigma) = \frac{1}{\sigma^{n+2}} \, f\left(\frac{r_{n+1} - \mu}{\sigma}\right) \prod_{i=0}^{n} \left\{ \frac{f\left(\frac{r_i - \mu}{\sigma}\right)}{1 - F\left(\frac{r_i - \mu}{\sigma}\right)} \right\}.$$

Upon transforming $(r_0, r_1, \ldots, r_{n+1})$ to $(\tilde{\mu}, \tilde{\sigma}, a_0, \ldots, a_{n-1})$, where a_i's are
the ancillary statistics defined by

$$a_i = \frac{r_i - \tilde{\mu}}{\tilde{\sigma}}, \qquad i = 0, 1, \ldots, n+1,$$

we may obtain the joint pdf of $(\tilde{\mu}, \tilde{\sigma}, a_0, a_1, \ldots, a_{n-1})$ as

$$C(\boldsymbol{a}, n) \frac{\tilde{\sigma}^n}{\sigma^{n+2}} \, f\left(a_{n+1} \frac{\tilde{\sigma}}{\sigma} + \frac{\tilde{\mu} - \mu}{\sigma}\right) \prod_{i=0}^{n} \left\{ \frac{f\left(a_i \frac{\tilde{\sigma}}{\sigma} + \frac{\tilde{\mu} - \mu}{\sigma}\right)}{1 - F\left(a_i \frac{\tilde{\sigma}}{\sigma} + \frac{\tilde{\mu} - \mu}{\sigma}\right)} \right\}.$$

$$(5.7.4)$$

Now upon transforming the variables $(\tilde{\mu}, \tilde{\sigma}, a_0, a_1, \ldots, a_{n-1})$ to $(Z_1, Z_2, Z_3,$
$a_0, a_1, \ldots, a_{n-2})$ in (5.7.4), we will obtain the joint pdf of $(Z_1, Z_2, Z_3, a_0, a_1,$
$\ldots, a_{n-2})$, from which the conditional joint pdf of (Z_1, Z_2, Z_3) given $\boldsymbol{a}$ can
be obtained to be

$$C'(\boldsymbol{a}, n) z_2^{n+1} f\left((z_3 + a_n)z_2 + z_1 z_2\right) \prod_{i=0}^{n-1} \left\{ \frac{f(a_i z_2 + z_1 z_2)}{1 - F(a_i z_2 + z_1 z_2)} \right\}.$$

$$(5.7.5)$$

Here, $Z_1 = (\tilde{\mu} - \mu)/\tilde{\sigma}$ and $Z_2 = \tilde{\sigma}/\sigma$ are pivotal quantities as defined earlier in (5.5.5), and $Z_3 = (R_{n+1} - R_n)/\tilde{\sigma}$ is the pivotal quantity to be used for prediction. From (5.7.5), we obtain the conditional prediction pdf of Z_3 as (Chan, 1998)

$$f(z_3 \mid \boldsymbol{a}) = C'(\boldsymbol{a}, n) \int_{-\infty}^{\infty} \int_0^{\infty} z_2^{n+1} f\left((z_3 + a_n)z_2 + z_1 z_2\right)$$
$$\times \prod_{i=0}^{n-1} \left\{ \frac{f(a_i z_2 + z_1 z_2)}{1 - F(a_i z_2 + z_1 z_2)} \right\} dz_2 \, dz_1, \quad 0 \le z_3 < \infty.$$

$$(5.7.6)$$

Example 5.7.3 (Extreme Value Distribution; Chan, 1998) Let us take $EV(\mu, \sigma)$ distribution considered earlier in Example 5.5.2. Since

$$f(x) = e^x \, e^{-e^x}, \qquad -\infty < x < \infty,$$

and

$$1 - F(x) = e^{-e^x}, \qquad -\infty < x < \infty,$$

upon substituting these expressions in (5.7.6) and then carrying out the integration, we will obtain the conditional marginal pdf of Z_3 given $\boldsymbol{a}$ as

$$f(z_3 \mid \boldsymbol{a}) = \frac{C''(\boldsymbol{a}, n)}{\{(n+1)z_3 + n\}^{n+1}}, \qquad 0 \le z_3 < \infty. \qquad (5.7.7)$$

From the fact that $\int_0^\infty f(z_3 \mid \boldsymbol{a}) dz_3 = 1$, we can find the normalizing constant $C''(\boldsymbol{a}, n) = (n+1)n^{n+1}$; thence, we obtain from (5.7.7) the conditional marginal pdf of Z_3 given $\boldsymbol{a}$ as

$$f(z_3 \mid \boldsymbol{a}) = (n+1) \left\{ \frac{n}{(n+1)z_3 + n} \right\}^{n+1}, \qquad z_3 \ge 0. \qquad (5.7.8)$$

From (5.7.8), we find the conditional cdf of Z_3 given $\boldsymbol{a}$ as

$$F(z_3 \mid \boldsymbol{a}) = 1 - \left\{ \frac{n}{(n+1)z_3 + n} \right\}^n, \qquad z_3 \ge 0,$$

and the p-th quantile of the conditional distribution of Z_3 as

$$z_{3,p} = \left(\frac{n}{n+1} \right) (1 - p)^{-1/n} - 1 + \frac{1}{n+1} .$$

We then have the $100(1-p)\%$ conditional prediction interval for R_{n+1} as

$$\left[R_n + \tilde{\sigma} \, z_{3,p/2}, \ R_n + \tilde{\sigma} \, z_{3,1-p/2} \right], \qquad (5.7.9)$$

where we may use the BLUE σ^* (or any other equivariant estimator) for $\tilde{\sigma}$. ○

5.7.3 Tolerance Region Prediction

The problem of tolerance region prediction has been discussed by Aitchison and Dunsmore (1975). For illustrating this approach, we shall use one- and two-parameter exponential distributions as done by Dunsmore (1983).

Example 5.7.4 (One-parameter Exponential Distribution; Dunsmore, 1983) For $\text{Exp}(\sigma)$ distribution, we have already seen that $R_0, R_1 - R_0, \ldots,$ $R_n - R_{n-1}$ are all i.i.d. $\text{Exp}(\sigma)$ random variables, and that R_n is a sufficient statistic for σ with its distribution as $\text{Gamma}(n + 1, \sigma)$. We may present predictions for the future record R_m in terms of the random variable $Y = R_m - R_n$. From the independence of record spacings, it is also evident that Y is distributed independently of R_n, and also that Y is distributed as $\text{Gamma}(m - n, \sigma)$. Then, from the framework presented in Aitchison and Dunsmore (1975), the regions

$$\left\{ \frac{1 - \beta(n+1, m-n; c)}{\beta(n+1, m-n; c)} R_n, \ \infty \right\} \quad \text{and}$$

$$\left\{ 0, \ \frac{1 - \beta(n+1, m-n; 1-c)}{\beta(n+1, m-n; 1-c)} R_n \right\} \quad (5.7.10)$$

are both tolerance regions for $Y = R_m - R_n$ with similar mean coverage c, where $\beta(n+1, m-n; c)$ denotes the c-th quantile of the $\text{Beta}(n+1, m-n)$ distribution. Similarly, the tolerance regions for Y which provide coverage c with guarantee d are

$$\left\{ \frac{\gamma(m-n, 1; 1-c)}{\gamma(m+1, 1; d)} R_n, \ \infty \right\} \quad \text{and} \quad \left\{ 0, \ \frac{\gamma(m-n, 1; c)}{\gamma(m+1, 1; 1-d)} R_n \right\},$$

$$(5.7.11)$$

where $\gamma(\ell, 1; c)$ denotes the c-th quantile of the $\text{Gamma}(\ell, 1)$ distribution. $\bigcirc$

Example 5.7.5 (Two-parameter Exponential Distribution; Dunsmore, 1983) For $\text{Exp}(\mu, \sigma)$ distribution, we have already seen that $(R_0, R_n - R_0)$ are jointly sufficient for (μ, σ) and are independently distributed as $\text{Exp}(\mu, \sigma)$ and $\text{Gamma}(n, \sigma)$. Once again, the random variable $Y = R_m - R_n$ is distributed as $\text{Gamma}(m-n, \sigma)$ independently of R_0 and R_n. Then, the regions

$$\left\{ \frac{1 - \beta(n, m-n; c)}{\beta(n, m-n; c)} (R_n - R_0), \ \infty \right\} \quad \text{and}$$

$$\left\{0, \ \frac{1 - \beta(n, m - n; 1 - c)}{\beta(n, m - n; 1 - c)} \ (R_n - R_0)\right\} \quad (5.7.12)$$

are both tolerance regions for $Y = R_m - R_n$ with similar mean coverage c. Moreover, the tolerance regions for Y which provide coverage c with guarantee d are

$$\left\{\frac{\gamma(m - n, 1; 1 - c)}{\gamma(n, 1; d)} \ (R_n - R_0), \ \infty\right\} \quad \text{and}$$

$$\left\{0, \ \frac{\gamma(m - n, 1; c)}{\gamma(n, 1; 1 - d)} \ (R_n - R_0)\right\}. \quad (5.7.13)$$

$\bigcirc$

5.7.4   Bayesian Prediction Intervals

A Bayesian approach may be adopted in order to derive the necessary predictive distribution; see Dunsmore (1983). Let θ (in the parameter space Θ) be a parameter with prior pdf $f(\theta)$. Suppose the given record value data is summarized by a sufficient statistic S that has pdf $f(s|\theta)$. Let $f(\theta|s)$ be the resulting posterior pdf of θ. The information about a future value Y can then be secured from the predictive density function

$$f(y|s) = \int_\theta f(y|\theta) f(\theta|s) \ d\theta. \quad (5.7.14)$$

Example 5.7.6 (One-parameter Exponential Distribution; Dunsmore, 1983) We noted earlier that $s = r_n$ is sufficient for θ in the case of one-parameter exponential distribution. For convenience, let us take the pdf as

$$f(x; \sigma) = \sigma \, e^{-\sigma x}, \qquad x \geq 0, \ \sigma > 0.$$

If we take the prior $f(\sigma)$ to be the conjugate Gamma(q, h), then the predictive distribution for $Y = R_m - R_n$ can be obtained from (5.7.14) as

$$f(y|r_n) = \frac{h_0^{q_0} \, y^{m-n-1}}{B(m - n, q_0)(y + h_0)^{q_0 + m - n}}, \qquad y > 0, \quad (5.7.15)$$

where $q_0 = q + m + 1$ and $h_0 = h + r_n$. Note that this is the beta density of the second kind (see Johnson, Kotz and Balakrishnan, 1995, p. 248) and will be denoted by $\text{Beta}_{\text{II}}(m - n, q_0, h_0)$. Instead, suppose we had started instead with the one-parameter exponential distribution with $1/\sigma$ in place of σ, then we would have taken $f(\sigma)$ to be the conjugate inverse gamma and proceeded similarly to derive the predictive distribution for Y. ○

Example 5.7.7 (Two-parameter Exponential Distribution; Dunsmore, 1983) For $\text{Exp}(\mu, \sigma)$ distribution, we saw earlier that R_0 and $R_n - R_0$ are jointly sufficient for (μ, σ). If we take the prior $f(\mu, \sigma)$ to be the conjugate exponential-gamma, Exp-Gamma(b, c, q, h), with pdf

$$f(\mu, \sigma) = c\sigma \, e^{-c\sigma(b-\mu)} h^q \sigma^{q-1} e^{-h\sigma} / \Gamma(q), \quad \mu < b, \ \sigma > 0,$$

then the predictive distribution for Y is once again $\text{Beta}_{\text{II}}(m - n, q_0, h_0)$, where $q_0 = q + m + \delta(c)$,

$$h_0 = \begin{cases} h + r_n - r_0 - c(r_0 - b) & \text{if } \ r_0 < b \\ h + r_n - r_0 + (r_0 - b) & \text{if } \ r_0 \geq b \end{cases}$$

and

$$\delta(c) = \begin{cases} 0 & \text{if } \ c = 0 \\ 1 & \text{if } \ c > 0 \end{cases} .$$

For point prediction, one may use the posterior mode $(m-n-1)h_0/(q_0+1)$ or the posterior mean $(m-n)h_0/(q_0-1)$. For interval prediction, one may use the intervals

$$\left\{ \frac{1 - \beta(q_0, m - n; k)}{\beta(q_0, m - n; k)} \, h_0, \ \infty \right\} \quad \text{and} \quad \left\{ 0, \ \frac{1 - \beta(q_0, m - n; 1 - k)}{\beta(q_0, m - n; 1 - k)} \, h_0 \right\}, \tag{5.7.16}$$

which provide Bayesian cover k.

It is useful to observe that, in the case of vague prior knowledge, the above point predictors for $Y = R_m - R_n$ simply become

$$\left(\frac{m - n - 1}{n + 1} \right) (r_n - r_0) \quad \text{and} \quad \left(\frac{m - n}{n - 1} \right) (r_n - r_0), \tag{5.7.17}$$

respectively. (Compare these with (5.6.2).) Further, the intervals in (5.7.16) become the mean coverage tolerance predictors in (5.7.12) with $k = c$. ○

5.8   ILLUSTRATIVE EXAMPLES

We will now consider several examples and illustrate the various inferential procedures developed in the preceding sections of this chapter.

Example 5.8.1 (Dunsmore, 1983; Balakrishnan and Chan, 1994) A rock crushing machine has to be reset if, at any operation, the size of rock being crushed is larger than any that has been crushed before. The data below are the sizes dealt with up to the third time that the machine has been reset:

$$9.3, \ 0.6, \ 24.4, \ 18.1, \ 6.6, \ 9.0, \ 14.3, \ 6.6, \ 13.0, \ 2.4, \ 5.6, \ 33.8.$$

If only the sizes at the operations when resetting was necessary had been observed, we would have observed the record values as

$$9.3, \ 24.4, \ 33.8.$$

Based on these observed values of the first three record values and assuming them to have come from an $\text{Exp}(\sigma)$ distribution, a 95% mean coverage tolerance region for the increase in size can be obtained to be $(0, 57.9)$ which provides $(33.8, 91.7)$ as an interval for the next record size of the rock. Similarly, with a vague prior, a 95% Bayesian prediction interval for the fourth record value turns out to be $(33.8, 91.7)$.

For the same record data, if a one-parameter Rayleigh distribution is assumed, a 90% prediction interval for the fourth record can be derived to be $(33.8, 49.6)$. Similarly, if a two-parameter Rayleigh distribution is assumed, a 90% prediction interval for the fourth record can be derived to be $(33.8, 69.1)$. $\bigcirc$

Example 5.8.2 (Balakrishnan and Chan, 1994) The first seven upper record values were simulated from a two-parameter Rayleigh distribution with location parameter $\mu = 50$ and scale parameter $\sigma = 10$. They are as follows:

$$66.24, \ 72.27, \ 78.07, \ 81.82, \ 86.33, \ 87.42, \ 90.05.$$

The BLUE's of μ and σ are computed in this case to be

$$
\begin{aligned}
\mu^* &= (1.2245 \times 66.24) + (0.2041 \times 72.27) + (0.1361 \times 78.07) \\
&\quad + (0.1020 \times 81.82) + (0.0817 \times 86.33) + (0.0680 \times 87.42) \\
&\quad - (0.8163 \times 90.05) \\
&= 54.54
\end{aligned}
$$

and

$$
\begin{aligned}
\sigma^* &= -(0.3332 \times 66.24) - (0.0555 \times 72.27) - (0.0370 \times 78.07) \\
&\quad - (0.0277 \times 81.82) - (0.0222 \times 86.33) - (0.0185 \times 87.42) \\
&\quad + (0.4942 \times 90.05) \\
&= 9.67.
\end{aligned}
$$

The 90% prediction interval for the eighth record then turns out to be (90.05, 97.14). ○

Example 5.8.3 (Normal Distribution; Balakrishnan and Chan, 1995) Let us consider the Venice sea-level data, giving the highest sea-levels in centimeters in Venice for the years 1931–1981, cited by Pirazzoli (1982). The upper record values for these largest sea-level measurements are

$$103, \ 121, \ 147, \ 151, \ 194,$$

with the last record of 194 observed in the year 1966.

Using the first four record values and assuming that they have come from $N(\mu, \sigma^2)$ distribution, the BLUEs of μ and σ are computed to be

$$
\begin{aligned}
\mu^* \ &= \ (0.7799 \times 103) + (0.3152 \times 121) + (0.2073 \times 147) \\
&\quad - (0.3025 \times 151) \\
&= \ 103.265
\end{aligned}
$$

and

$$
\begin{aligned}
\sigma^* \ &= \ -(0.4069 \times 103) - (0.1460 \times 121) - (0.0926 \times 147) \\
&\quad + (0.6454 \times 151) \\
&= \ 24.267.
\end{aligned}
$$

Using these estimates and the pivotal quantity $(R_5 - \mu^*)/\sigma^*$, we can conclude at 10% level of significance that the highest sea-level observation of 194 in the year 1966 to be a spurious value (or an outlier). ○

Remark 5.8.1 Though the above analysis based on the normal assumption may seem to be a naive one, it is worth mentioning that Smith (1986) carried out an elaborate analysis by admitting a linear trend in the location of a generalized extreme value model for the entire data set. Interestingly, Smith (1986, p. 40) pointed out that the very large observation of 194 corresponding to the flood of 1966 did not fit in under any statistical model considered.

Example 5.8.4 (Normal Distribution; Balakrishnan and Chan, 1998) Roberts (1979) has given the one-hour mean concentration of sulfur dioxide (in pphm) from Long Beach, California, for the years 1956 to 1974. From these data, the upper record values for the month of October turn out to be

$$26, \ 27, \ 40, \ 41.$$

Using these record values and assuming that they have come from $N(\mu, \sigma^2)$ population, we determine the BLUE's of μ and σ to be

$$
\begin{aligned}
\mu^* &= (0.7799 \times 26) + (0.3152 \times 27) + (0.2073 \times 40) - (0.3025 \times 41) \\
&= 24.677
\end{aligned}
$$

and

$$
\begin{aligned}
\sigma^* &= -(0.4069 \times 26) - (0.1460 \times 27) - (0.0926 \times 40) + (0.6454 \times 41) \\
&= 8.236;
\end{aligned}
$$

the standard errors of these estimates are given by

$$ SE(\mu^*) = 8.236\sqrt{0.9475} = 8.017 $$

and

$$ SE(\sigma^*) = 8.236\sqrt{0.227} = 3.887. $$

Also, the 90% prediction interval for the next upper record value is obtained as $(41, 51.917)$. ○

Example 5.8.5 (Logistic Distribution; Balakrishnan, Ahsanullah and Chan, 1995) For the Venice sea-level data considered earlier in Example 5.8.3, let us now assume $L(\mu, \sigma^2)$ distribution and determine the corresponding BLUE's of μ and σ from the first four record values as

$$
\begin{aligned}
\mu^* &= (0.8413 \times 103) + (0.1734 \times 121) + (0.2101 \times 147) \\
&\quad - (0.2248 \times 151) \\
&= 104.575
\end{aligned}
$$

and

$$
\begin{aligned}
\sigma^* &= -(0.2737 \times 103) + (0.0456 \times 121) - (0.0265 \times 147) \\
&\quad + (0.2546 \times 151) \\
&= 11.876.
\end{aligned}
$$

Thus, the BLUE's of the mean and standard deviation are 104.575 and 21.542 (why?). Observe the closeness of these estimates to the corresponding values determined in Example 5.8.3 based on the normal distribution. Once again, upon using the above estimates and the pivotal quantity $(R_5 - \mu^*)/\sigma^*$, one can also conclude at 10% level of significance that the record value of 194 in the year 1966 to be an outlier. ○

Example 5.8.6 (Extreme Value Distribution; Chan, 1998) For the data presented presented earlier in Example 5.8.4, let us consider the log-record values

$$3.258, \ 3.296, \ 3.689, \ 3.714$$

and assume them to have come from an extreme value distribution (so that the original record data are assumed to have come from a two-parameter Weibull distribution). Then the BLUE's of μ and σ are given by

$$
\begin{aligned}
\mu^* &= \frac{1.2561}{3} (3.258 + 3.296 + 3.689) + (1 - 1.2561)3.714 \\
&= 3.338
\end{aligned}
$$

and

$$
\begin{aligned}
\sigma^* &= 3.714 - \frac{1}{3} (3.258 + 3.296 + 3.689) \\
&= 0.300.
\end{aligned}
$$

After finding that

$$P(Z_1 \le -0.981|\boldsymbol{a}) = 0.05 \text{ and } P(Z_1 \le 3.811|\boldsymbol{a}) = 0.95,$$

where $Z_1 = (\mu^* - \mu)/\sigma^*$, an exact 90% conditional confidence interval for the location parameter μ is obtained to be $(2.195, 3.632)$. Similarly, after finding that

$$P(0.206 \le Z_2 \le 2.408|\boldsymbol{a}) = 0.950,$$

where $Z_2 = \sigma^*/\sigma$, an exact 95% conditional confidence interval for the scale parameter σ is obtained to be $(0.124, 1.451)$.

Next, let us consider the conditional prediction interval for the next record. For this purpose, we set $n = 3$ in (5.7.8), in order to obtain the conditional marginal pdf of the predictive pivotal quantity $Z_3 = (R_4 - R_3)/\sigma^*$. For the 90% prediction interval, we find

$$z_{3,0.9} = \frac{0.75}{(0.1)^{1/3}} - 1 + 0.25 = 0.8658.$$

Consequently, the 90% conditional prediction interval for R_4 turns out to be

$$(R_3, \ R_3 + 0.8658\sigma^*) = (3.714, \ 3.9737).$$

Upon exponentiating this interval, we obtain the 90% conditional prediction interval for the record R_4 in the original scale (mean concentration of sulfur dioxide in pphm) as (41, 53.183). Note that this prediction interval is quite close to the 90% unconditional prediction interval of (41, 51.917) determined earlier in Example 5.8.4 under the assumption of $N(\mu, \sigma^2)$ distribution for the original record data. ○

Example 5.8.7 (Gumbel Distribution; Ahsanullah, 1990) For the annual maxima of the one-hour mean concentration of sulfur dioxide (in pphm) data from Long Beach, California, mentioned earlier in Example 5.8.4, Roberts (1979) fitted a Gumbel(μ, σ) distribution and obtained the corresponding point estimates as 31.5 and 12.346, respectively. From the data on the annual maxima, one gets the lower record values as

$$47, 41, 32, 27, 20, 18.$$

Assuming that these lower record values have come from a Gumbel distribution, we may determine the BLUE's of μ and σ as

$$
\begin{aligned}
\mu^* &= 0.3412(47 + 41 + 32 + 27 + 20) - (0.7061 \times 18) \\
&= 44.3
\end{aligned}
$$

and

$$
\begin{aligned}
\sigma^* &= 0.2(47 + 41 + 32 + 27 + 20) - 18 \\
&= 15.4.
\end{aligned}
$$

Similarly, the BLIE's of μ and σ may be computed to be 48.7 and 12.8, and the MLE's of μ and σ as 51.7 and 12.8, respectively.

The standard errors of the BLUE's can be computed to be

$$\mathrm{SE}(\mu^*) = \sigma^* \sqrt{V_1} = 15.4\sqrt{0.763} = 13.45$$

and

$$\mathrm{SE}(\sigma^*) = \sigma^* \sqrt{V_2} = 15.4\sqrt{0.2} = 6.89.$$

Then, by determining the exact percentage points of the pivotal quantities $(\mu^* - \mu)/\sigma^*$ and σ^*/σ (of the former through numerical integration), Sultan and Balakrishnan (1997a) obtained the exact 90% confidence intervals for μ and σ to be (24.0, 90.1) and (8.4, 39.1), respectively. ○

5.9   INFERENCE WITH RECORDS AND INTER-RECORD TIMES

Consider an experiment in which only record-breaking values are observed; here, we shall assume that the data available is of the form R'_0, Δ_1, R'_1, $\Delta_2, \ldots$, where, as before, $R'_0, R'_1, \ldots$ are successive lower records and Δ_1, $\Delta_2, \ldots$ are the inter-record times (or the number of trials needed to obtain new records). Under this form of data, and assuming a specific parametric family of distributions for the underlying sequence of random variables, one may develop inferential procedures along the lines similar to those presented in the preceding sections. Samaniego and Whitaker (1986), for example, considered the exponential distribution and discussed the maximum likelihood estimation of the scale parameter in this framework under both inverse sampling and fixed sampling schemes.

For illustration, let us assume the underlying variables X_i's to be distributed as $\text{Exp}(\sigma)$ and that the observed data is $(r'_0, \delta_1, r'_1, \delta_2, \ldots, r'_n)$. In this case, the likelihood is readily obtained to be

$$
\begin{aligned}
L &= \prod_{i=0}^{n} f(r'_i; \sigma)\{1 - F(r'_i; \sigma)\}^{\delta_{i+1}-1} \\
&= \frac{1}{\sigma^{n+1}}\, e^{-\frac{1}{\sigma}\sum_{i=0}^{n}\delta_{i+1}r'_i}, \quad 0 \le r'_0 < r'_1 < \cdots < r'_n,
\end{aligned}
\qquad (5.9.1)
$$

where $\delta_{n+1} \equiv 1$. The MLE of σ is then

$$
\hat{\sigma} = \frac{1}{n+1}\sum_{i=0}^{n}\Delta_{i+1}R'_i.
\qquad (5.9.2)
$$

It can be easily verified that $\hat{\sigma}$ is a complete sufficient statistic and is also the MVUE of σ. It is worth noting that the above likelihood function is very similar in form to the likelihood function based on a Type-II progressively right censored sample; see, for example, Balakrishnan and Cohen (1991).

Instead of an inverse sampling scheme as discussed just now, we may consider the estimation problem under a fixed sample. Suppose the data collected is from the original variables $X_1, X_2, \ldots, X_n$, where the sample size n is fixed. Of course, the number of lower record values that occur in this sample is random. In this case, the data available may be represented as $r'_0, \delta_1, r'_1, \delta_2, \ldots, r'_N, \delta_{N+1}$, where $\delta_{N+1} = n - \sum_{i=1}^{N}\delta_i$ with $N \in \{0, 1, \ldots, n-1\}$ being random. For the case of $\text{Exp}(\sigma)$ distribution, the MLE of σ can be shown to be

$$\hat{\sigma} = \frac{1}{N+1} \sum_{i=0}^{N} \Delta_{i+1} R_i'. \tag{5.9.3}$$

Although this estimator is identical in form to the MLE presented earlier in (5.9.2) for the case of inverse sampling, the statistical properties and distribution theory are very different. Once again, it should be noted that this is analogous to the difference between the distributional properties of the MLE based on Type-I and Type-II progressively right censored samples from the exponential distribution; see Balakrishnan and Cohen (1991). For some asymptotic properties (as $n \to \infty$) of the MLE $\hat{\sigma}$, one may refer to Samaniego and Whitaker (1986).

One can also discuss nonparametric inferential methods in this context. For example, by assuming that the underlying sequence of variables X_i's are i.i.d. with cdf $F(x)$, and that the observed data is of the form $(r_0', \delta_1, r_1', \delta_2, \ldots, r_N', \delta_{N+1})$, we can discuss the MLE of the cdf F. Clearly, the MLE $\hat{F}$ then should maximize the likelihood function

$$L = \prod_{i=0}^{N} \{F(r_i'; \sigma) - F(r_i' - ; \sigma)\}\{1 - F(r_i'; \sigma)\}^{\delta_{i+1}-1},$$

$$r_0' < r_1' < \cdots < r_N', \tag{5.9.4}$$

where $N \in \{0, 1, \ldots, n-1\}$ is random, and $\delta_{N+1} = n - \sum_{i=1}^{N} \delta_i$. Samaniego and Whitaker (1988) have then derived the nonparametric MLE of the survival function $1 - F$, which is similar to the Kaplan-Meier estimator. Considerable amount of further work has been carried out in this direction (using kernel density estimation methods) by Gulati and Padgett (1992, 1994a,b,c,d, 1997).

Mention should also be made to the works of Smith (1988), Smith and Miller (1986) and Tryfos and Blackmore (1985) in which the estimation of parameters as well as predictions have been discussed based on data concerning best performances in Olympic and marathon events. Note that though their articles mention records, the data they have considered involve only observations on best performances (in the event of interest) every year and not records in the sense we have seen so far.

5.10   DISTRIBUTION-FREE TESTS IN TIME-SERIES USING RECORDS

Following the foundational work of Chandler (1952), Foster and Stuart (1954) used the idea of records to propose two simple statistics for distribution-free tests of the randomness of a series of observations. Having made a series of observations, let S and D denote the sum and the difference of the numbers of upper and lower record values in the series. Intuitively, it is clear that the difference D will provide a test for trend in location since, if there is a trend in location, the time-series will produce more records in one direction than in the other. Similarly, it is also evident that the sum S will provide a test for trend in the dispersion of the series since, if there is an increasing (decreasing) dispersion trend, the time-series will produce many (few) records.

Under the null hypothesis of randomness (i.e., no trend in the time-series), Foster and Stuart (1954) discussed the joint distribution of S and D. They are uncorrelated for any sample size and are asymptotically independently normally distributed. Tables 5.10.1 and 5.10.2, taken from Foster and Stuart (1954), give the cdf's of the statistics S and D for some selected values of n. It should be mentioned that here the time-series observed is taken to be $X_0, X_1, X_2, \ldots, X_n$, with X_0 not being taken to be a record value. Note that if we treat the initial value X_0 as a record, then only a minor change is needed in Table 5.10.1. From Table 5.10.2, we observe that the normal approximation for the distribution of D works remarkably well, even at $n = 5$. But for the distribution of S, the approximation is close in the upper tail, even at $n = 14$, but not so in the lower tail. The mean and standard deviation of S and D that are needed for the normal approximation are presented in Table 5.10.3, taken from Foster and Stuart (1954), for $n = 9(5)99$.

Table 5.10.1. Cdf of S for $2 \leq n \leq 14$, with Normal Approximation for $n = 14$

					s						
n	1	2	3	4	5	6	7	8	9	10	11
2	.333	1.0000									
3	.167	.667	1.0000								
4	.100	.467	.867	1.0000							
5	.067	.345	.733	.956	1.0000						
6	.048	.265	.622	.892	.987	1.0000					
7	.036	.211	.533	.825	.964	.9968	1.0000				
8	.028	.172	.461	.760	.933	.989	.9993	1.0000			
9	.022	.143	.403	.700	.898	.978	.9974	.9999	1.0000		
10	.018	.121	.356	.646	.862	.964	.9938	.9994	1.0000		
11	.015	.104	.317	.598	.826	.947	.989	.9985	.9999	1.0000	
12	.013	.090	.284	.555	.791	.928	.982	.9970	.9997	1.0000	
13	.011	.079	.256	.516	.757	.908	.975	.9949	.9993	.9999	1.0000
14	.010	.070	.233	.481	.725	.888	.966	.9922	.9987	.9998	1.0000
Normal approxi-mation, $n = 14$	0.20	0.080	.228	.464	.715	.890	.970	.9945	.9993	.9999	1.0000

Table 5.10.2. Cdf of D for $2 \leq n \leq 5$, with Normal Approximation for $n = 5$. Only the Positive Half of the Symmetrical Distribution is Presented

			n		Normal Approximation,
d	2	3	4	5	$n = 5$
0	.667	.625	.617	.611	.615
1	.833	.833	.817	.804	.811
2	1.0000	.958	.942	.929	.929
3		1.0000	.9917	.985	.980
4			1.0000	.9986	.9959
5				1.0000	.9994

Table 5.10.3. Mean and Standard Deviation of S, and Standard Deviation of D, for $n = 9(5)99$

n	$E(S)$	S.D.(S)	S.D.(D)
9	3.858	1.288	1.964
14	4.636	1.521	2.153
19	5.195	1.677	2.279
24	5.632	1.791	2.373
29	5.990	1.882	2.447
34	6.294	1.956	2.509
39	6.557	2.019	2.561
44	6.790	2.072	2.606
49	6.998	2.121	2.645
54	7.187	2.163	2.681
59	7.360	2.201	2.713
64	7.519	2.236	2.742
69	7.666	2.268	2.769
74	7.803	2.297	2.793
79	7.931	2.324	2.816
84	8.051	2.349	2.837
89	8.165	2.373	2.857
94	8.273	2.395	2.876
99	8.375	2.416	2.894

These tests for trend are, however, not invariant under a reversal of the direction of the time variable, since an observation which is a record in one direction is not necessarily a record in the other. If we now denote S' and D' for the corresponding statistics obtained by counting backwards, then Foster and Stuart (1954) also considered the "round-trip" tests defined by

$$D^* = D - D' \qquad \text{and} \qquad S^* = S - S'.$$

As before, D^* may be used in a test for trend in location and S^* may be used in a test for trend in the dispersion. Foster and Stuart (1954) carried out a limited power study to examine the performance of the tests based on D and D^* using simulations from a normal distribution with a positive linear trend in the mean and a constant variance. A more extensive empirical power study in the same setting was carried out by Foster and Teichroew (1955); see also, Iiyama, Nishimura and Sibuya (1995). While examining the asymptotic relative efficiencies of various tests of randomness against normal regression (such as Spearman's rank correlation test, Kendall's rank correlation test, rank serial correlation test, Brown-Mood median test, and sign tests), Stuart (1954, 1956, 1957) showed that the D-test based on records has an asymptotic relative efficiency of zero compared with the regression coefficient test and some other tests as well. However, Diersen and Trenkler (1996) have shown that these records tests have good efficiencies in the case of some non-normal distributions such as uniform and exponential. With the intention of improving the performance of the records tests, Diersen and Trenkler (1996) also considered breaking the time-series into several parts, finding the statistics D and D^* in each part, and using the sum of these statistics as a test statistic for trend in location. By considering the randomness as the null hypothesis and a polynomial trend in location as the alternative hypothesis, these authors have examined the asymptotic relative efficiencies of the records tests based on a split series. Due to the fact that the probability that an observation exceeds the actual record decreases with the number of observations taken, Diersen and Trenkler (1996) proposed tests in which optimal weights and linear weights are assigned to records, in order to assign more weight to records occurring near the end of the series.

Example 5.10.1 Consider the data presented in Table 5.10.4, taken from Foster and Stuart (1954), giving total annual rainfall (in inches) at Oxford, England, for the years 1858–1952.

From these data, we have $n = 94$ and also find

$$D = 0, \ S = 8, \ D' = -1, \ S' = 7, \ D^* = 1 \text{ and } S^* = 1.$$

While we have the observed value of D to be 0, which is exactly the same as the expected value of D, we also note from Table 5.10.3 that the observed value of S being 8 is very close to the expected value of S given by 8.273. We, therefore, can conclude that the time-series data on annual rainfall at Oxford do not show a trend in either the mean or the variance.

○

Table 5.10.4. Total Annual Rainfall (in Inches) at Oxford, England

Year	Rainfall	From top	From bottom	Year	Rainfall	From top	From bottom
1858	20.77			1906	24.01		
1859	25.30	R_1		1907	26.77		
1860	30.41	R_2		1908	23.57		
1861	22.59			1909	27.50		
1862	26.72			1910	28.82		
1863	21.91			1911	21.09		
1864	17.87	R'_1		1912	32.43		
1865	28.07			1913	25.11		
1866	30.17			1914	29.51		
1867	26.56			1915	31.45		
1868	25.88			1916	31.55		
1869	25.94			1917	24.92		
1870	17.20	R'_2		1918	27.13		
1871	20.71			1919	26.30		
1872	29.21			1920	25.46		
1873	22.96			1921	14.94	R'_4	R'_4
1874	20.97			1922	26.38		
1875	32.42	R_3		1923	26.29		
1876	31.77			1924	33.70		
1877	29.53			1925	26.64		
1878	26.43			1926	27.38		
1979	30.79			1927	34.84		R_2
1880	30.91			1928	22.65		
1881	26.05			1929	22.43		
1882	31.48			1930	26.91		
1883	26.63			1931	27.23		
1884	18.89			1932	26.48		
1885	25.75			1933	19.30		R'_3
1886	31.94			1934	21.92		
1887	19.09			1935	30.30		
1888	27.29			1936	25.71		
1889	23.53			1937	27.55		
1890	17.80			1938	22.67		
1891	27.66			1939	30.43		
1892	20.50			1940	25.75		
1893	17.65			1941	25.53		
1894	28.54			1942	26.26		
1895	22.59			1943	20.22		
1896	23.68			1944	21.95		
1897	26.36			1945	22.38		
1898	19.22			1946	26.96		
1899	21.03			1947	19.93		R'_2
1900	23.61			1948	28.64		
1901	22.27			1949	22.72		R'_1
1902	16.66	R'_3		1950	28.84		
1903	35.95	R_4	R_3	1951	33.70		R_1
1904	22.66			1952	25.87		
1905	20.92						

EXERCISES

1. (a) For $N(\mu, \sigma^2)$ distribution, use (5.2.21) and (5.2.22) to develop a computer program to numerically determine the MLE's of μ and σ based on the observed values of the upper records $R_0, R_1, \ldots, R_n$.

   (b) For the data in Example 5.8.4, use your program to numerically determine the MLE's of μ and σ. Compare these with the values of BLUE's of μ and σ given by 24.677 and 8.236, respectively, and comment.

   (c) For the data in Example 5.8.3, by considering the first four upper record values only, compute the MLE's of μ and σ. Do the same by considering all five upper record values. Compare the results and comment.

2. (a) For $L(\mu, \sigma^2)$ distribution, use (5.2.23) and (5.2.24) to develop a computer program to numerically determine the MLE's of μ and σ based on the observed values of the upper records $R_0, R_1, \ldots, R_n$.

   (b) For the data in Example 5.8.5, by considering the first four upper record values only, compute the MLE's of μ and σ. Do the same by considering all five upper record values. Compare the results and comment.

3. Prove Lemma 5.3.1.   [Graybill, 1983, p. 198]

4. Assume that the record values $R_0, R_1, \ldots, R_n$ are observed from the Uniform$(0, \sigma)$ distribution.

   (a) Show that the random variables $R_0, R_1 - \frac{1}{2} R_0, R_2 - \frac{1}{2} R_1, \ldots,$ $R_n - \frac{1}{2} R_{n-1}$ are all uncorrelated.

   (b) Show that $E(R_i - \frac{1}{2} R_{i-1}) = \frac{1}{2}$ and $\text{Var}(R_i - \frac{1}{2} R_{i-1}) = \frac{1}{4 \times 3^{i+1}}$ for $i = 0, 1, \ldots, n$ (with $R_{-1} \equiv 0$).

   (c) Then, upon taking the BLUE of σ to be $\sigma^* = \sum_{i=0}^{n} c_i (R_i - \frac{1}{2} R_{i-1})$, show that

   $$E(\sigma^*) = \frac{\sigma}{2} \sum_{i=0}^{n} c_i \quad \text{and} \quad \text{Var}(\sigma^*) = \frac{\sigma^2}{4} \sum_{i=0}^{n} c_i^2 / 3^{i+1}.$$

   (d) Using these expressions and adopting the Lagrangian method, show that the optimal solution is given by $c_i = 4 \times 3^i / (3^{n+1} - 1)$

for $i = 0, 1, \ldots, n$; thence, show that the BLUE σ^* is as given in (5.3.7) and its variance to be as presented in (5.3.8).

5. For the scale-parameter Weibull distribution considered in Section 5.3, carry out the necessary algebraic steps, in order to show that the BLUE of σ is as given in (5.3.13) and its variance is as in (5.3.14).

[Balakrishnan and Chan, 1994]

6. For $EV(\mu, \sigma)$ distribution considered in Section 5.3, using Lemma 5.3.1, derive directly the BLUE's of μ and σ and show them to be the same as presented in Section 5.3.

[Ahsanullah, 1990; Sultan and Balakrishnan, 1997c]

7. For Uniform$(\mu, \mu + \sigma)$ distribution considered in Section 5.3, use the direct Lagrangian method to derive the BLUE's of μ and σ. Show that they are the same as presented in (5.3.25) and (5.3.26), respectively, thus showing that these estimators are trace-efficient and also determinant-efficient linear unbiased estimators.

8. For the two-parameter Rayleigh and Weibull distributions considered in Section 5.3, derive explicit expressions for the BLUE's of μ and σ and the variances and covariance of these estimators.

[Nagaraja, 1984; Balakrishnan and Chan, 1994]

9. (a) Based on the BLUE's derived in the last exercise, propose some pivotal quantities for the construction of confidence intervals for μ (when σ is known) and σ (when μ is unknown).

(b) After deriving expressions for moments and product moments (of order up to four) of record values from Rayleigh and Weibull distributions, compute the mean, variance and the coefficients of skewness and kurtosis of your pivotal quantities (since they will be linear functions of record values).

(c) Using the computed values of the mean, variance and the coefficients of skewness and kurtosis, develop Edgeworth series approximations for the distributions of the two pivotal quantities.

(d) For selected values of n (say, up to 8), determine the lower and upper 1%, 2.5% and 5% points from the Edgeworth series approximations.

[Sultan and Balakrishnan, 1997a,b]

10. First seven upper record values were simulated from a logistic distribution with location parameter $\mu = 4$ and scale parameter $\sigma = 3$. They are as follows:

$$5.024,\ 7.795,\ 10.192,\ 17.850,\ 18.988,\ 22.008,\ 25.777.$$

Based on these record data, compute the BLUE's of μ and σ and their standard errors. [Balakrishnan, Ahsanullah and Chan, 1995].

11. For $N(\mu, \sigma^2)$ and $L(\mu, \sigma^2)$ distributions, Tables 5.3.1 and 5.3.2 present the coefficients a_i and b_i of the BLUE's μ^* and σ^* and the variances and covariance of these estimators. Upon using these tables and (5.4.1), construct the corresponding tables for the BLIE's $\tilde{\mu}^*$ and $\tilde{\sigma}^*$.

12. Show that the statistic Z_3 in (5.5.3) has a $F_{2,2n}$ distribution.

13. For $EV(\mu, \sigma)$ distribution, starting from (5.5.15), show that the conditional marginal distributions of the pivotal quantities Z_1, Z_2 and Z_p are as presented in (5.5.17), (5.5.16) and (5.5.18), respectively.

[Chan, 1998]

14. (a) For Uniform$(\mu, \mu + \sigma)$ distribution, by repeatedly applying the argument used to derive the BLUP of R_{n+1} in Section 5.6, derive the BLUP of R_m for $m \geq n + 2$.

    (b) In this case, proceed directly from the formula in (5.6.1) to show that the BLUP of R_{n+1} is as presented in (5.6.9).

    (c) Proceed similarly to derive the BLUP of R_m, where $m \geq n + 2$.

15. (a) For $EV(\mu, \sigma)$ distribution, by repeatedly applying the argument used to derive the BLUP of R_{n+1} in Section 5.6, derive the BLUP of R_m for $m \geq n + 2$.

    (b) In this case, proceed directly from the formula in (5.6.1) to show that the BLUP of R_{n+1} is as presented in (5.6.12).

    (c) Proceed similarly to derive the BLUP of R_m, where $m \geq n + 2$.

16. From (5.6.13), derive the BLIP $\tilde{R}_m^*$ for Uniform$(\mu, \mu + \sigma)$ distribution.

17. From (5.6.13), derive the BLIP $\tilde{R}_m^*$ for $EV(\mu, \sigma)$ distribution.

18. Consider the two-parameter power function distribution with pdf

$$f(x; \mu, \sigma) = \frac{\nu}{\sigma} \left(\frac{x - \mu}{\sigma} \right)^{\nu-1}, \quad \mu < x < \mu + \sigma,$$

where the shape parameter ν is assumed to be known.

(a) Derive the BLUE's and BLIE's of the parameters μ and σ based on the record values.

(b) Can you derive the MLE's of the parameters μ and σ, or do they have to be numerically determined?

(c) Derive the BLUP's and BLIP's of future records.

19. Consider the two-parameter Pareto distribution with pdf

$$f(x; \mu, \sigma) = \frac{\nu}{\sigma} \left(\frac{x - \mu}{\sigma} \right)^{-\nu-1}, \quad \mu + \sigma < x < \infty,$$

where the shape parameter ν is assumed to be known.

(a) Derive the BLUE's and BLIE's of the parameters μ and σ based on the record values.

(b) Can you derive the MLE's of the parameters μ and σ, or do they have to be numerically determined?

(c) Derive the BLUP's and BLIP's of future records.

20. Consider the generalized extreme value distribution with cdf

$$F(x; \mu, \sigma) = exp \left[- \left\{ 1 - k \left(\frac{x - \mu}{\sigma} \right) \right\}^{1/k} \right],$$

where the shape parameter k is assumed to be known. The support of the distribution is $x < \mu + (\sigma/k)$ when $k > 0$, $x > \mu + (\sigma/k)$ when $k < 0$, and $-\infty < x < \infty$ when $k = 0$ (in which case the distribution is $EV(\mu, \sigma)$).

(a) Derive the BLUE's and BLIE's of μ and σ.

(b) Derive the BLUP's and BLIP's for the future record R_m.

(c) Propose some pivotal quantities for the construction of confidence intervals for μ and σ, and prediction intervals of R_{n+1}.

(d) Can you find the necessary percentage points of these pivotal quantities?

21. Consider the generalized Pareto distribution with pdf

$$f(x; \mu, \sigma) = \frac{1}{\sigma} \left\{ 1 + k \left(\frac{x - \mu}{\sigma} \right) \right\}^{-(1+k^{-1})},$$

where the shape parameter k is assumed to be known. The support of the distribution is $x \geq \mu$ when $k > 0$, $\mu < x < \mu - (\sigma/k)$ when $k < 0$, and $\mu \leq x < \infty$ when $k = 0$ (in which case the distribution is $\text{Exp}(\mu, \sigma)$).

(a) Derive the BLUE's and BLIE's of μ and σ.

(b) Derive the BLUP's and BLIP's for the future record R_m.

(c) Propose some pivotal quantities for the construction of confidence intervals for μ and σ, and prediction intervals of R_{n+1}.

(d) Can you find the necessary percentage points of these pivotal quantities?

22. For the extreme value distribution, starting from (5.7.6), show that the conditional marginal pdf of the pivotal quantity Z_3 is as given in (5.7.8). [Chan, 1998]

23. For the two-parameter exponential distribution, derive the maximum likelihood estimators under the two sampling schemes discussed in Section 5.9.

24. For the two-parameter Weibull distribution with pdf

$$f(x; \sigma, c) = \frac{cx^{c-1}}{\sigma^c} e^{-(x/\sigma)^c}, \quad x > 0, \ \sigma > 0, \ c > 0,$$

discuss the maximum likelihood estimation of the parameters σ and c under the two sampling schemes discussed in Section 5.9.

[Hoinkes and Padgett, 1994]

25. For the statistic S defined in Section 5.10 as the sum of the numbers of upper and lower record values in the series $X_0, X_1, \ldots, X_n$ (with X_0 not being taken to be a record value), show that $P(S = 1) = 2/\{n(n+1)\}$. [Foster and Stuart, 1954]

26. Show that the statistics S and D defined in Section 5.10 are uncorrelated for any n. Show also that they are asymptotically independently normally distributed.

[Foster and Stuart, 1954]

27. The following data represent the amount of annual (Jan. 1 – Dec. 31) rainfall in inches recorded at Los Angeles Civic Center during the 100-year period from 1890 until 1989:

12.69	12.84	18.72	21.96	7.51	12.55	11.80	14.28
4.83	8.69	11.30	11.96	13.12	14.77	11.88	19.19
21.46	15.30	13.74	23.92	4.89	17.85	9.78	17.17
23.21	16.67	23.29	8.45	17.49	8.82	11.18	19.85
15.27	6.25	8.11	8.94	18.56	18.63	8.69	8.32
13.02	18.93	10.72	18.76	14.67	14.49	18.24	17.97
27.16	12.06	20.26	31.28	7.40	22.57	17.45	12.78
16.22	4.13	7.59	10.63	7.38	14.33	24.95	4.08
13.69	11.89	13.62	13.24	17.49	6.23	9.57	5.83
15.37	12.31	7.98	26.81	12.91	23.65	7.58	26.32
16.54	9.26	6.54	17.45	16.69	10.70	11.01	14.97
30.57	17.00	26.33	10.92	14.41	34.04	8.90	8.92
18.00	9.11	11.57	4.56				

(a) Identify the values of R_i and R_i' when counting from the top as well as from the bottom.

(b) Determine the values of the statistics D, S, D', S', D^* and S^*.

(c) Do the data show any trend in either the mean or the variance?

[Arnold, Balakrishnan and Nagaraja, 1992, p. 257]

28. For the data presented in Table 5.10.4, assume $N(\mu, \sigma^2)$ distribution and do the following:

(a) Identify the upper record values observed.

(b) Using these record data, determine the BLUE's of μ and σ and their standard errors.

(c) Determine the BLIE's of μ and σ and their estimated mean square errors.

(d) Determine the MLE's of μ and σ and compare these with the BLUE's.

(e) Based on the BLUE's of μ and σ, construct a 95% prediction interval for the next upper record value.

(f) Now assuming $L(\mu, \sigma^2)$ for the record data, obtain the corresponding results. Compare and comment.

29. Let N_n denote the number of records in the sequence $X_0, X_1, \ldots, X_n$ of i.i.d. random variables from distribution F.

   (a) Show that

   $$f_{N_n}(m + 1) = \frac{n}{n + 1} f_{N_{n-1}}(m) + \frac{1}{n + 1} f_{N_{n-1}}(m - 1)$$

   for $m = 1, 2, \ldots, n$.

   (b) Show that the pmf $f_{N_n}(m)$, for fixed n, is a concave function.

   (c) If n is assumed to be unknown, show that $2^{N_n} - 2$ is an unbiased estimator of n? What is its variance?

   (d) Derive the MLE of n.   [Moreno Rebollo et al., 1996]

CHAPTER 6

GENERAL RECORD MODELS

6.1 INTRODUCTION

All our discussion thus far was restricted to the situation where we were observing a sequence of i.i.d. random variables. There too, mostly we dealt with the classical record model, where the population cdf is assumed to be continuous. In that setting we discussed the distribution theory for the record values and for the various statistics associated with the record counting process. We also investigated various characterization results and surveyed the inferential procedures that involve record value data. While the results were fascinating, it must be admitted that the i.i.d. model is too simplistic to explain the stochastic nature of record breaking phenomena abundant in real life. The search for a more reasonable model continues. Such a model is needed to explain a higher incidence of new records than predicted by the classical model. A case in point is the frequent breaking of sports records. The i.i.d. model surely cannot be used here.

In this chapter we explore some important record models in which the observation process fails to consist of i.i.d. random variables. We begin with setups in which the observations are not identically distributed while being independent, and conclude the chapter with a brief discussion of models with dependent sequences of random variables.

In Section 6.2 we introduce a model proposed by Yang (1975). It assumes that we are observing the maxima of a geometrically increasing population. Finite-sample properties of record indicators for the Yang model resemble those for the classical model. However, the limit distribution of the inter-record times turns out to be different here. A generalization of the Yang model – to be called the F^α *record model* – is discussed in Section 6.3. Following the work of Nevzorov, and Ballerini and Resnick, we establish several finite-sample and asymptotic results for that model. The asymptotic results depend on the value of the limiting record breaking rate and, thus, we find it convenient to introduce two types of F^α record models depending on whether it is positive or zero.

Yang's model did not explain the breaking of Olympic records very well. Ballerini and Resnick (1985) introduced the concept of improvement in the population according to a linear trend. We present some interesting results for this *linear drift record model* in Section 6.4. Pfeifer (1982) suggested a variation of the classical record model in which the population distribution changes after every record event. In Section 6.5 we discuss some basic properties of record statistics from that model. Section 6.6 is devoted to characterization results pertaining to the various record models introduced in this chapter. In the last section we elaborate on three models with dependent observations. First we consider records from Markov sequences and note that some of the nice features of the classical record model do hold for records in this case as well. It is followed by a discussion of record statistics from exchangeable sequences and sequences of dependent random variables having *Archimedean copula* type dependence structure. We conclude the chapter with the discussion of a general record model for dependent data introduced very recently by Hofmann (1997). It encompasses several of the F^α type record models.

6.2 GEOMETRICALLY INCREASING POPULATIONS

As we noted in Section 2.5, if we are observing an i.i.d. sequence of continuous random variables, the records occur more and more infrequently – more precisely, $\log \Delta_n \approx n$ for large n. But, in the Olympics and other athletic competitions, records are broken at a much faster rate. How could we explain this phenomenon? For what stochastic models do the observed record data appear plausible? Of course, we will have to dispense with the assumption that the observed sequence of values are i.i.d. One simple modification to the classical record model suggested by Yang (1975) assumes that the observations, while being independent, are non-identically distributed in a special way. In this section we discuss some basic distributional results for the Yang model. It is a special case of a more general setup to be called the F^α record model, being discussed in the next section.

Let $\{Y_n, n \geq 1\}$ represent an infinite sequence of independent continuous random variables. Suppose Y_n represents the maximum of $\alpha(n)$ i.i.d. random variables, each having continuous cdf F. Thus, F_n, the cdf of Y_n, is given by $F_n(y) = \{F(y)\}^{\alpha(n)}$. In the Olympic games and other international athletic events, $\alpha(n)$ may represent the size of the population of athletes at the time of the nth event. As in the classical record model, let $I_n, n \geq 1$, denote the sequence of record indicator random variables. Then, we have the following generalization of a basic fact established in Section 2.5.

Lemma 6.2.1 (Nevzorov, 1985) *The record indicators $I_j, j \geq 1$, are independent and I_j is Ber(p_j), where*

$$P(I_j = 1) \equiv p_j = \frac{\alpha(j)}{S(j)} \qquad (6.2.1)$$

and $S(j) = \sum_{i=1}^{j} \alpha(i)$.

Proof. Our proof follows closely the informal arguments presented in Section 2.5 for the classical record model. Suppose, for a fixed n, we want to show that $I_1, \ldots, I_n$ are independent. To prove the independence of these Bernoulli random variables, it suffices to show that, for an arbitrary set of integers $1 \leq j_1 < j_2 < \cdots < j_k \leq n$, $P(I_{j_1} = 1, I_{j_2} = 1, \ldots, I_{j_k} = 1)$ factors into the product of associated marginal probabilities. Without loss of generality, we may take F to be the standard uniform cdf. Upon conditioning on the values of $Y_{j_1}, Y_{j_2}, \ldots, Y_{j_k}$, and proceeding as in Section 2.5 we obtain

$$P(I_{j_1} = 1, I_{j_2} = 1, \ldots, I_{j_k} = 1) = \frac{\alpha(j_1)}{S(j_1)} \frac{\alpha(j_2)}{S(j_2)} \cdots \frac{\alpha(j_k)}{S(j_k)}$$
$$= p_{j_1} p_{j_2} \cdots p_{j_k}, \qquad (6.2.2)$$

where the p_j's are given by (6.2.1). On putting $k = 1$ in (6.2.2) it follows that I_j is Ber(p_j) for all $j \geq 1$. Thus, the lemma is established. $\bigcirc$

The above fact is very helpful in the study of the finite-sample as well as asymptotic properties of various record counting processes. As our first application, let us examine the distribution of the inter-record time Δ_n. Other applications will be discussed in Section 6.3 in a more general setup. Of special interest here is the population that is increasing at a geometric rate θ. Consider

$$P(\Delta_n > k) = \sum_{j=n}^{\infty} P(I_{j+1} = 0, \ldots, I_{j+k} = 0, T_{n-1} = j)$$

$$= \sum_{j=n}^{\infty} (1 - p_{j+1}) \cdots (1 - p_{j+k}) P(T_{n-1} = j),$$

where we have used the mutual independence of the event $\{T_{n-1} = j\}$ ($\equiv \{I_1 + \cdots + I_{j-1} = n-1, I_j = 1\}$) and the random variables $I_{j+1}, \ldots, I_{j+k}$. Thus, we obtain

$$P(\Delta_n > k) = \sum_{j=n}^{\infty} \frac{S(j)}{S(j+k)} \, P(T_{n-1} = j). \qquad (6.2.3)$$

To represent the geometrically increasing population, let us take $\alpha(j) = c(1 + \theta)^{j-1}$, $j \geq 1$, where θ is the growth rate. (To be more realistic, one could choose $\alpha(j)$ to be an integer closest to this value, but the final conclusion, stated below in Theorem 6.2.2, will be no different.) Then, as $j \to \infty$, p_j in (6.2.1) approaches $\theta(1 + \theta)^{-1} \equiv p(> 0)$, say. In that case, first we show that T_{n-1} is a proper random variable, or in other words, $P(T_{n-1} < \infty)$ is 1. For this purpose, let the event $A_j = \{I_j = 1\}$ and note that $P(A_j) = p_j$. From Lemma 6.2.1 we know that the A_j's are independent. Since $p_j \to p > 0$, $\sum_{j=1}^{\infty} p_j$ diverges, and therefore from Borel-Cantelli lemma, $P(A_j$ infinitely often$) = 1$. This means, an $(n-1)$th record will be established with probability 1 which indicates that $P(T_{n-1} < \infty)$ is 1. We now consider

$$(1 - p)^k - \frac{S(j)}{S(j+k)} = \frac{1}{(1+\theta)^k} - \frac{(1+\theta)^j - 1}{(1+\theta)^{j+k} - 1}$$

$$= \frac{S(k)}{S(j+k)} \frac{1}{(1+\theta)^k} .$$

Since $S(j)$ is non-negative and increases with j, we may write, for $j \geq n$,

$$0 < (1 - p)^k - \frac{S(j)}{S(j+k)} < \frac{S(k)}{S(n+k)} \frac{1}{(1+\theta)^k} .$$

As the above upper bound is less than $(1 + \theta)^{-(n+k)}$, from (6.2.3) we conclude that

$$0 < \{(1 - p)^k - P(\Delta_n > k)\} < \frac{1}{(1+\theta)^{n+k}} . \qquad (6.2.4)$$

Hence, we have proved the following result:

Theorem 6.2.1 (Yang, 1975) *Let* $\{Y_n, n \geq 1\}$ *be a sequence of indepen-dent random variables such that the cdf of* Y_n *is of the form* $\{F(y)\}^{\alpha(n)}$, *where* F *is a continuous cdf. If the* $\alpha(n)$*'s increase at a geometric rate such that* $\alpha(n) = c(1 + \theta)^{n-1}$, $n \geq 1$, *then* Δ_n *converges in distribution to a* $Geo(p)$ *random variable, with* $p = \theta/(1 + \theta)$.

Note that (6.2.4) gives a quick and powerful bound on $P(\Delta_n > k)$. While the limit distribution of Δ_n here is non-degenerate without any transformation or normalization, this was not the case in the classical record model.

Yang examined the applicability of the above result for modeling the breaking of records in the Olympics. On noting that, since 1900, the world population has been doubling every 36 years, he estimated that the growth rate θ in a span of four years representing the time between successive Olympic games to be $(2^{1/9} - 1)$ or 0.08. This means, if the model were true, Δ_n would be approximately geometric with $p = 0.074$, giving us an expected waiting time between successive records of 13.5 games. But, in almost all sport events, records have been broken more often, indicating that the model does not give a satisfactory fit to the data. Probable causes for the discrepancy are the improved facilities and training techniques and possibly the fact that the athletic population has grown at a faster rate than the general world population.

6.3   THE F^α RECORD MODEL

Yang's investigation of the properties of record statistics from a geomet-rically increasing population was pursued in a more general setting in the 1980s. For such models, Nevzorov, in a series of papers, carried out a thorough study of their finite-sample, as well as asymptotic properties. We draw heavily upon his work in the presentation of the results of this section. Formally, the F^α *record model* consists of an infinite sequence $\{Y_n, n \geq 1\}$ of independent random variables, where the cdf of Y_n has the form $F_n(y) = \{F(y)\}^{\alpha(n)}$, $\alpha(n) > 0$, and the associated record statistics. Here the cdf F is assumed to be continuous and the $\alpha(n)$'s need not be integers. Nevzorov (1990) refers to this model as the F^α scheme.

The Yang model, discussed in Section 6.2, was an F^α (record) model that assumed $\alpha(n)$'s to be geometrically increasing. On taking $\alpha(n) = 1$ for all n, we obtain the classical record model. Thus, the F^α model can accommodate a variety of situations associated with a sequence of independent possibly non-identically distributed random variables. Yet, it retains several interesting features of the classical record model, primarily due to the independence of the record indicators discussed in Lemma 6.2.1. However, the asymptotic properties of the record counting statistics depend on the rate of growth of the $\alpha(n)$'s. Thus, we find it convenient to distinguish between the following two cases identified by the limit behavior of the record rate $p_n = \alpha(n)/S(n)$ (defined in (6.2.1)).

(i) F^α *Model 1*, where $p_n \to p$, $0 < p < 1$.

(ii) F^α *Model 2*, where $p_n \to 0$, and $S(n) \to \infty$.

The first model is asymptotically close to the Yang model and the latter one includes the classical record model.

6.3.1   Finite-sample Properties

Theorem 6.3.1 (Ballerini and Resnick, 1987a) *In the F^α model, the record indicators $I_1, \ldots, I_n$ and M_n are mutually independent, where $M_n = \max(Y_1, \ldots, Y_n)$.*

Proof. (Hofmann, personal communication) The mutual independence of the I_j's has been established in Lemma 6.2.1. To prove the mutual independence of M_n and the Bernoulli random variables, observe that, for y in the support of F,

$$
\begin{aligned}
&P(I_1 = i_1, \ldots, I_n = i_n \mid M_n \leq y) \\
&= \ P(I_1 = i_1, \ldots, I_n = i_n \mid Y_i \leq y, \ 1 \leq i \leq n).
\end{aligned}
$$

Let F_0 be the cdf obtained by truncating F at y from above. That is, $F_0(x) = F(x)/F(y)$, $x \leq y$. Then the right side above represents the joint probability associated with the record indicator variables from the F_0^α model. As we know from Lemma 6.2.1 that in an F^α model, the stochastic structure of the I_j's is free of F, the above conditional probability is nothing but $p_{i_1} p_{i_2} \cdots p_{i_n}$, where $p_j = P(I_j = 1) = \alpha(j)/S(j)$. $\bigcirc$

While proving the above theorem, Ballerini and Resnick used basic properties of extremal processes and embedding techniques. For the F^α model, just as in the classical case, the independence of the record indicators yields simple and elegant proofs of several finite-sample and asymptotic properties of T_n, Δ_n and N_n. We will explore them next. While N_n is always a proper random variable, T_n and consequently Δ_n need not be. The following result, adapted from Nevzorov (1985), provides a necessary and sufficient condition for T_n to be a proper random variable.

Lemma 6.3.1 *The waiting time for the first nontrivial record, T_1, is a proper random variable (a) iff the series $\sum_{j=1}^{\infty} \alpha(j)$ is divergent (b) iff T_n is proper for any n.*

Proof. Observe that $P(T_1 > j) = P(I_1 = 1, I_2 = 0, \ldots, I_j = 0) = \alpha(1)/S(j)$. Thus $P(T_1 > j) \to 0$ as $j \to \infty$ iff $S(j) \to \infty$, which means (a) is established.

To prove (b), first note that if T_n is proper, evidently T_1 is also proper. Next, note that $\{T_n > j\} = \{N_j \le n\}$ and the associated probability monotonically decreases as j increases. Hence, as $j \to \infty$, $P\{T_n > j\} \to P\{N_\infty \le n\} = P(\text{At most } n \ A_j\text{'s occur})$ where $A_j = \{I_j = 1\}$. We will now claim that this probability is zero if T_1 is proper.

If T_1 is a proper random variable, we may conclude from part (a) that $\sum_{j=1}^{\infty} \alpha(j)$ diverges. Now since $p_j = \alpha(j)/S(j)$, from Abel-Dini test (see, e.g., Knopp, 1956, p. 125), it follows that if $\sum_{j=1}^{\infty} \alpha(j)$ diverges, so does $\sum_{j=1}^{\infty} p_j$. Since the events A_j are independent and $P(A_j) = p_j$, in view of the Borel-Cantelli lemma we notice that $P(A_j\text{'s occur infinitely often}) = 1$. This means the probability of the event that at most n of the A_j's occur is 0. ○

As $\sum_{j=1}^{\infty} \alpha(j)$ diverges iff $\sum_{j=1}^{\infty} p_j$ diverges, the above lemma implies that the divergence of $\sum_{j=1}^{\infty} p_j$ is necessary and sufficient for T_n to be a proper random variable for any n. For both the F^α models, the series diverges, and consequently, the T_n's and Δ_n's are all proper random variables for these schemes.

Finite-sample properties of the sequence $\{T_n, \ n \ge 1\}$ are similar to those for the classical model. For example, the T_n's form a Markov chain with stationary transition probabilities. On using Theorem 6.3.1 we obtain

$$
\begin{aligned}
P(T_n = j \mid T_{n-1} = i) &= P(I_{i+1} = 0, \ldots, I_{j-1} = 0, I_j = 1) \\
&= \frac{S(i)\alpha(j)}{S(j-1)S(j)}, \qquad j > i. \qquad (6.3.1)
\end{aligned}
$$

When the record accrual rate p_n is a constant for $n \ge 2$, the transition probability in (6.3.1) will depend only on $(j - i)$. This happens when $\alpha(n) = \theta(1+\theta)^{n-2}\alpha(1)$, where $\theta = p/(1-p)$, p being the constant record accrual rate. In that case, Δ_n's are all i.i.d. Geo(p) random variables. It is no surprise that for the Yang model (6.2.4) holds.

The martingale properties of record times arising from the classical record model continue to hold for the F^α models. Below we present a result that generalizes the result in Exercise 2.4.

Theorem 6.3.2 (Nevzorov, 1990) *In the F^α model suppose $S(n)$ diverges. Then the sequence of random variables $\{A(T_n) - (n+1), \; n \geq 1\}$ is a martingale, where $A(n) = \sum_{j=1}^{n} p_j$.*

Proof. From (6.2.1) and (6.3.1), we may write

$$
\begin{aligned}
E(A(T_n) \mid T_{n-1} = i) &= S(i) \sum_{j=i+1}^{\infty} A(j) \left\{ \frac{1}{S(j-1)} - \frac{1}{S(j)} \right\} \\
&= S(i) \left\{ \sum_{j=i+1}^{\infty} \frac{A(j) - A(j-1)}{S(j-1)} + \frac{A(i)}{S(i)} \right\} \\
&= A(i) + S(i) \sum_{j=i+1}^{\infty} \frac{\alpha(j)}{S(j-1)S(j)} \\
&= A(i) + S(i) \sum_{j=i+1}^{\infty} \left\{ \frac{1}{S(j-1)} - \frac{1}{S(j)} \right\} \\
&= A(i) + 1, \qquad\qquad\qquad\qquad\qquad (6.3.2)
\end{aligned}
$$

since $S(n) \to \infty$. From (6.3.2) it follows that $E(A(T_n)) = E(A(T_{n-1})) + 1$, $n \geq 1$ and consequently $E(A(T_n)) = n + 1$. Now define $W_n = A(T_n) - (n+1)$, for $n \geq 1$. We then have $E(W_n) = 0$ and

$$
\begin{aligned}
E(W_n | W_{n-1} = w) &= E(A(T_n) - (n+1) | A(T_{n-1}) - n = w) \\
&= E\{A(T_n) | T_{n-1} = A^{-1}(n+w)\} - (n+1) \\
&= w
\end{aligned}
$$

on using (6.3.2). This confirms that the W_n sequence is a martingale. ○

For another martingale sequence, see Exercise 6.1.

6.3.2 Asymptotic Properties

The random variable N_n is always proper. Its asymptotic properties depend on the limiting behavior of the sequences $E(N_n) \equiv \sum_{j=1}^{n} p_j = A(n)$ and $\mathrm{Var}(N_n) \equiv A(n) - \sum_{j=1}^{n} p_j^2 = B(n)$, say. It is easily seen from the strong law of large numbers that if $\{A(n)/n\} \to p$ as $n \to \infty$, $N_n/n \to p$, a. s. If $B(n) \to \infty$, $\{N_n - A(n)\}/\sqrt{B(n)} \overset{d}{\to} N(0,1)$ as $n \to \infty$. If, further, $\sum_{j=1}^{n} p_j^2 = o(A(n))$, then $B(n) \approx A(n)$ and thus $\sqrt{A(n)}$ can be used as the scaling factor.

The asymptotic properties of $A(n)$ are closely tied to those of $S(n)$. We have seen in the proof of Lemma 6.3.1 that if $S(n)$ diverges, so does

$A(n)$. If $S(n)$ converges, by the comparison test, it follows that $A(n)$ also converges. In fact,

$$
\begin{aligned}
A(n) &= \sum_{j=1}^{n} p_j \\
&= \sum_{j=1}^{n} \frac{S(n) - S(n-1)}{S(n)} \\
&\approx \int_{1}^{n} \frac{dS(x)}{S(x)} \\
&\approx \log(S(n)),
\end{aligned}
\tag{6.3.3}
$$

where we have extended the domain of $S(n)$ to $(0, \infty)$. Hence, in some cases one can replace $A(n)$ by $\log S(n)$ while normalizing N_n. Also note that from Theorem 6.3.2 and (6.3.3) it follows that $E(\log(S(T_n))) \approx n$.

If in the F^α Model 1, we assume $A(n) = np + o(\sqrt{n})$ and $\sum_{j=1}^{n}(p_j - p)^2 = o(\sqrt{n})$, then, $(N_n - np)/\sqrt{np(1-p)} \xrightarrow{d} N(0,1)$ (see Ballerini and Resnick, 1987a; Nevzorov, 1995). This result is applicable to the Yang model discussed in Section 6.2. Nevzorov also shows that in F^α Model 2, $(N_n - \log S(n))/\sqrt{\log S(n)}$ is asymptotically standard normal if $\sum_{j=1}^{n} p_j^2 = o(\sqrt{A(n)})$. This generalizes (2.5.11).

As pointed out in Chapter 2, one can use the equivalence of the events $\{N_n \geq k\}$ and $\{T_k < n\}$ and obtain parallel asymptotic results for the sequence of record times. We now examine the asymptotic properties of T_n through the study of the limiting joint distribution of inter-record times. The following result, applicable to F^α Model 1, generalizes Theorem 6.2.1.

Theorem 6.3.3 (Ballerini and Resnick, 1987a) *In the F^α model, let*

$$
\lim_{n \to \infty} \frac{\alpha(n)}{S(n)} = p, \qquad 0 < p < 1.
\tag{6.3.4}
$$

Then, as $n \to \infty$, $\Delta_n, \Delta_{n+1}, \ldots$ behave asymptotically like i.i.d. $\mathrm{Geo}(p)$ random variables.

Proof. To simplify the details we consider the joint distribution of Δ_n and Δ_{n+1} only. First note that if (6.3.4) holds, $S(n) \to \infty$ as $n \to \infty$, and consequently T_{n-1} is a proper random variable. Further, since $S(n-1)/S(n) \to (1-p)$ as $n \to \infty$, in view of (6.3.4) we may conclude that

$$
\frac{S(n)}{S(n+i-1)} \frac{\alpha(n+i)}{S(n+i)} = \left\{ \prod_{m=1}^{i-1} \frac{S(n+m-1)}{S(n+m)} \right\} \frac{\alpha(n+i)}{S(n+i)}
$$
$$
\to (1-p)^{i-1} p.
\tag{6.3.5}
$$

Now, consider

$$P(\Delta_n = i, \Delta_{n+1} = j)$$

$$= \sum_{k=n}^{\infty} P(T_{n-1} = k, I_{k+1} = 0, \ldots, I_{k+i} = 1, I_{k+i+1} = 0, \ldots, I_{k+i+j} = 1)$$

$$= \sum_{k=n}^{\infty} P(T_{n-1} = k) \frac{S(k)}{S(k+i-1)} \frac{\alpha(k+i)}{S(k+i)} \frac{S(k+i)}{S(k+i+j-1)}$$

$$\times \frac{\alpha(k+i+j)}{S(k+i+j)} \qquad (6.3.6)$$

on using Theorem 6.3.1. From (6.3.5), we can conclude that, for fixed i, j, for any $\varepsilon > 0$, there exists an $N(\varepsilon)$ such that, for all $k \geq n \geq N(\varepsilon)$,

$$\left| \frac{S(k)}{S(k+i-1)} \frac{\alpha(k+i)}{S(k+i)} \frac{S(k+i)}{S(k+i+j-1)} \frac{\alpha(k+i+j)}{S(k+i+j)} \right.$$

$$\left. - (1-p)^{i-1} p(1-p)^{j-1} p \right| < \varepsilon.$$

Since T_{n-1} is a proper random variable, the above inequality and (6.3.6) together imply that $|P(\Delta_n = i, \Delta_{n+1} = j) - (1-p)^{i-1} p(1-p)^{j-1} p| < \varepsilon$ for all $n \geq N(\varepsilon)$. In other words, Δ_n and Δ_{n+1} are asymptotically i.i.d. Geo(p) random variables. $\bigcirc$

The above theorem implies that, for a geometrically increasing population with a growth rate of θ, the inter-record times are asymptotically independent Geo(p) random variables, where $p = \theta/(1 + \theta)$. This is an improvement over Theorem 6.2.2 in that, we now are able to establish the asymptotic independence of the inter-record times as well.

The exact distribution of R_n is quite involved for the F^α model. In fact, no closed-form expression for its pdf exists. However, some attempts have been made toward the determination of the asymptotic distribution of R_n. We now elaborate on the work of Ballerini and Resnick (1987a) and Nevzorov (1995). The former paper provides a limit result with random norming constants for the F^α Model 1, while the latter obtains several limit results for the F^α model assuming F is exponential. Nevzorov also discusses the class of limit distributions for the general F^α Models 1 and 2.

Let $a(n)$ and $b(n) > 0$ be such that

$$F^n(a(n) + b(n)x) \to G(x), \qquad (6.3.7)$$

as $n \to \infty$ for all real x, where G is one of the three extreme value distributions (see (2.3.21–2.3.23) for their canonical forms.) Then we have the following result:

Theorem 6.3.4 (Ballerini and Resnick, 1987a) *For an F^α model, suppose (6.3.7) holds and $S(n) \to \infty$ as $n \to \infty$. Then, for all x,*

$$P\left(\frac{R_n - a(S(T_n))}{b(S(T_n))} \le x\right) \to G(x). \qquad (6.3.8)$$

Proof. The assumption $S(n) \to \infty$ ensures that T_n is a proper random variable for all n. Now consider

$$
\begin{aligned}
P(R_n \le y) &= \sum_{i=n+1}^{\infty} P(M_i \le y, \, T_n = i) \\
&= \sum_{i=n+1}^{\infty} P(M_i \le y, \, N_{i-1} = n, \, I_i = 1) \\
&= \sum_{i=n+1}^{\infty} P(M_i \le y) P(N_{i-1} = n, \, I_i = 1),
\end{aligned}
$$

from Theorem 6.3.1. Thus,

$$
\begin{aligned}
P(R_n \le y) &= \sum_{i=n+1}^{\infty} \{F(y)\}^{S(i)} P(T_n = i) \\
&= E\{F(y)\}^{S(T_n)}.
\end{aligned}
$$

Since $T_n \to \infty$ almost surely as $n \to \infty$, so does $S(T_n)$. So, from (6.3.7) we can conclude that $\{F(a(S(T_n)) + b(S(T_n))x)\}^{S(T_n)} \to G(x)$ for all x. Hence, on using dominated convergence theorem, we conclude that

$$
\begin{aligned}
P\left(\frac{R_n - a(S(T_n))}{b(S(T_n))} \le x\right) &= E\{F(a(S(T_n)) + b(S(T_n))x\}^{S(T_n)} \\
&\to G(x). \qquad \bigcirc
\end{aligned}
$$

When F is standard exponential, it is known that (6.3.7) holds with $a(n) = \log n$ and $b(n) = 1$, and $G(x) = \exp(-\exp(-x)) \equiv G_3(x)$ given by (2.3.23). Hence (6.3.8) implies that $\{R_n - \log S(T_n)\}$ converges in distribution to the Gumbel distribution. Theorem 6.3.1 implies that M_i and the event $\{T_n = i\}$ are independent. This fact can be used to find non-random norming constants that yield limiting normal distribution for R_n. This was considered by Nevzorov (1995). For the F^α Model 2, assuming F is exponential and

$$\sum_{j=1}^{n} p_j^2 = o\left(\left\{\sum_{j=1}^{n} p_j\right\}^{1/2}\right), \qquad (6.3.9)$$

he shows that

$$P\left(R_n - \log S(T_n) \le x, \log S(T_n) - n \le y\sqrt{n}\right) \to G_3(x)\Phi(y),$$

for all real x and y. As a consequence, we have

$$\frac{R_n - n}{\sqrt{n}} = \frac{R_n - \log S(T_n)}{\sqrt{n}} + \frac{\log S(T_n) - n}{\sqrt{n}}$$

$$\xrightarrow{d} N(0,1), \tag{6.3.10}$$

just as in the case of the classical model. See Exercise 6.4 for a similar result for the F^α Model 1.

The conclusion reached in (6.3.10) has important implications. We can use Theorem 2.3.1 to obtain the limit distribution of R_n for a more general F. To be precise, for the F^α Model 2, if F satisfies (2.3.15) and (6.3.9) holds, then

$$\frac{R_n - \psi_F(n)}{\psi_F(n + \sqrt{n}) - \psi_F(n)} \xrightarrow{d} N(0,1),$$

where $\psi_F(x)$ is the inverse hazard function defined in (2.3.14).

6.4   LINEAR DRIFT RECORD MODEL

As we noted at the end of Section 6.2, Olympic records are being broken at a rate faster than predicted by the model that assumes a geometrically increasing population. Thus, to explain the frequent record breaking phenomenon, one may take into account an improvement over time in the population performance. Stochastic modelers have used a linear trend component to explain this improvement and we will discuss such models now.

Let $\{X_n, \ n \ge 1\}$ be a sequence of i.i.d. random variables having continuous cdf F. Suppose $Y_n = X_n + cn$, where $c > 0$ represents the rate of improvement. The record statistics for the Y_n sequence are of interest now. We will call this model the *linear drift record model*.

Let us now take $F = G_3$, the Gumbel cdf, to obtain

$$
\begin{aligned}
P(Y_n \le y) &= F(y - cn) \\
&= \{\exp(-\exp(-y))\}^{\exp(cn)} \\
&= \{F(y)\}^{\alpha(n)}, \tag{6.4.1}
\end{aligned}
$$

where $\alpha(n) = \exp(cn)$. From (6.4.1), it is evident that we now have an F^α model with $F = G_3$. In fact, since

$$
\begin{aligned}
p_n &\equiv \frac{\alpha(n)}{S(n)} \\
&= (1 - e^{-c}) \frac{e^{cn}}{e^{cn} - 1} \\
&\to (1 - e^{-c}),
\end{aligned}
$$

we have an F^α Model 1 with p, the limiting record rate, being $(1 - e^{-c})$. Therefore we can use the asymptotic results developed for that model to obtain limit laws for the record counting statistics for this *linear drift Gumbel record model*, as it was referred to by Ballerini (1987).

For the general linear drift model, let us consider the probability of a record at time n. It is given by

$$
\begin{aligned}
p_n &= P(X_j + cj < X_n + cn \ \ j = 1, \ldots, n-1) \\
&= P(X_j < X_n + c(n - j), \ j = 1, \ldots, n-1) \\
&= \int_{-\infty}^{\infty} \prod_{j=1}^{n-1} F(x + cj) \, dF(x) \\
&= \int_{-\infty}^{\infty} G_n(x) \, dF(x), \qquad\qquad (6.4.2)
\end{aligned}
$$

where

$$
G_n(x) = \prod_{j=1}^{n-1} F(x + cj). \qquad\qquad (6.4.3)
$$

Clearly $G_n(x)$ decreases and, consequently, from (6.4.2) it follows that p_n decreases to p, the limiting record rate. Note that for the linear drift model, except when F is a Gumbel cdf, the record indicators are dependent, and the p_n's depend on F. As we have seen above, for the Gumbel case $p = 1 - e^{-c}$. When $F = \Phi$ and $c = 1$, $p \approx 0.725$ (Ballerini and Resnick, 1985). The right tail behavior of F affects the value of the limiting record accrual rate in an interesting way.

Theorem 6.4.1 (Nagaraja, 1994a) *In the linear drift record model with a positive drift $(c > 0)$ the limiting record accrual rate p is positive iff*

$$
\int_0^{\infty} \{1 - F(x)\} dx < \infty. \qquad\qquad (6.4.4)
$$

Proof. If $F^{-1}(1)$ is finite, obviously (6.4.4) holds. If we take $x_0 = F^{-1}(1) - c$, $G_n(x) = 1$ for all $x \geq x_0$ for all n and from (6.4.2) we can conclude that $p \geq (1 - F(x_0)) > 0$.

If $F^{-1}(1) = \infty$, we observe the following equivalence relations:

$$\int_0^\infty \{1 - F(x)\}dx < \infty \quad \Leftrightarrow \quad E(X_1^+) < \infty$$

$$\Leftrightarrow \quad E(X_1/c)^+ < \infty$$

$$\Leftrightarrow \quad \sum_{j=1}^\infty P(X_1 > cj) < \infty$$

$$\Leftrightarrow \quad \sum_{j=1}^\infty P(X_j > cj) < \infty, \qquad (6.4.5)$$

where $x^+ = \max(0, x)$. (The last equivalence holds since the X_j's are identically distributed.) Thus, if (6.4.4) holds, from the Borel-Cantelli lemma we can conclude that $P(X_j - cj > 0$ infinitely often$) = 0$. This means that $\max_{1 \leq j < \infty}(X_j - cj)$ is a proper random variable whose cdf is given by $G_\infty(x) = \lim G_n(x)$. Since G_∞ is a proper cdf, there exists an x_0 such that $G_\infty(x_0) > 0$. From (6.4.2), we note that $p \geq G_\infty(x_0)(1 - F(x_0)) > 0$.

To prove the converse when $F^{-1}(1) = \infty$, let us assume $p > 0$. Then there exists an x_0 such that $G_\infty(x_0) > 0$. From (6.4.3) we then see that $\prod_{j=1}^\infty \{F(x_0+cj)\} > 0$. We now observe the following equivalence relations:

$$\prod_{j=1}^\infty \{F(x_0 + cj)\} > 0 \quad \Leftrightarrow \quad \sum_{j=1}^\infty \{1 - F(x_0 + cj)\} < \infty$$

$$\Leftrightarrow \quad \sum_{j=1}^\infty \{1 - F(cj)\} < \infty.$$

We now appeal to (6.4.5) to conclude that (6.4.4) indeed holds. ◯

The above result implies that the limiting record rate is zero when F is, for example, a standard Pareto cdf.

Ballerini and Resnick (1985) have studied the limiting properties of N_n and below we present one of their results.

Theorem 6.4.2 (Ballerini and Resnick, 1985) *In the linear drift record model, $E(N_n/n - p)^2 \to 0$ as $n \to \infty$.*

Proof. Since $p_n \to p$, as $n \to \infty$, $E(N_n/n) = \left(\sum_{i=1}^n p_i\right)/n \to p$. As

$$E\left(\frac{N_n}{n} - p\right)^2 = \frac{1}{n^2} E(N_n^2) + p^2 - 2pE\left(\frac{N_n}{n}\right),$$

it is enough if we show that $\frac{EN_n^2}{n^2} \to p^2$. For this, first we consider

$$
\begin{aligned}
E(I_i I_{i+m}) \\
= \ & P[X_j < X_i + (j-i)c, \ j = 1, \ldots, (i-1); \ X_j < X_{i+m} + (j-i-m)c, \\
& \quad j = i+1, \ldots, (i+m-1); \ X_i < X_{i+m} + mc] \\
= \ & \int_{-\infty}^{\infty} \left\{ \int_{x-cm}^{\infty} G_m(y) \, dF(y) \right\} G_i(x) \, dF(x),
\end{aligned} \tag{6.4.6}
$$

where the G_j's are defined in (6.4.3). Therefore,

$$
\begin{aligned}
\lim_{m \to \infty} E(I_i I_{i+m}) & = \int_{-\infty}^{\infty} \int_{-\infty}^{\infty} G_\infty(y) \, dF(y) G_i(x) \, dF(x) \\
& = p \int_{-\infty}^{\infty} G_i(x) \, dF(x),
\end{aligned}
$$

and hence

$$
\lim_{i \to \infty} \lim_{m \to \infty} E(I_i I_{i+m}) = p^2. \tag{6.4.7}
$$

Next, we note that

$$
\begin{aligned}
E\left\{ \frac{N_n^2}{n^2} \right\} & = \frac{1}{n^2} \sum_{i=1}^{n} \sum_{j=1}^{n} E(I_i I_j) \\
& = \frac{\sum_{i=1}^{n} p_i}{n^2} + \frac{2}{n^2} \sum_{i=1}^{n-1} \sum_{m=1}^{n-i} E(I_i I_{i+m}) \\
& \to \ 0 + p^2,
\end{aligned}
$$

from (6.4.7). $\qquad\qquad\qquad\qquad\qquad\qquad\qquad\qquad\qquad\qquad\quad\bigcirc$

From the above result we may conclude that N_n/n is a mean squared error consistent estimator of p. Ballerini and Resnick establish the almost sure convergence of N_n/n also. Further, using the asymptotic theory for stationary sequences, they show that if $E(X_1^+)^2$ is finite (this implies $p > 0$), $\sqrt{n} \left(\frac{N_n}{n} - p \right)$ is asymptotically normally distributed. They also provide an expression for the variance of the limit distribution. In contrast, for the classical model, $p = 0$ and we have seen in Chapter 2 that $(N_n - \log n)/\sqrt{\log n}$ is asymptotically standard normal. de Haan and Verkade (1987) discuss the extreme value theory for the sequence of observations $Y_n = X_n + ca(n)$, $c > 0$, where the $a(n)$'s are chosen to satisfy (6.3.7). Introduction of such a trend provides asymptotic results for the record statistics that are closer to the ones obtained for the classical model. Some of their results are explored in Exercise 6.7.

Smith (1988) has considered some generalizations of the linear drift model for describing annual best performance (Y_n) in mile and marathon races where one observes $\min(Y_1, \ldots, Y_n)$, $n \geq 1$. Note that here, as in Section 5.9, the data is made up of the sequences of lower records (R'_n) and record times (T'_n). He suggests the model $Y_n = \sigma X_n + c_n$, $\sigma > 0$ unknown, where the trend component c_n has either linear or quadratic, or an exponential-decay form in n. Further, Smith takes the cdf of X_n (F) to be one of the following: Φ, the generalized extreme value cdf corresponding to the limit distribution of the sample minimum (whose three possible forms are given in (4.3.5)) and a special member of this family, namely the extreme value cdf given by (2.7.12). The maximum likelihood method is used to fit these nine models to the series of athletic records. Smith concludes that the model in which c_n is linear and F is normal does reasonably well, at least for the short-term forecasting of future annual best performances in these races. Recently, Feuerverger and Hall (1996) have observed that for the linear drift model, the maximum likelihood estimators of location and scale coefficients are quite sensitive to misspecification of F. See also Carlin and Gelfand (1993).

6.5   THE PFEIFER MODEL

The F^α record model assumed that the distribution changes with each observation. Pfeifer (1982) introduced an interesting variation of the classical model wherein the successive (upper) record values constitute a Markov chain with nonstationary transition distribution given by

$$P(R_n > y \mid R_{n-1} = x) = \frac{1 - F_n(y)}{1 - F_n(x)}, \quad y \geq x. \tag{6.5.1}$$

Such a dependence structure for the record value sequence can be produced as follows: Suppose we have a double array of independent random variables $\{X_{01}, X_{nj}; \ n, j \geq 1\}$ such that X_{nj} has cdf F_n, $n \geq 0$. Now take $R_0 = X_{01}$ and define $\Delta_n = \min\{j : X_{nj} > R_{n-1}\}$ and $R_n = X_{n, \Delta_n}$ for $n \geq 1$. We call this setting the *Pfeifer record model*. When the F_n's are continuous, in order that all these record statistics are well defined, it is necessary to assume that the $F_n^{-1}(1)$'s form a nondecreasing sequence. If an F_n is discrete such that $P(X_{nj} = F_n^{-1}(1))$ is positive, then one needs to assume that $F_{n+1}^{-1}(1) > F_n^{-1}(1)$. To keep our discussion simple, we will assume hereinafter that the F_n's are all continuous cdf's with nondecreasing upper limits. Note that while the X_{nj}'s are independent, the record generating observations are not.

Pfeifer presents an example of a shock model where the above distributional assumptions are plausible. Suppose a system with initial strength X_{01} is subjected to a sequence of shocks of random magnitudes represented by the X_{nj}'s. The system will fail as soon as a shock exceeding R_{n-1} arrives. For safety reasons the system is modified such that it can withstand a shock of magnitude equal to that of the last fatal shock, namely R_n. Further, some new safety factors are built in so that the cdf of the magnitude of subsequent shocks changes to F_n. Then R_n represents the strength of the system on modification after the nth failure. If the distributions associated with F_n's are stochastically decreasing, we obtain a shock model with an increasing safety feature.

The Pfeifer model shares several interesting features with the classical model. For example, from the definition, it is clear that $\{(R_n, \Delta_n), \ n \geq 1\}$ is a Markov chain and the transitions are determined by the following conditional probability:

$$P(R_n > y, \Delta_n = j \mid R_{n-1} = x, \Delta_{n-1} = i)$$
$$= \{F_n(x)\}^j \{1 - F_n(y)\}, \qquad x < y, \ i, j = 1, 2, \dots . \quad (6.5.2)$$

From (6.5.2) it follows that, given $R_{n-1} = x$, Δ_n and Δ_{n-1} are independent and Δ_n is $\text{Geo}(1 - F_n(x))$. More generally, conditioned on the sequence $\{R_n, \ n \geq 0\}$, the Δ_n's are independent, Δ_n being $\text{Geo}(1 - F_n(R_{n-1}))$ for $n \geq 1$.

Pfeifer (1984) has explored the asymptotic properties of record counting statistics for his model assuming that the F_n's are nondecreasing in n. The following result, adapted from his work, is a direct generalization of a result due to Shorrock (1972a) for the classical model.

Theorem 6.5.1 *Let $\{F_n, \ n \geq 0\}$ be a sequence of continuous cdf's nondecreasing in n but having supports with a common upper limit. Let H_n denote the hazard function of F_n. Then*

$$\varlimsup_{n \to \infty} \left| \frac{\log \Delta_n - H_n(R_{n-1})}{\log n} \right| = 1 \ a.s. \quad (6.5.3)$$

Proof. As we noted above, conditioned on the R_n's, the Δ_n's are independent and Δ_n is $\text{Geo}(1 - F_n(R_{n-1}))$. From Lemma 2 of Shorrock (1972a) (see Exercise 6.10), it then follows that (6.5.3) holds if we can show that $\sum_{n=1}^{\infty}(1 - F_n(R_{n-1})) < \infty$ a.s. This is confirmed if we can show that

$$\sum_{n=1}^{\infty} E(1 - F_n(R_{n-1})) < \infty. \quad (6.5.4)$$

Now, from (6.5.1), we have

$$E(E(1 - F_n(R_n)) \mid R_{n-1} = x) = \int_x^\infty \frac{1 - F_n(y)}{1 - F_n(x)} \, dF_n(y)$$

$$= \frac{1}{2} (1 - F_n(x)),$$

and hence

$$E(1 - F_n(R_n)) = E(1 - F_n(R_{n-1}))/2. \qquad (6.5.5)$$

If the F_n's are nondecreasing in n, we have $F_n(R_{n-1}) \geq F_{n-1}(R_{n-1})$ a.s. Thus,

$$E(1 - F_n(R_{n-1})) \leq E(1 - F_{n-1}(R_{n-1}))$$
$$= E(1 - F_{n-1}(R_{n-2}))/2,$$

from (6.5.5). Therefore, on induction, we may conclude that

$$E(1 - F_n(R_{n-1})) < \left(\frac{1}{2}\right)^{n-1} E(1 - F_1(R_0)).$$

Clearly, this inequality implies that (6.5.4) holds. ○

We now explore some consequences of the above theorem. Following Pfeifer (1984), let us assume

$$1 - F_i(x) = (1 - F(x))^{\lambda_i}, \qquad i \geq 0, \qquad (6.5.6)$$

where F is continuous and the λ_i's form a nondecreasing sequence of positive numbers. If (6.5.6) holds, the failure rates of the F_i's are proportional and are nondecreasing in i. Since F is continuous, the distribution of Δ_n depends only on the λ_i's and hence, as was done in the case of classical model (the case in which $\lambda_i \equiv 1$), without loss of generality, let us take F to be standard exponential. Then, from (6.5.1) we may conclude that $R_n - R_{n-1}$ and R_n are independent and the former random variable has $\text{Exp}(1/\lambda_n)$ distribution. Thus, we may write

$$R_n \stackrel{d}{=} \sum_{i=0}^n (X_i^*/\lambda_i), \qquad (6.5.7)$$

where the X_i^*'s are i.i.d. standard exponential random variables. The asymptotic behavior of R_n depends on the limiting behavior of $\text{Var}(R_n) = \sum_{i=0}^n \lambda_i^{-2}$. In fact, we have the following:

(i) If $\sum_{i=0}^{\infty} \lambda_i^{-2} < \infty$, there exists a random variable W such that

$$R_n - \sum_{i=0}^{n} \lambda_i^{-1} \equiv R_n - E(R_n) \to W \text{ a.s.} \tag{6.5.8}$$

(ii) If $\sum_{i=0}^{\infty} \lambda_i^{-2}$ is divergent,

$$\frac{R_n - E(R_n)}{\sqrt{\text{Var}(R_n)}} \xrightarrow{d} N(0,1). \tag{6.5.9}$$

Since $H_n(x) = \lambda_n x$, from (6.5.3) we obtain

$$\varlimsup_{n\to\infty} \frac{\lambda_n}{\log n} \left| \frac{\log \Delta_n}{\lambda_n} - R_{n-1} \right| = 1 \text{ a.s.}$$

When combined with appropriate assumptions on the growth rate of λ_n, this fact can be used to obtain limit results for Δ_n. For example, if $(\log n / \lambda_n) \to 0$, we can replace R_n by $(\log \Delta_n)/\lambda_n$ in (6.5.8) and (6.5.9). We now present an interesting special case, which shows that the limiting behavior can differ substantially from the asymptotic normality of $\log \Delta_n$ encountered in the classical model.

Theorem 6.5.2 (Pfeifer, 1984) *In the Pfeifer model, suppose the F_n's satisfy (6.5.6) with $\lambda_n = n + m$, where m is a fixed positive integer. Then,*

$$\frac{\log \Delta_n}{n} - \log n \xrightarrow{d} R'_{m-1}, \tag{6.5.10}$$

where R'_{m-1} behaves like the $(m-1)$th lower record value from the Gumbel cdf (given by (2.3.23)).

Proof. First note that since $(\log n / \lambda_n) \to 0$, $\{(\log \Delta_n)(n + m)^{-1}\} - R_{n-1} \to 0$ a.s. Next, we observe that on setting $\lambda_n = n + m$ in (6.5.7), R_{n-1} may be viewed as the upper mth order statistic from a random sample of size $n + m - 1$. Hence it is known that (see, e.g., Arnold, Balakrishnan and Nagaraja, 1992, pp. 207, 220), $(R_{n-1} - \log(n + m - 1)) \xrightarrow{d} R'_{m-1}$. Thus, we may write

$$\frac{\log \Delta_n}{n + m} - \log(n + m - 1) \xrightarrow{d} R'_{m-1}.$$

On using the fact that $\log(n) = o(n)$, we can then establish that (6.5.10) indeed holds. $\qquad \bigcirc$

When $\lambda_n = n + m$, $\sum_{i=0}^{\infty} \lambda_i^{-2}$ is convergent, and $E(R_{n-1}) \approx \gamma + \log n$, and consequently we could have used (6.5.8) to prove that $(\log \Delta_n/n) - \log n$ converges almost surely to a random variable W. But we could not have identified its distribution. Incidentally, combining (6.5.8) and Theorem 6.5.2, we may conclude that the convergence in (6.5.10) occurs in fact with probability 1.

As discussed in Exercise 2.13 and Section 4.7, Kamps (1995a) has considered properties of the sequence of random variables called generalized order statistics. These random variables possess the Markov property (6.5.1) where the F_i's are of the form (6.5.6) with $\lambda_i = k + (n - i)c$, $0 \le i \le n$. Here k and n are positive integers and c is an arbitrary real number. When we choose $c = 0$, these random variables behave like Type 2 k-records from the classical model and when $c = 1$ and $k = 1$ they behave like order statistics from a random sample of size $(n + 1)$. Thus, the stochastic structures of order statistics from a random sample and of records from the classical model coincide with that of records from special Pfeifer models. Consequently, Kamps' work unifies several distributional results known for order statistics and record values.

6.6 CHARACTERIZATIONS

Here we present a few characterizations associated with the record models introduced in this chapter. First, we obtain a necessary and sufficient condition for a record model with independent observations to be an F^α model. Next we obtain a characterization of the exponential distribution within the F^α model. We also provide a characterization of the Gumbel distribution in the linear drift record model. Finally, in the Pfeifer model, we obtain characterizations of the exponential and geometric-tail distributions using the independence of R_{n-1} and $R_n - R_{n-1}$. These results parallel similar characterizations considered in Chapter 4.

6.6.1 F^α and Linear Drift Record Models

In Theorem 6.3.1 we noted that, for the F^α record model, the record indicators are mutually independent and are independent of the maximum of the available observations. We now establish the converse result under the basic assumption that the observation sequence consists of independent continuous random variables. We thus provide a characterization of the F^α model.

Theorem 6.6.1 *Let Y_j, $1 \leq j \leq n$, be independent continuous random variables such that $F_j^{-1}(0)$ is a constant where F_j is the cdf of Y_j. If $M_n = \max(Y_1, \ldots, Y_n)$ and the record indicators I_j, $2 \leq j \leq n$, are independent, then there exist positive constants $\alpha(j)$ such that $F_j(y) = \{F_1(y)\}^{\alpha(j)}$, $2 \leq j \leq n$.*

Proof. Let $F_i^{-1}(0) = y_0$ and define $G_i(y) = \prod_{j=1}^{i} F_j(y)$, for $1 \leq i \leq n$. Then, for all $y > y_0$, and $2 \leq j \leq n$,

$$
\begin{aligned}
P(M_n \leq y, \ I_j = 1) &= \frac{G_n(y)}{G_j(y)} \int_{y_0}^{y} G_{j-1}(x) \, dF_j(x) \\
&= P(M_n \leq y) \frac{1}{G_j(y)} \int_{y_0}^{y} G_{j-1}(x) \, dF_j(x).
\end{aligned}
$$

Consequently, the independence assumption implies

$$
\int_{y_0}^{y} G_{j-1}(x) \, dF_j(x) = p_j G_j(y), \quad y > y_0, \tag{6.6.1}
$$

where $p_j \equiv P(I_j = 1) = \int_{y_0}^{\infty} G_{j-1}(x) \, dF_j(x)$. Since the F_i's have common lower limit, $G_{j-1}(y) > 0$ for all $y > y_0$ and hence $p_j > 0$. Further, $p_j < 1$ since in a right neighborhood of y_0, $G_{i-1}(y) < 1$ and F_j has points of increase there. On differentiating (6.6.1) we get

$$
(1 - p_j) G_{j-1}(y) \, dF_j(y) = p_j F_j(y) \, dG_{j-1}(y), \quad y > y_0. \tag{6.6.2}
$$

Since $0 < p_j < 1$, (6.6.2) can be expressed as

$$
\frac{dF_j(y)}{F_j(y)} = \beta(j) \frac{dG_{j-1}(y)}{G_{j-1}(y)}, \quad y > y_0, \ 2 \leq j \leq n,
$$

where $\beta(j) = p_j/(1-p_j)$ is finite and positive. On integration over (y, ∞), the above equation implies $\log F_j(y) = \beta(j) \log G_{j-1}(y)$ for all $y > y_0$. On using this relationship iteratively for $j = 2, \ldots, n$, we may conclude that the theorem holds. The $\alpha(j)$'s can be computed recursively using the relation $\beta(j) = \alpha(j)/\sum_{i=1}^{j-1} \alpha(i)$, $j \geq 2$, where $\alpha(1) \equiv 1$. $\quad \bigcirc$

Note that M_n represents the last record value in the given sequence or, more precisely, it is nothing but $R_{(N_n-1)}$. For some variations of the above characterization, see Exercises 6.13 and 6.14. We now present a characterization of the exponential distribution within the F^α model.

Theorem 6.6.2 (Nevzorov, 1986) *In the F^α model, let $\alpha(2) = 1$ and F be absolutely continuous with support $\mathbf{R}^+$. Suppose its pdf f satisfies the condition*

$$f(0) \equiv \lim_{x \to 0+} \frac{F(x)}{x} = \lambda, \tag{6.6.3}$$

where λ is finite and positive. Then conditions (a) and (b) below are equivalent:

(a) *The random variables R_0 and $R_1 - R_0$ are independent.*

(b)   (i) *F is an $Exp(1/\lambda)$ cdf,*

      (ii) *$\alpha(j) = 1$ for $j \geq 3$.*

   Proof. For all $y \geq 0$ and x, a point of increase of F, $K(x, y) = P(R_1 - R_0 > y \mid R_0 = x)$ is given by

$$
\begin{aligned}
K(x, y) \;=\;\; & \{1 - F(x + y)\} + F(x)\{1 - F^{\alpha(3)}(x + y)\} \\
& + F(x) \sum_{i=3}^{\infty} F^{\alpha(3) + \cdots + \alpha(i)}(x)\{1 - F^{\alpha(i+1)}(x + y)\}.
\end{aligned}
$$
$$\tag{6.6.4}$$

If the two conditions in (b) hold, $K(x, y)$ in (6.6.4) simplifies to $\exp(-\lambda y)$ indicating the independence claimed in (a).

   To prove the converse, let us note that if the conditional probability $K(x, y)$ is free of x, it coincides with the quantity obtained by letting x go to 0. In other words, from (6.6.4) we obtain $1 - F(y) = K(x, y)$ or

$$
\begin{aligned}
1 - F(y) \;=\;\; & \{1 - F(x + y)\} + F(x)\{1 - F^{\alpha(3)}(x + y)\} \\
& + F(x) \sum_{i=3}^{\infty} F^{\alpha(3) + \cdots + \alpha(i)}(x)\{1 - F^{\alpha(i+1)}(x + y)\}
\end{aligned}
$$
$$\tag{6.6.5}$$

for all $y \geq 0$ and all x that are points of increase of F. From (6.6.5) we may conclude that if F is increasing at x, F is also increasing at $x + y$ for any $y > 0$. Hence we may conclude that (6.6.5) holds for all $x, y \geq 0$. On rearranging it, we obtain

$$
\begin{aligned}
F(x + y) - F(y) \;=\;\; & F(x)\{1 - F^{\alpha(3)}(x + y)\} \\
& + F(x) \sum_{i=3}^{\infty} F^{\alpha(3) + \cdots + \alpha(i)}(x)\{1 - F^{\alpha(i+1)}(x + y)\}.
\end{aligned}
$$
$$\tag{6.6.6}$$

On dividing both sides of (6.6.6) by $F(x)$ and letting $x \to 0+$, the R.H.S. yields $1 - F^{\alpha(3)}(y)$, whereas from the L.H.S. we obtain

$$\lim_{x \to 0+} \{F(x+y) - F(y)\}\{F(x)\}^{-1}.$$

Thus, from (6.6.3) we conclude that

$$
\begin{aligned}
f(y) &= \lim_{x \to 0+} \frac{F(x+y) - F(y)}{x} \\
&= \lim_{x \to 0+} \frac{F(x+y) - F(y)}{F(x)} \frac{F(x)}{x} \\
&= \{1 - F^{\alpha(3)}(y)\} f(0), \qquad y \geq 0. \qquad (6.6.7)
\end{aligned}
$$

Now differentiate both sides of (6.6.6) with respect to y and recall $\alpha(2) = 1$ to observe

$$f(y) - f(x+y) = f(x+y) \left\{ \sum_{i=2}^{\infty} \alpha(i+1) F^{\alpha(2)+\cdots+\alpha(i)}(x) \right\} \{F^{\alpha(i+1)-1}(x+y)\}.$$

On letting y go to 0, and rearranging, the above equation yields

$$
\begin{aligned}
f(0) &= f(x) \left\{ 1 + \sum_{j=3}^{\infty} \alpha(j) F^{\alpha(2)+\cdots+\alpha(j)-1}(x) \right\} \\
&= f(0)\{1 - F^{\alpha(3)}(x)\} \left\{ 1 + \sum_{j=3}^{\infty} \alpha(j) F^{\alpha(3)+\cdots+\alpha(j)}(x) \right\}, \qquad x \geq 0,
\end{aligned}
$$

on using (6.6.7). Now put $u = \{F(x)\}^{\alpha(3)}$ and notice that we can rewrite this as

$$\sum_{i=0}^{\infty} u^i = 1 + \alpha(3)u + \sum_{j=4}^{\infty} \alpha(j) u^{1+\beta(4)+\cdots+\beta(j)}, \qquad 0 \leq u < 1,$$

where $\beta(j) = \alpha(j)/\alpha(3)$, $j \geq 4$. On comparing the coefficients of successive powers of u, we are led to the conclusion that $\alpha(j) = 1$, $j \geq 3$; that is, the condition (b) (ii) of the theorem holds. Since $\alpha(3) = 1$, (6.6.7) implies that the failure rate of F is a constant, being λ. In other words, (b) (i) also holds. ○

Nevzorov (1995) actually proved the exponential characterization without the assumption of absolute continuity and the condition given by (6.6.3). These assumptions have simplified our proof. It is interesting to note that we said nothing about $\alpha(1)$ – in fact, Y_1 can have an arbitrary distribution and still $R_1 - R_0$ and R_0 are independent as long as $F_2, F_3 \ldots$ are all $\text{Exp}(1/\lambda)$ cdf's.

The next result is a characterization of the Gumbel cdf in a linear drift record model. It is a refinement of a result due to Ballerini (1987).

Theorem 6.6.3 (Nagaraja, 1994a) *Let X_1, X_2 be i.i.d. continuous random variables with cdf F and define $Y_n = X_n + nc$, $n = 1, 2$, $c > 0$. If $M_2 = \max(Y_1, Y_2)$ and I_2, the indicator random variable corresponding to the event $\{Y_2 > Y_1\}$, are independent for all c, then F is a Gumbel cdf.*

Proof. Let us denote the cdf of Y_n by F_n and note that $F_n(y) = F(y - nc)$, $n = 1, 2$. Further let $F^{-1}(0) = a$, $F^{-1}(1) = b$, and $p(c) = P(I_2 = 1) = P(X_1 < X_2 + c)$. Since $c > 0$, $p(c) > 0$ (in fact, it exceeds 0.5), and for c small, say $< (b - a)$, $p(c) < P(X_1 < X_2 + b - a) = 1$. Now, let us fix a c for which $0 < p(c) < 1$ and follow the proof of Theorem 6.6.1. The independence assumption implies (recall (6.6.1))

$$\int_{-\infty}^{y} F_1(x) dF_2(x) = p(c) F_1(y) F_2(y), \qquad y > a + 2c.$$

On differentiation and rearrangement, the above equation leads us to the differential equation

$$\frac{dF_2(y)}{F_2(y)} = \frac{p(c)}{1 - p(c)} \frac{dF_1(y)}{F_1(y)}, \qquad y > a + 2c,$$

which can be expressed as

$$\frac{dF(x)}{F(x)} = \alpha(c) \frac{dF(x + c)}{F(x + c)}, \qquad x > a,$$

where $\alpha(c) = p(c)\{1 - p(c)\}^{-1} > 1$. Integrating both sides of the above equation over (x, ∞), we may conclude that

$$- \log F(x) = -\alpha(c) \log\{F(x + c)\}, \tag{6.6.8}$$

$x > a$. If a were finite, since F is continuous, by letting $x \to a+$, we observe that (6.6.8) holds for $x = a$ as well. Now (6.6.8) implies $F(a + c)$ is also 0, which contradicts the assumption that $F^{-1}(0) = a$. Thus, a has to be $-\infty$. Arguing similarly we conclude that $b = +\infty$. Further, (6.6.8) implies that F is strictly increasing and thus $0 < p(c) < 1$ for all $c > 0$. Hence (6.6.8) holds for all real x and $c > 0$. By defining $\alpha(-c) = 1/\alpha(c)$, we see that the equation holds for all c.

Now with $g_1(x) = -\log F(x)$, (6.6.8) can be expressed as $g_1(x) = \alpha(c)g_1(x + c)$, for all real x and c. On putting $x = 0$, we notice that $g_1(c) = g_1(0)/\alpha(c)$ and consequently we may conclude that $\alpha(x)$ satisfies the CFE (recall (4.4.1)), $\alpha(x + c) = \alpha(x)\alpha(c)$ for all x and c. Thus, $\alpha(x) = \exp(\gamma x)$ for some $\gamma > 0$, since the function is increasing. In other words, we have $-\log F(x) = \{-\log F(0)\} \exp(-\gamma x)$, for all x. This implies that F is the G_3 in (2.3.23), but for a possible change of location and scale. ○

6.6.2   The Pfeifer Model

For the Pfeifer model, the dependence structure of the R_n's remains the same as in the classical model, except for the lack of stationarity. Hence several of the characterizations developed in Chapter 4 can be generalized to this setting. To be precise, any characterization that made use of (2.6.3) (or (4.2.2)) has a parallel result that would involve (6.5.1). While we present just two such results here, it is worth pointing out that, since the cdf of the record generating sequence may change with n, some interesting twists do exist in such characterizations.

We now consider an exponential characterization that can be viewed as a generalization of a result (due to Tata, 1969) for the classical model. From (6.5.1) it is evident that for $n \geq 1$, if F_n is an exponential cdf, R_{n-1} and $R_n - R_{n-1}$ are independent and the record spacing has cdf F_n. A converse to this fact is given below.

Theorem 6.6.4 (Pfeifer, 1982) *For $0 \leq j \leq n - 2$, let F_j be a continuous cdf with $F_j^{-1}(0) = 0$. Further, assume that F_{n-1} is strictly increasing in $\mathbf{R}^+$. Then, if R_{n-1} and $R_n - R_{n-1}$ are independent, F_n is an exponential cdf.*

Proof. For $x \geq 0$, from (6.5.1) we have,

$$P(R_n - R_{n-1} > x \mid R_{n-1} = y) = \frac{1 - F_n(x + y)}{1 - F_n(y)} \quad \text{a.s. } F_{R_{n-1}}(y). \quad (6.6.9)$$

Under our assumptions, $F_{R_{n-1}}$ is strictly increasing in $\mathbf{R}^+$ and hence for any given x, (6.6.9) holds in a dense subset of $\mathbf{R}^+$. By the right continuity of the cdf's, the independence assumption implies

$$1 - G_n(x) = \frac{1 - F_n(x + y)}{1 - F_n(y)}, \quad \text{for all } x, y \text{ in } \mathbf{R}^+, \quad (6.6.10)$$

where G_n is the cdf of $R_n - R_{n-1}$. On putting $y = 0$ in (6.6.10), we note that $G_n \equiv F_n$ and consequently we see (guess what!) a CFE (4.4.1) with

$g_0(x) = 1 - F_n(x)$. Since this g_0 is clearly bounded, we may conclude that F_n as well as G_n are exponential cdf's. ○

If F_{n-1} is not strictly increasing, F_n need not be an exponential cdf. (See Exercise 6.15 for an example.)

In the discrete case, results that characterize the geometric-tail distributions in the i.i.d. model (see Section 4.6) can be easily generalized to include the Pfeifer model. The following simple result illustrates the fact that these generalizations are quite direct. For another similar result, see Exercise 6.17. Recall that a cdf F whose support is the integers has a geometric-tail distribution, and we write F is GeoT(j, p), if $f(x) = p\{1 - F(x-)\}$, $x \geq j$.

Theorem 6.6.5 *Let $F_0, \ldots, F_n$ be cdf's having supports on the set of non-negative integers such that $f_i(i) > 0$, and $F_i^{-1}(1) = \infty$, $i = 0, \ldots, n$. If R_{n-1} and the event $\{R_n - R_{n-1} = 1\}$ are independent, then F_n is a GeoT(n, p) cdf with $p = f_n(n)\{1 - F_n(n-1)\}^{-1}$ $(= P(R_n - R_{n-1} = 1))$.*

Proof. Consider the conditional probability

$$
\begin{aligned}
&P(R_n - R_{n-1} = 1 \mid R_{n-1} = y) \\
&= P(R_n = y + 1 \mid R_{n-1} = y) \\
&= \frac{f_n(y + 1)}{1 - F_n(y)} \\
&= P(R_n - R_{n-1} = 1), \text{ from the assumption of independence} \\
&= p, \text{ say.}
\end{aligned}
$$

So, we have $f_n(y) = p(1 - F_n(y - 1))$, $y \geq n$. In other words, F_n is the cdf of a GeoT(n, p) distribution. Further, in view of the independence assumption, $p = P(R_n - R_{n-1} = 1 \mid R_{n-1} = n-1) = f_n(n)\{1 - F_n(n-1)\}^{-1}$ is strictly between 0 and 1. ○

6.7 RECORDS FROM DEPENDENT SEQUENCES

Thus far we have explored models in which the record generating process consisted of independent random variables. An exception was the Pfeifer model, where records were generated from a dependent sequence determined by the record history. We now introduce some other dependent record models.

First we observe that some of the results discussed in Sections 6.3 and 6.4 were actually proved in the literature for special dependent sequences. Ballerini and Resnick (1987a) established Theorem 6.3.1 for a sequence of random variables, for which $\{M_n, n \geq 1\}$ can be embedded in some *extremal process* (see Resnick, 1987, p. 179 for a definition). For the linear drift record model, assuming the X_n's to be strictly stationary, Ballerini and Resnick (1987b) proved various limit theorems for the record statistics. From their work it follwos that Theorem 6.4.2 holds even when $\{X_n\}$ is an ergodic sequence. Exercise 6.22 presents another limit result for a dependent linear drift record model.

We will discuss Markov sequences and the sequences of exchangeable random variables below in some detail. We will also elaborate on a dependent model possessing an Archimedean copula property and a random power model called *Random F^α model*. There are papers scattered in the literature that deal with dependent sequences. Some of these show that Rényi's limit results for the record counting statistics from the classical model do hold for the particular dependent sequence considered. For example, Guthrie and Holmes (1975) show that the limit results discussed in Section 2.5 do hold for the sequence *defined on a finite recurrent Markov chain*. Deken (1978) obtains limit laws for the record counting statistics observed from a *scheduled maxima sequence*. We do not plan to pursue such special models here.

6.7.1   Markov Sequences

An important class of dependent random variables is generated by the assumption of Markovian dependence. Biondini and Siddiqui (1975) conducted an extensive study of the dependence structure of the record statistics generated from a Markov sequence of random variables. They considered discrete as well as continuous marginal distributions and also the case where the transition distributions are nonstationary. As to be expected, the details get quite involved. Subsequently, Adke (1993) considered record statistics generated from a stationary Markov sequence. We follow his work in our treatment of the topic. Let us suppose $\{Y_n, n \geq 1\}$ forms a Markov process with stationary transition distribution given by $F(y|x) = P(Y_{n+1} \leq y | Y_n = x)$. In order that the record statistics are well defined, in addition to the common assumption that the upper limit of the state space is not an atom, we need to assume that the Markov chain is irreducible.

Theorem 6.7.1 (Adke, 1993) *Suppose the Y_n's form a Markov process with stationary transition cdf $F(y|x)$. Then $\{(R_n, T_n), n \geq 0\}$ is also a Markov process. For $y > x$, and $i, j \geq 1$, its transition distribution is given by*

$$P(R_{n+1} \leq y, \ T_{n+1} = i + j \mid R_n = x, \ T_n = i)$$

$$= \begin{cases} F(y|x) - F(x|x), \quad j=1 \\[2mm] \int_{y_1=-\infty}^{x} \cdots \int_{y_{j-1}=-\infty}^{x} \{F(y|y_{j-1}) - F(x|y_{j-1})\} \qquad (6.7.1) \\[2mm] \qquad \times \ \prod_{m=1}^{j-2} dF(y_{m+1}|y_m) dF(y_1|x), \ j \geq 2. \end{cases}$$

Proof. For $y_0 < y_1 < \cdots < y_{n-1} < x$ and positive integers $i_1 < i_2 < \cdots < i_{n-1} < i, \ j \geq 2$,

$$P(R_{n+1} \leq y, \ T_{n+1} = i + j \mid R_0 = y_0, \ldots, R_{n-1} = y_{n-1}, \ R_n = x;$$
$$T_0 \equiv 1, \ldots, T_{n-1} = i_{n-1}, \ T_n = i)$$
$$= P(x < Y_{i+j} \leq y, \ Y_{i+1} \leq x, \ldots, Y_{i+j-1} \leq x \mid R_0, \ldots, R_{n-1},$$
$$T_0, \ldots, T_{n-1}, \ T_n = i, \ Y_i = x),$$

where the conditioning event depends only on the values of $Y_m, \ m \leq n$. Hence from the Markov property of the Y_n's, we may write

$$P(R_{n+1} \leq y, \ T_{n+1} = i + j \mid R_0, \ldots, R_{n-1}, \ R_n = x;$$
$$T_0, \ldots, T_{n-1}, \ T_n = i)$$
$$= P(x < Y_{i+j} \leq y, \ Y_{i+1} \leq x, \ldots, Y_{i+j-1} \leq x \mid Y_i = x).$$
$$(6.7.2)$$

Note that the right side above also represents $P(R_{n+1} \leq y, \ T_{n+1} = i + j \mid R_n = x, \ T_n = i)$. This concludes the proof that (R_n, T_n) forms a Markov chain. The transition probabilities can be determined from the right side expression in (6.7.2). It can be evaluated by conditioning on the values of $Y_{i+j}, \ldots, Y_{i+j-1}$ and using the stationary Markov property of the Y_n's. The resulting quantity is the multiple integral given in (6.7.1).

When $j = 1$, there are no integrals to evaluate and one obtains
$$P(R_{n+1} \leq y, \ T_{n+1} = i + 1 \mid R_n = x, \ T_n = i) = P(x < Y_{i+1} \leq y \mid Y_i = x)$$
$$= F(y|x) - F(x|x). \qquad \bigcirc$$

Biondini and Siddiqui (1975) describe the dependence structure of the records through the transition pdf, assuming the absolute continuity of

$F(y|x)$ and of the initial distribution. They obtain an integral equation satisfied by the pdf that involves an auxiliary function. Explicit solutions of these equations appear to be intractable, except for the classical model.

From (6.7.1) we notice that $P(R_{n+1} \leq y, T_{n+1} = i+j \mid R_n = x, T_n = i)$ depends explicitly on x, y, and j, but is free of the value of i. Let us denote the conditional probability by $Q(x, y, j)$. Now, consider

$$
\begin{aligned}
& P(R_{n+1} \leq y, \ \Delta_{n+1} = j \mid R_n = x, \ T_n = i) \\
= \ & P(R_{n+1} \leq y, \ T_{n+1} = i + j \mid R_n = x, \ T_n = i) \\
\equiv \ & Q(x, y, j) \\
= \ & P(R_{n+1} \leq y, \ \Delta_{n+1} = j \mid R_n = x).
\end{aligned}
\tag{6.7.3}
$$

Hence, we may write

$$
\begin{aligned}
P(R_{n+1} \leq y \mid R_n = x) \ &= \ \sum_{j=1}^{\infty} P(R_{n+1} \leq y, \ \Delta_{n+1} = j \mid R_n = x) \\
&= \ \sum_{j=1}^{\infty} Q(x, y, j).
\end{aligned}
\tag{6.7.4}
$$

Since (R_n, T_n) is a Markov process, in view of (6.7.3) we note that the expression in (6.7.4) also represents $P(R_{n+1} \leq y \mid R_n = x, \ R_{n-1} = y_{n-1}, \dots, R_0 = y_0)$ for any $y_0 < \cdots < y_{n-1} < x < y$. In other words, the R_n's themselves form a Markov chain, just as in the case of the classical record model. However, as (6.7.4) shows, the transition probabilities are more involved here.

For the classical model, it is known that marginally, the T_n's form a Markov chain. Adke points out that the T_n's here need not be a Markov chain. However, there do exist Markov sequences Y_n's for which the bivariate process $\{(R_n, T_n), \ n \geq 0\}$ has the same structure as the one produced by the classical model. Exercises 6.19 and 6.20 provide two such examples. These situations illustrate the possibility that the record statistics data given by $\{(R_n, T_n), \ n \geq 0\}$ may fail to identify the intrinsic record generating phenomenon. Thus one faces with the problem of nonidentifiability while modelling the record generating phenomenon on the basis of record statistics data.

6.7.2 Exchangeable Observations

Another frequently studied dependent model assumes that Y_n's form an infinite sequence of exchangeable random variables. In this case, the record indicators behave just like those in the classical model. Thus, all the finite-sample and asymptotic results for the record counting statistics remain the same. What about the record values?

de Finetti's theorem (see, e.g., Galambos, 1987, p. 184) serves as a useful tool in the investigation of the distribution of R_n. Since an exchangeable infinite sequence can be represented as a mixture of sequences of i.i.d. random variables, we can visualize a random variable Z with cdf F_Z and a family of cdf's $\{F(y|z), -\infty < z < \infty\}$, such that for every n, the joint cdf of $Y_1, \ldots, Y_n$ can be expressed as

$$F_n(y_1, \ldots, y_n) = \int_{-\infty}^{\infty} \prod_{i=1}^{n} F(y_i|z) dF_Z(z). \tag{6.7.5}$$

In view of the conditional independence property exhibited in (6.7.5), we can use the known finite-sample results for the classical record model and obtain similar results for exchangeable random variables. For example, in the continuous case,

$$P(R_n > y) = \int_{-\infty}^{\infty} (1 - F(y|z)) \sum_{i=0}^{n} \frac{\{-\log(1 - F(y|z))\}^i}{i!} \, dF_Z(z).$$

A particularly simple example of an exchangeable sequence was discussed by Nayak and Inginshetty (1995). Following their work, let us assume $\{Y_n, n \geq 1\}$ form a stationary Gaussian sequence with standard normal marginals and common correlation ρ, $0 < \rho < 1$. Then, it is well known that we can represent the Y_n's as $Y_n = \sqrt{\rho} \, Z + \sqrt{1 - \rho} \, X_n$, $n \geq 1$, where Z and the X_n's are all i.i.d. standard normal random variables. Thus, we obtain

$$R_n(Y) = \sqrt{\rho} \, Z + \sqrt{1 - \rho} \, R_n(X), \tag{6.7.6}$$

where $R_n(Y)$ and $R_n(X)$ represent the nth upper record values from the Y and X sequences, respectively. We know from (2.3.27) that $\sqrt{2} \, (R_n(X) - \Phi^{-1}(1 - e^{-n})) \overset{d}{\to} N(0, 1)$. Now from (6.7.6) we may conclude that

$$\begin{aligned}
\sqrt{2} \, & \{R_n(Y) - \sqrt{1 - \rho} \, \Phi^{-1}(1 - e^{-n})\} \\
&= \sqrt{2\rho} \, Z + \sqrt{1 - \rho} \, \sqrt{2} \, (R_n(X) - \Phi^{-1}(1 - e^{-n})) \\
&\overset{d}{\to} N(0, 1 + \rho),
\end{aligned}$$

since Z and $R_n(X)$ are independent.

6.7.3 Dependent Models Based on Archimedean Copula

Ballerini (1994) introduced an F^α type model, where the Y_n's are dependent, but the independence result established in Theorem 6.3.1 still holds. He assumed that the finite dimensional joint

$$P(Y_1 \le y_1, ..., Y_n \le y_n) = C\left(\sum_{i=1}^{n} C^{-1}(F_i(y_i))\right), n \ge 1, \qquad (6.7.7)$$

where C is a completely monotone function mapping $[0, \infty)$ onto $(0, 1]$ with $C(0) = 1$. Here C can be thought of as the Laplace transform of a positive random variable (see Feller, 1965, p. 415).

When $F_i = F^{\alpha(i)}$ and (6.7.7) holds, we have an F^α model, where the observations are dependent. Assuming $C(s) = \exp(-s^{1/\theta})$, $\theta \ge 1$, Ballerini showed that for this model, just as in Theorem 6.3.1, $I_1, \ldots, I_n$ and M_n are mutually independent. The Bernoulli parameter associated with I_j is given by

$$p_j = \frac{\{\alpha(j)\}^\theta}{\sum_{i=1}^{j}\{\alpha(i)\}^\theta}. \qquad (6.7.8)$$

Observe that Ballerini's assumptions imply that for some γ_j, $C^{-1}(F_j(y)) = \gamma_j C^{-1}(F(y))$, for $1 \le j \le n$. This is the key criterion that ensures the independence of the record indicators and M_n. In fact, it leads to a characterization in the family of dependent observations having the Archimedean copula property.

Theorem 6.7.2 (Nevzorova, Nevzorov and Balakrishnan, 1997) *Let Y_j, $1 \le j \le n$, be continuous random variables having the joint cdf given by (6.7.7). As usual, let I_j be the record indicator corresponding to Y_j and $M_j = \max(Y_1, \ldots, Y_j)$.*

(a) If

$$C^{-1}(F_j(y)) = \gamma_j C^{-1}(F_1(y)), \quad 2 \le j \le n, \qquad (6.7.9)$$

where $\gamma_j > 0$, then the random variables $I_1, \ldots, I_n$ and M_n are mutually independent.

(b) Assume that there exists an interval (c, d) such that $0 < F_j(c) < F_j(d) < 1$. If I_j and M_j are independent for all $j = 2, 3, ..., n$, then (6.7.9) holds for some $\gamma_j > 0$.

We refer you to the original source for a proof. As C in (6.7.7) can be chosen as the Laplace transform of an arbitrary positive random variable, the above result provides a rich class of dependent random variables outside even the F^α model setup, for which the record indicators and the observed maximum are all mutually independent. In fact, one can find examples outside the class of dependent models based on Archimedean copula where the record indicators are independent (see Hofmann, 1997). Thus, the finite-sample and asymptotic results developed for record counting statistics in the context of the fixed F^α model that are based on the independence of record indicators do directly extend to more general dependent models.

6.7.4   A Random Power Record Model

Thus far, we have assumed that the $\alpha(n)$'s in the F^α model are constants. This is not quite realistic in many cases. For example, the best performance in a particular athletic event depends on the possibly random number of contestants. Hence, it is of interest to study the behavior of record statistics from models in which the α's are random variables. Very recently, Hofmann (1997) has introduced one such model and called it *Random F^α model*. It has the following structure. Let $\alpha(n)$'s form a sequence of possibly dependent positive random variables. Given the $\alpha(n)$'s, the observations Y_n's are conditionally independent and Y_n has cdf $F^{\alpha(n)}$, where F is a continuous cdf. Thus, the Y_n's are in general neither independent nor are identically distributed. This random power model includes as special cases the fixed F^α model discussed in Section 6.3, as well as the models introduced in Sections 6.7.2 and 6.7.3. This extremely versatile model produces quite general situations with independent record indicators.

Example 6.7.1 (Random F^α Model) For $n \geq 0$, let $\beta(n)$ represent the nth upper record value from a Pareto(1,1) distribution. Let $\alpha(1) = \beta(0)$ and for $n > 1$ let $\alpha(n)$ represent the record spacing $\beta(n-1) - \beta(n-2)$. Given the sequence $\{\alpha(n), n \geq 1\}$, assume that the Y_n's are conditionally independent and Y_n has cdf $F^{\alpha(n)}$. Here as $\beta(n) = \sum_{i=1}^{n+1} \alpha(i)$ approaches infinity with probability 1, records are broken infinitely often with probability 1. Now define $\pi(n) = \alpha(n)/\beta(n)$. Note that $E\pi(n) = P(I_n = 1)$, the probability that Y_n is a record value. In view of the representation for the Pareto records given in (2.4.9), it follows that the $\pi(n)$'s are independent random variables. On using conditioning argument it follows that

$$P(I_1 = 1, \ldots, I_n = 1) = E\{\pi(1) \cdots \pi(n)\}$$

and the right side above clearly factorizes. This suggests that the corresponding record indicators are mutually independent. Note, however, that the Y_n's are dependent and marginally do not have an F^α structure. ◯

Hofmann (1997) provides several distributional results for the record values and record counting statistics for the random F^α model. Some of these are similar to the ones discussed in Section 6.3 for the fixed α case. We will not pursue the details here.

EXERCISES

1. Assume that we have an F^α model. Let $T_{n(k)}$ be the waiting time for the nth Type 2 k-record.

    (a) For all $k \geq 1$ and $\beta < k$, show that $\{W_n, n \geq 0\}$ is a martingale, where
    $$W_n = \left(\frac{k - \beta}{k}\right)^{n+1} \frac{\Gamma(T_{(n+1)(k)} + 1)}{\Gamma(T_{n(k)} - \beta + 1)}.$$

    (b) If $\alpha > 0$, for n large show that
    $$E\left[\left\{T_{(n-1)(k)}\right\}^{k-\alpha}\right]$$
    $$= \frac{\Gamma(k)}{\Gamma(\alpha)}\left\{\left(\frac{k}{\alpha}\right)^n + \frac{(\alpha - k)(\alpha - k + 1)}{2\alpha}\left(\frac{k}{\alpha + 1}\right)^n\right.$$
    $$\left. + O\left(\left(\frac{k}{\alpha + 2}\right)^n\right)\right\}.$$

    [Nevzorov, 1990]

2. In the F^α model, if $p_n \to p$ for some $0 \leq p \leq 1$ as $n \to \infty$, show that, with probability 1,

    (a) $\dfrac{N_n}{n} \to p$ 	 (b) $\dfrac{T_n}{n} \to \dfrac{1}{p}$.

    [Ballerini and Resnick, 1987a]

3. Let $S(t)$ be a continuous function increasing in $(0, \infty)$ such that $S(n) = \alpha(1) + \cdots + \alpha(n)$, where the $\alpha(i)$'s are the coefficients of an F^α model. Suppose $S(t)$ is a *regularly varying* function with exponent $\gamma > 0$, as $t \to \infty$; that is, $\{S(tx)/S(t)\} \to x^\gamma$ for all $x > 0$.

   (a) For a fixed m, as $n \to \infty$, show that the m random variables $\{T_{i+1}/T_i\}$, $i = n, n+1, \ldots, n+m$, are asymptotically i.i.d. such that
   $$P\left\{\frac{T_{n+1}}{T_n} > x\right\} \to x^{-\gamma}, \qquad x \geq 1.$$
   In other words, they are asymptotically independent Pareto random variables.

   (b) Find the limit distributions of $\{\Delta_n/T_n\}$ and of $\{\Delta_{n+1}/\Delta_n\}$.

   [Nevzorov, 1985]

4. In the F^α Model 1, take F to be the standard exponential cdf.

   (a) Assuming that $\sum_{n=1}^{\infty} |p_n - p|$ is convergent, show that
   $$\frac{R_n - na}{b\sqrt{n}} \xrightarrow{d} N(0, 1),$$
   where $a = -\log(1 - p)/p$, and $b = a\sqrt{(1 - p)}$.

   (b) Now suppose $\sum_{i=1}^{n}(p_i - p)$ and $\sum_{i=1}^{n}(p_i - p)^2$ are both of $o(\sqrt{n})$. Prove that
   $$P(R_n - \log S(T_n) \leq x, \ \log S(T_n) \leq na + \sqrt{n}\,by) \to G_3(x)\Phi(y)$$
   as $n \to \infty$, where a and b are given in part (a).

   [Nevzorov, 1995]

5. In the F^α Model 1, assume $\sum_{i=1}^{n}(p_i - p)$ and $\sum_{i=1}^{n}(p_i - p)^2$ are both of $o(\sqrt{n})$ and establish the following:
   $$\frac{T_n - n/p}{\sqrt{n(1 - p)/p^2}} \xrightarrow{d} N(0, 1).$$

   [Ballerini and Resnick, 1987a; Nevzorov, 1995]

6. In the F^α model suppose $S(n) \to \infty$ as $n \to \infty$.

   (a) If $\max\{a(1), \ldots, a(n)\} = o(S(n))$, show that $\{N_n/\log(S(n))\}$ converges to 1 almost surely.

   (b) Show that the condition in (a) holds iff $p_n \to 0$. (Thus, the result in (a) holds for the F^α Model 2.)

   [Weissman, 1995]

7. Let $\{X_n, \ n \geq 1\}$ be a sequence of i.i.d. random variables with cdf F. Define $Y_n = X_n + c\,a(n)$, $c \geq 0$, and let N_n denote the number of records among $Y_1, \ldots, Y_n$.

   (a) Let $F = G_3$ (given by (2.3.23)), and $a(n) = \log n$. Prove that $\{N_n/\log n\}$ converges almost surely to $c + 1$.

   (b) Suppose F is in the domain of maximal attraction of G_3 such that $\{F(a(n) + b(n)x)\}^n \to G_3(x)$ as $n \to \infty$. Show that

   $$E\left(\frac{N_n}{\log n} - (c+1)\right)^2 \to 0.$$

   (c) Let $F(x) = G_1(x; 1) = \exp(-1/x)$, $x > 0$, and take $a(n) = n$. Establish the following:

   (i) If $c = 1$, $E\left\{\frac{N_n}{(\log n)^2}\right\} \to \frac{1}{2}$ and $\mathrm{Var}\left\{\frac{N_n}{(\log n)^2}\right\} \to \frac{1}{6}$.

   (ii) If $0 < c < 1$, $E\left\{\frac{N_n}{\log n}\right\} \to \frac{1}{1-c}$.

   [de Haan and Verkade, 1987]

8. Let $\{Y_n, \ n \geq 1\}$ be a sequence of independent random variables where Y_n has continuous cdf F_n such that for a fixed positive integer m, $F_{mi+j} = F_j$, $1 \leq j \leq m$, $i \geq 1$. Such a situation arises when m athletes participate in a contest performing trials in a sequence (e.g., as in long jump). If N_n denotes the number of records among $Y_1, \ldots, Y_n$, then prove that

   (a) $\frac{N_{mn} - \log n}{\sqrt{\log n}} \xrightarrow{d} N(0, 1)$.

   (b) $\sup_x \left| P\left(\frac{N_{mn} - \log n}{\sqrt{\log n}} \leq x\right) - \Phi(x) \right| \leq c_0 \left(\frac{m^2}{\log n}\right)^{1/4}$, for some constant c_0.

   [Balabekyan and Nevzorov, 1986]

9. In the Pfeifer record model, let $F_n(x) = \{F(x)\}^{\alpha(n)}$, where $\alpha(n) > 0$ and F is a continuous cdf. This can be viewed as a Pfeifer type F^α record model.

   (a) Show that

   $$P(T_n = i_n \mid T_1 = i_1, \ldots, T_{n-1} = i_{n-1})$$
   $$= \alpha(n) \, \frac{\beta(n-1) + \alpha(n-1)i_{n-1}}{\{\beta(n) + \alpha(n-1)(i_n-1)\}\{\beta(n) + \alpha(n)i_n\}} \, ,$$

   where $\beta(n) = \sum_{j=1}^{n-1}(\alpha(j-1) - \alpha(j))i_{j-1} + \alpha(n-1)i_{n-1}$.

   (b) When can the sequence $\{T_n, \ n \geq 1\}$ be a Markov chain?

   [Pfeifer, 1982]

10. Let $\{W_n, \ n \geq 1\}$ be a sequence of independent geometric random variables, W_n being $\mathrm{Geo}(p_n)$. If $\sum_{n=1}^{\infty} p_n$ is convergent, show that

   $$\overline{\lim_{n \to \infty}} \left| \frac{\log W_n p_n}{\log n} \right| = 1 \text{ almost surely.}$$

   [Shorrock, 1972a]

11. In the Pfeifer model, assume that F_i is $\mathrm{Exp}(1/\lambda_i)$, where the λ_i's are nondecreasing for $i \geq 1$. Show that (6.5.8) and (6.5.9) hold.

   [Pfeifer, 1984]

12. Let k and n be positive integers and c be an arbitrary real number. For $i \geq 0$, let $1 - F_i(x) = (1 - F(x))^{\lambda_i}$, $\lambda_i = k + (n-i)c$, $0 \leq i \leq n$, where F is an absolutely continuous cdf. Now suppose $\{R_n, \ n \geq 0\}$ forms the sequence of records from the Pfeifer model with these F_i's.

   (a) For $i < n$, show that (i) if R_i is IFR, so is R_{i+1} and (ii) R_i is DFR if R_{i+1} is.

   (b) For every $c \geq 0$, and $k \geq 1$, and any cdf F, there exist n and i such that R_i does not possess DFR property.

   (c) Discuss the consequences of the results in (a) and (b) on the reliability properties of k-records in the classical model and of order statistics of a random sample of size $n + 1$ from the cdf F.   [Kamps, 1994; 1995a]

13. Let the cdf $F_1, \ldots, F_n$ of independent random variables $Y_1, \ldots, Y_n$ be continuous and $0 < F_i(c) < F_i(d) < 1$, $1 \leq i \leq n$ for some real c and d. Let $M_i = \max(Y_1, \ldots, Y_i)$, and define $\boldsymbol{M}_n = (M_1, \ldots, M_n)$ and $\boldsymbol{I}_n = (I_1, \ldots, I_n)$ where the I_j's are the record indicators. Now, suppose $\boldsymbol{M}_n$ and $\boldsymbol{I}_n$ are independent.

   (a) Show that there exist positive constants $\alpha(j)$ such that $F_j(y) = \{F_1(y)\}^{\alpha(j)}$, for $2 \leq j \leq n$.

   (b) Establish that the random variables $I_2, \ldots, I_n$ are independent.

   [Nevzorov, 1990]

14. Let $Y_1, \ldots, Y_n$ be absolutely continuous independent random variables with F_j (f_j) being the cdf (pdf) of Y_i. Suppose there exists an interval such that $\prod_{j=1}^{n-1} f_j(y) \neq 0$ and let f_n be an arbitrary pdf. Suppose the vector $\boldsymbol{I}_{n-1}$ (defined in Exercise 6.13 above) and I_n are independent for any f_n. Prove that

   (a) $F_j(y)$ is necessarily of the form $F_j(y) = \{F_1(y)\}^{\alpha(j)}$, for some positive constants $\alpha(j)$, for $2 \leq j \leq n-1$.

   (b) $I_1, \ldots, I_{n-1}$ are mutually independent.

   [Nevzorov, 1986]

15. Suppose we are observing records from a Pfeifer model. Let F_0 be an arbitrary cdf with support on the non-negative integers and let

$$F_1(x) = \begin{cases} 1 - \exp\{-([x] + x)\}, & x \geq 0 \\ 0, & x < 0. \end{cases}$$

Here, as usual, $[x]$ stands for the integer part of x. Show that R_0 and $R_1 - R_0$ are independent. [Pfeifer, 1982]

16. In the Pfeifer model, suppose F_0 is an exponential cdf. Show that if $R_0, R_1 - R_0, \ldots, R_n - R_{n-1}$ are independent, $F_1, \ldots, F_n$ are all exponential cdf's. The characterization fails if F_0 is not assumed to be exponential! [Pfeifer, 1982]

17. Consider the Pfeifer model where the F_n's are cdf's of unbounded discrete random variables with support being a subset of the positive integers. Under appropriate assumptions (to be stated by you) show that if $E(R_n - R_n|R_{n-1})$ is a constant, then F_n has a geometric tail.

[Ahsanullah and Holland, 1987; Nagaraja, Sen and Srivastava, 1989]

18. As usual, let I_j denote the record indicator associated with Y_j, $j \geq 1$. Sibuya and Nishimura introduce a record model where that the I_j's are independent and I_j is a $\mathrm{Ber}(p_j)$ random variable with $p_j = \theta/(\tau + \theta + j - 1)$ for some $\theta > 0$ and $\tau \geq 0$. They carry out inference on the parameters τ and θ, and using this model, consider prediction of the number of future records from some weather related and athletic data. Note that now we have some flexibility in choosing the record breaking rate but unless $\tau = 0$, I_1 is nondegenerate. (What are the implications of this?)

(a) Show that the pgf of N_n is given by

$$E(s^{N_n}) = \frac{(\tau + \theta s)^{(n)}}{(\tau + \theta)^{(n)}},$$

where $x^{(n)} = x(x+1) \cdots (x+n-1)$. The associated probabilities correspond to a generalized form of Stirling numbers of the first kind. Compare the above pgf with that of N_n for the classical model, obtained in (2.5.8).

(b) Determine $E(N_n)$ and $\mathrm{Var}(N_n)$.

[Sibuya and Nishimura, 1997]

19. Let $\{X_n,\ n \geq 1\}$ be a sequence of i.i.d. random variables with cdf F. Let Y_1 have cdf F_1 and be independent of the X_n sequence. Now, for $n \geq 1$, define the sequence $Y_{n+1} = k \max(Y_n, X_n)$, where $0 < k < 1$.

(a) Show that $Q(x, y, j) \equiv P(R_{n+1} \leq y,\ T_{n+1} = i + j \mid R_n = x,\ T_n = i)$ is given by $\{F(x/k)\}^{j-1}\{F(y/k) - F(x/k)\}$, $x < y$.

(b) If we choose $F_1(y) = F(x/k)$, the bivariate sequence $\{(R_n, T_n),\ n \geq 0\}$ above has the same distributional structure as the corresponding sequence obtained from a classical model with cdf F_1. What are the implications of this finding?　　[Adke, 1993]

20. In Exercise 6.19, if the Y_n's are determined from the relation $Y_{n+1} = \max(Y_n, X_n) - c$, $c > 0$, find $Q(x, y, j)$.　　[Adke, 1993]

21. Let $\{Y_n,\ n \geq 1\}$ form an infinite sequence of exchangeable random variables such that Y_n given $Z = z$ is exponentially distributed with mean z^{-1}, and $f_Z(z) = ze^{-z}$, $z > 0$.

(a) What is the marginal distribution of Y_n?

(b) Find $E(R_n)$.

(c) Compare your answer with $E(R_n)$ you would get on assuming that the Y_i's are i.i.d. standard exponential random variables.

(d) Comment on the characterization given in Theorem 4.2.1.

[Arnold and Balakrishnan, 1989, p. 147]

22. Let $\{X_n,\ n = 0, \pm 1, \pm 2, \ldots\}$ be a doubly infinite strictly stationary sequence in the linear drift record model $Y_n = X_n + cn,\ n \geq 1$, $c > 0$. If $\{X_n\}$ is ergodic, show that $\frac{N_n}{n} \to p$ almost surely, where $p = P(X_j > \max_{1 \leq i < \infty}(X_{j-i} - ci))$.

[Ballerini and Resnick, 1987b]

23. Consider a sequence of exchangeable random variables $\{Y_n,\ n \geq 1\}$ defined in the following manner: Let X_n's be i.i.d. continuous and V_n's be i.i.d. Ber(p) random variables. Let Z be a continuous random variable, and assume all the random variables involved are mutually independent. Now define $Y_n = pX_n + ZV_n$ and let $R_n(X)\ (R_n(Y))$ denote the nth record value from the $X_n\ (Y_n)$ sequence. Finally, assume that , as $n \to \infty$, $(R_n(X) - a_n)/b_n \overset{d}{\to} W$, a nondegenerate random variable, and $b_n \to b$, finite and positive. Determine the limit distribution of $(R_n(Y) - pa_n)/pb_n$. [Nayak, 1985]

24. Let $Y_1, \ldots, Y_n$ be continuous random variables having the joint cdf given by (6.7.7). Let I_j be the record indicator corresponding to Y_j and $M_j = \max(Y_1, \ldots, Y_j),\ 1 \leq j \leq n$.

(a) Take $F_j = F^{\alpha(j)}$ and $C(s) = \exp(-s^{1/\theta})$, $\theta \geq 1$. Show that I_1, ..., I_n and M_n are mutually independent and that the success parameter associated with I_j is given by (6.7.8).

(b) Give an example where the Y_j's are dependent and F_j cannot be expressed as $F^{\alpha(j)}$ and yet I_1, ..., I_n and M_n are mutually independent.

[Ballerini, 1994; Nevzorova, Nevzorov and Balakrishnan, 1997]

25. We say two continuous random variables Y_i and Y_j are ordered in the sense of likelihood ratio and write $Y_i \overset{lr}{\leq} Y_j$ if $f_j(y)/f_i(y)$ is nondecreasing in y, where $f_i(f_j)$ is the pdf of $Y_i(Y_j)$. Suppose $Y_1, \ldots, Y_n$ are independent continuous random variables such that $Y_1 \overset{lr}{\leq} Y_2 \overset{lr}{\leq} \ldots \overset{lr}{\leq} Y_n$. Let $N_1(n)$ be the number of upper records

among $Y_1, \ldots, Y_n$ and $N_2(n)$ be the number of upper records when the data is presented in reverse order beginning with Y_n. Show that $N_1(n)$ is stochastically larger than $N_2(n)$; that is,

$$P\{N_1(n) \leq k\} \leq P\{N_2(n) \leq k\} \text{ for all } k.$$

[Haiman and Nevzorov, 1996]

26. Let $\{X_n, n \geq 1\}$ be a sequence of i.i.d. random variables with absolutely continuous cdf F and pdf f. For a fixed real δ, define $T_0(\delta) \equiv 1$ and $T_n(\delta) = \min\{j : X_j > X_{T_{n-1}(\delta)} + \delta\}$ for $n \geq 1$. Further let $R_n(\delta) = X_{T_n(\delta)}$. These are δ–$exceedance$ $record$ statistics. (Clearly, when $\delta = 0$, one obtains the classical model.)

   (a) Find the joint pdf of $R_j(\delta)$, $0 \leq j \leq n$.

   (b) Assuming F is standard exponential, show that $R_n - n\delta$ is a Gamma$(n + 1, 1)$ random variable.

   (c) For a positive integer j, find an expression for $P(T_1(\delta) > j)$ in terms of F and f.

   (d) Evaluate the expression in (c) above assuming F is a standard exponential cdf.

[Balakrishnan, Balasubramanian and Panchapakesan, 1997]

CHAPTER 7

RANDOM AND POINT PROCESS RECORD MODELS

7.1 INTRODUCTION

In this chapter we introduce another important variation of the classical model. Instead of assuming the availability of an infinite sequence of observations, we assume here that the sequence is of random length. We call this a *random record model*. Such a situation arises naturally when the observations arrive at time points determined by an independent point process observed over a finite time. We call such a setup a *point process record model* (PPRM). These extensions to the classical model add new twists to the distribution theory for the record counting statistics. Yet, one obtains nice results in quite a few of these situations.

In Section 7.2 we introduce the random record model and present the basic distribution theory for record statistics. In Section 7.3 we provide an introduction to PPRM and discuss some applications of that model. Section 7.4 explores the case where the pacing process P is a Poisson process. A comparison with the classical model is also provided there. In Section 7.5 we examine the behavior of the record arrival process when P is a renewal process. We consider birth process based models in Section 7.6. We present two distinct PPRM's for which records arrive according to a homogeneous Poisson process. This alerts us of nonidentifiability issues we face while modeling the record generating process on the basis of record arrivals. We consider two other PPRM's in Section 7.7. In the last section we present an elaborate discussion of the role of record arrival process in finding optimal solution to the classical *secretary problem*.

Throughout this chapter we assume that the sequence of i.i.d. observations is independent of the number of available observations and of the pacing process **P**. Some of our results can be modified to handle situations when the observations are independent but not identically distributed. See Bunge (1989), Bunge and Nagaraja (1991) and Hofmann (1997) for some work in that direction. What happens if **P** and the observations are dependent? Westcott (1977) considers situations that incorporate two simple models for dependence. We do not pursue these generalizations.

7.2 BASIC RANDOM RECORD MODEL

Let $\{X_i, \ i \geq 1\}$ be a sequence of i.i.d. random variables with cdf F. Let us assume that X_1 is always observed and N will represent the additional random number of observed X_i's. Assume that N is a non-negative integer valued random variable independent of the X_i-sequence. Suppose now that records are observed from the sequence of random length given by $\{X_1, \ldots, X_{N+1}\}$. We will call this *random record model* and present several interesting features of the distributions of various record statistics arising from this scheme. To begin with, note that, with positive probability, the first record will not be broken. In fact,

$$P\{X_1 = \max(X_1, \ldots, X_{N+1})\} = E\left(\frac{1}{N+1}\right)$$

is always positive, unless N is infinite with probability 1 or when we have a classical model.

7.2.1 Joint Distribution of Record Values

As R_n may not be observable with positive probability, the distribution theory for the successive record values $R_1, \ldots, R_n$ would have to incorporate the event that ensures their presence. Accordingly, let M denote the number of nontrivial records.

To make the exposition simpler, let us assume F is standard uniform. Let us denote $P\{N = j\}$ by p_j and introduce $\pi(s) = \sum_{j=0}^{\infty} p_j s^j$, the pgf of N. We now consider the joint distribution of $R_0, \ldots, R_n$. Define

$$P_n = P\{M \geq n, R_0 \in ds_0, \ R_1 \in ds_1, \ldots, R_n \in ds_n\},$$

where the s_i's are strictly increasing and $\{R_i \in ds_i\}$ represents the event $\{s_i < R_i \leq s_i + ds_i\}$. We will now find an expression for P_n.

Let $m = n - 1$ and define $\boldsymbol{s} = (s_0, ..., s_m)$ and $\partial(\boldsymbol{s}) = \prod_{j=0}^{m} ds_j$. For at least n nontrivial records to occur, it is necessary that $N \geq n$. Thus, for $k \geq n$,

$$P\{M \geq n, R_0 \in ds_0, \ldots, R_n \in ds_n \mid N = k\}$$
$$= \sum_{0 \leq i_0 + i_1 + \cdots + i_m \leq k - n} ds_0 s_0^{i_0} ds_1 s_1^{i_1} \cdots ds_m s_m^{i_m} ds_n + o(\partial(\boldsymbol{s})), \quad (7.2.1)$$

where the i_j's are non-negative integers and $o(\partial(\boldsymbol{s})) \to 0$ as $\partial(\boldsymbol{s}) \to 0$. The above expression holds, even when $k < n$, where we interpret the summation over an empty set to be 0. On applying the law of total probability we obtain

$$\begin{aligned}
P_n &= \sum_{k=0}^{\infty} p_k \sum_{0 \leq i_0 + i_1 + \cdots + i_m \leq k - n} ds_0 s_0^{i_0} ds_1 s_1^{i_1} \cdots ds_m s_m^{i_m} ds_n \\
&\qquad\qquad + o(\partial(\boldsymbol{s})) \\
&= \prod_{j=0}^{n} ds_j g(\boldsymbol{s}) + o(\partial(\boldsymbol{s})), \quad\quad (7.2.2)
\end{aligned}$$

where

$$g(\boldsymbol{s}) = \sum_{k=0}^{\infty} p_k \sum_{0 \leq i_0 + \cdots + i_m \leq k - m - 1} s_0^{i_0} \cdots s_m^{i_m}. \quad\quad (7.2.3)$$

While $g(\boldsymbol{s})$ has a nice algebraic form in terms of the probabilities associated with N, it also has an interesting probabilistic interpretation. For this purpose, let Z_j, $0 \leq j \leq m$, be independent random variables, Z_j being Geo$(1 - s_j)$, and let $W = \sum_{j=0}^{m}(Z_j - 1)$. Then for $w \geq 0$,

$$P(W \leq w) = \sum_{0 \leq i_0 + \cdots + i_m \leq w} (1 - s_0) s_0^{i_0} \cdots (1 - s_m) s_m^{i_m}.$$

Hence, if N is independent of W,

$$\begin{aligned}
P(N - W \geq n) &= \sum_{k=0}^{\infty} p_k P(W \leq k - n) \\
&= \prod_{j=0}^{m}(1 - s_j) g(\boldsymbol{s}) \quad\quad (7.2.4)
\end{aligned}$$

on using (7.2.3). Therefore, we have

$$P_n = \left\{ \prod_{j=0}^{n-1} \frac{ds_j}{1 - s_j} ds_n \right\} P(N - W \geq n) + o(\partial(\boldsymbol{s})),$$

where $P(N - W \geq n)$ is given by (7.2.4). The first factor of the leading term above represents the joint pdf of standard uniform record values from the classical model (recall (2.3.9)). When N is a proper random variable, we have to discount the probability by the factor $P\{N - W \geq n\}$. However, note that the computation of this probability requires a nontrivial effort and it depends on the values of the observed records.

Let us now obtain an explicit finite-series expression for $g(\boldsymbol{s})$ in terms of the $\pi(s_i)$'s.

Lemma 7.2.1 *Suppose $s_i \in [0, 1)$, $0 \leq i \leq m$, are all distinct and $g(\boldsymbol{s})$ is given by (7.2.3). Then, for $k \geq 0$,*

$$\sum_{i_0 + \cdots + i_m \leq k - m - 1} s_0^{i_0} \cdots s_m^{i_m} = \prod_{i=0}^{m} \frac{1}{(1 - s_i)} - \sum_{i=0}^{m} \frac{\rho_i s_i^k}{1 - s_i}, \qquad (7.2.5)$$

and

$$g(\boldsymbol{s}) = \prod_{i=0}^{m} \frac{1}{(1 - s_i)} - \sum_{i=0}^{m} \frac{\rho_i \pi(s_i)}{(1 - s_i)}, \qquad (7.2.6)$$

where

$$\rho_i = \prod_{\substack{j=0 \\ j \neq i}}^{m} \frac{1}{(s_i - s_j)} \qquad i = 0, \ldots, m. \qquad (7.2.7)$$

Proof. We begin by considering the partial fraction expansion for the product $\prod_{j=0}^{m}(s - s_j)^{-1}$ where $s \geq 1 > \max_{0 \leq j \leq m} s_j$. From Feller (1968, p. 276) it follows that

$$\prod_{j=0}^{m} \frac{1}{s - s_j} = \sum_{i=0}^{m} \frac{\rho_i}{s - s_i}, \qquad (7.2.8)$$

where the ρ_i's are given by (7.2.7). With $\theta = s^{-1}$, the L.H.S. of (7.2.8) can be expressed as $\theta^{m+1} \prod_{j=0}^{m}(1 - s_j\theta)^{-1}$. On using geometric series expansion, we then obtain

$$\prod_{j=0}^{m} \frac{1}{s - s_j} = \theta^{m+1} \prod_{i=0}^{m} \left(\sum_{k=0}^{\infty} s_i^k \theta^k \right).$$

On expanding the expression on the R.H.S. of (7.2.8) on similar lines, we get

$$\sum_{i=0}^{m} \frac{\rho_i}{s - s_i} = \theta \sum_{i=0}^{m} \rho_i \left(\sum_{k=0}^{\infty} s_i^k \theta^k \right).$$

Consequently, (7.2.8) implies

$$\theta^m \prod_{i=0}^{m} \left(\sum_{k=0}^{\infty} s_i^k \theta^k \right) = \sum_{i=0}^{m} \rho_i \left(\sum_{k=0}^{\infty} s_i^k \theta^k \right),$$

for all θ in $(0,1]$. On equating the coefficients of θ^t we observe that

$$\sum_{i_0 + \cdots + i_m = t - m} s_0^{i_0} \cdots s_m^{i_m} = \sum_{i=0}^{m} \rho_i s_i^t, \quad t \geq 0, \tag{7.2.9}$$

where, for $t < m$, the summation over the empty set $\{i_0 + \cdots + i_m = t - m\}$ is interpreted as 0. Consequently for all $k \geq 1$,

$$\sum_{i_0 + \cdots + i_m \leq k - m - 1} s_0^{i_0} \cdots s_m^{i_m} = \sum_{i=0}^{m} \rho_i \sum_{t=0}^{k-1} s_i^t,$$

$$= \sum_{i=0}^{m} \rho_i \frac{1 - s_i^k}{1 - s_i}$$

$$= \sum_{i=0}^{m} \rho_i \frac{1}{1 - s_i} - \sum_{i=0}^{m} \rho_i \frac{s_i^k}{1 - s_i}.$$

The above relation holds for $k = 0$ as well. On taking $s = 1$ in (7.2.8) we note that the first term on the R.H.S. above is $\prod_{j=0}^{m}(1 - s_j)^{-1}$. This establishes (7.2.5).

To establish (7.2.6), first observe that, in view of (7.2.3) and (7.2.5),

$$g(s) = \sum_{k=0}^{\infty} p_k \left\{ \prod_{i=0}^{m} \frac{1}{(1 - s_i)} - \sum_{i=0}^{m} \frac{\rho_i s_i^k}{1 - s_i} \right\},$$

and recall that $\pi(s) = \sum_{k=0}^{\infty} p_k s^k$. $\quad\bigcirc$

On combining (7.2.2) and (7.2.6) we have, for $0 < s_0 < \ldots < s_n < 1$,

$$P_n = ds_n \prod_{i=0}^{n-1} \frac{ds_i}{(1 - s_i)} - \prod_{j=0}^{n} ds_j \left\{ \sum_{i=0}^{m} \frac{\rho_i \pi(s_i)}{(1 - s_i)} \right\} + o(\partial(s)). \tag{7.2.10}$$

For continuous F, the $s_i(= F(r_i))$'s are distinct with probability 1 and, without loss of generality, we can take them to be strictly increasing in $(0,1)$. Consequently, (7.2.10) leads to the following general result:

Theorem 7.2.1 (Bunge and Nagaraja, 1991) *When F is absolutely continuous, the joint likelihood of $R_0, \ldots, R_n$ and the event $\{M \geq n\}$ is given by*

$$p_n(r_0, \quad \ldots, \quad r_n; M \geq n)$$
$$= \quad f(r_n) \prod_{i=0}^{n-1} \frac{f(r_i)}{(1 - F(r_i))} - \{\prod_{j=0}^{n} f(r_j)\} \sum_{i=0}^{n-1} \frac{\rho_i \pi(F(r_i))}{(1 - F(r_i))} \, ,$$

$$(7.2.11)$$

where $r_0 < \ldots < r_n$, and the ρ_i's are obtained from (7.2.7) by taking $s_i = F(r_i)$ and $m = n - 1$.

In (7.2.11) the first term on the right represents the likelihood for the first $n + 1$ records in the classical model. The second term represents the reduction due to the uncertainties about the number of available observations. The above result holds even when F is discrete as long as the upper limit is not one of its atoms. We can use Theorem 7.2.1 to evaluate $P(M \geq n)$ and to obtain the conditional joint pdf of $R_0, \ldots, R_n$ given, $\{M \geq n\}$.

When $\pi(s)$ is convenient to work with, (7.2.11) sometimes results in a compact expression for $p_n(\cdot)$.

Example 7.2.1 (Negative Binomial Model) Let $N + r$ be negative binomial with parameters r and p. Then the pgf of N is given by $\pi(s) = p^r(1 - qs)^{-r}$ and

$$\frac{\pi(s)}{1 - s} = \frac{1}{1 - s} - \frac{q}{p} \sum_{j=1}^{r} \left(\frac{p}{1 - qs}\right)^j.$$

Hence the sum in the second term on the R. H. S.— of (7.2.11) can be expressed as

$$\sum_{i=0}^{n-1} \frac{\rho_i \pi(s_i)}{(1 - s_i)} = \sum_{i=0}^{n-1} \frac{\rho_i}{(1 - s_i)} - \frac{q}{p} \sum_{i=0}^{n-1} \rho_i \sum_{j=1}^{r} \left(\frac{p}{1 - qs_i}\right)^j.$$

From (7.2.8), we observe that the first term on the right above is the product $\prod_{i=0}^{n-1}(1 - s_i)^{-1}$. Consequently, when $N + r$ is NBin(r, p), (7.2.11) reduces to

$$p_n(r_0, \ldots, r_n; M \geq n) = \{\prod_{k=0}^{n} f(r_k)\} \frac{q}{p} \sum_{i=0}^{n-1} \rho_i \sum_{j=1}^{r} \left(\frac{p}{1 - qs_i}\right)^j, \quad (7.2.12)$$

where the ρ_i's are given by (7.2.7) and $m = n-1$. If $r = 1$, or when $N+1$ is geometric, the expression for $p_n(\cdot)$ in (7.2.12) simplifies substantially. For this purpose, we consider the partial fraction expansion for $\prod_{j=0}^m (s - qs_j)^{-1}$ as done in (7.2.8). Note that

$$\prod_{i=0}^m \frac{1}{(s - qs_i)} = \sum_{i=0}^m \frac{\rho_i^*}{s - qs_i}, \qquad (7.2.13)$$

where the ρ_i^* can be obtained from the expression for ρ_i in (7.2.7) by replacing s_i by qs_i. Consequently, $\rho_i = q^m \rho_i^*$. Hence, when $N + 1$ is Geo(p),

$$\begin{aligned}
p_n(r_0, \ldots, r_n; M \geq n) &= \prod_{j=0}^n f(r_j) q^{m+1} \sum_{i=0}^m \frac{\rho_i^*}{1 - qs_i} \\
&= f(r_n) \prod_{i=0}^{n-1} \frac{qf(r_i)}{1 - qF(r_i)},
\end{aligned}$$

on using (7.2.13) with $s = 1$. If q were chosen to be 1, N will be infinity with probability 1 and $p_n(\cdot)$ above reduces to the joint pdf of the first $n+1$ records in the classical model. ○

7.2.2 Dependence Structure of Record Values and Record Counts

We can use Theorem 7.2.2 to study the dependence structure of record values and counts. It is interesting to compare the results thus obtained with the corresponding ones from the classical model. In Section 2.6 we have seen that many of the record statistics from the classical model possess the Markov property.

From (7.2.11) we note that

$$p_n(r_0, \ldots, r_n; M \geq n) = \prod_{i=0}^{n-1} \frac{f(r_i)}{1 - F(r_i)} f(r_n)\{1 - \varepsilon_{n-1}\},$$

and keep in mind that ε_{n-1} depends on $s_0, \ldots, s_{n-1}$ but not on s_n. (Recall $s_i = F(r_i)$.) Then,

$$\begin{aligned}
P\{M \geq n, R_0 \in dr_0, &\ldots, R_{n-1} \in dr_{n-1}\} \\
&\approx \prod_{i=0}^{n-1} \frac{f(r_i)dr_i}{1 - F(r_i)} \{1 - \varepsilon_{n-1}\} \int_{r_n = r_{n-1}}^\infty f(r_n)dr_n \\
&= \prod_{i=0}^{n-2} \frac{f(r_i)dr_i}{1 - F(r_i)} f(r_{n-1})dr_{n-1}\{1 - \varepsilon_{n-1}\}.
\end{aligned}$$

Consequently

$$P\{M \geq n, R_n \in dr_n \mid M \geq n, r_0, \dots, r_{n-1}\}$$
$$= \frac{f(r_n)dr_n}{1 - F(r_{n-1})} + o(dr_n) , \quad r_n > r_{n-1},$$

which is the same as in the classical model (see (2.6.3)). On the other hand,

$$p_n(r_0, \dots, r_{n-1}; M \geq n-1) = \prod_{i=0}^{n-2} \frac{f(r_i)}{1 - F(r_i)} \, f(r_{n-1})\{1 - \varepsilon_{n-2}\},$$

and so

$$P\{M \geq n, R_n \in dr_n \mid M \geq n-1, r_0, \dots, r_{n-1}\}$$
$$= \frac{f(r_n)}{1 - F(r_{n-1})} \left\{ \frac{1 - \varepsilon_{n-1}}{1 - \varepsilon_{n-2}} \right\} + o(dr_n),$$

which cannot be reduced further.

For the classical model, we observed that (see (2.6.4)), given the record values, the inter-record times (Δ_i's) are conditionally independent and Δ_i is a $\mathrm{Geo}(1 - s_{i-1})$ random variable. To be precise, for the classical model,

$$P\{\Delta_1 = i_1, \dots, \Delta_n = i_n \mid r_0, \dots, r_n\} = \prod_{j=0}^{n-1} (1 - s_j) s_j^{i_{j+1}-1}.$$

For the random record model, since

$$P\{\Delta_1 = i_1, \dots, \Delta_n = i_n, R_0 \in dr_0, \dots, R_n \in dr_n, M \geq n\}$$
$$= \left\{ \prod_{j=0}^{n-1} s_j^{i_{j+1}-1} ds_j \right\} ds_n P\{N \geq i_1 + \dots + i_n\} + o(dr_0 \dots dr_n),$$

we have

$$P\{\Delta_1 = i_1, \dots, \Delta_n = i_n \mid M \geq n, r_0, \dots, r_n\}$$
$$= \frac{\{\prod_{j=0}^{n-1} s_j^{i_{j+1}-1} f(r_j)\} f(r_n) P\{N \geq i_1 + \dots + i_n\}}{p_n(r_0, \dots, r_n; M \geq n)}.$$

Note that this expression can be factored iff $P\{N \geq i_1 + \dots + i_n\}$ can be factored. But $P\{N \geq i_1 + \dots + i_n\}$ can be factored iff N has $\mathrm{GeoT}(n, 1-c)$ distribution for some $0 < c < 1$. (Show it!) In that case, $P(N \geq i) = P(N \geq n)c^{i-n}$ for all $i \geq n$ and

$$P\{\Delta_1 = i_1, \dots, \Delta_n = i_n \mid M \geq n, r_0, \dots, r_n\} \propto \prod_{j=0}^{n-1} (cs_j)^{i_{j+1}-1}.$$

So, we have proved the following result:

Theorem 7.2.2 (Bunge and Nagaraja, 1991) *In the basic random record model, given the existence and values of $R_0, \ldots, R_n$, the inter-record times Δ_j, $1 \le j \le n$ are conditionally independent iff N has $GeoT(n, 1 - c)$ distribution for some $0 < c < 1$. Further, Δ_j is $Geo(1 - cs_{j-1})$, where $s_j = F(r_j)$.*

7.2.3   Number of Records

We can use the arguments of Section 7.2.2 to obtain the pdf of M, the number of nontrivial records. Now, without loss of generality, take F to be standard uniform and recall (7.2.1) and (7.2.9) to observe

$$
\begin{aligned}
P\{M &= m, R_0 \in ds_0, \ldots, R_m \in ds_m \mid N = k\} \\
&= \sum_{i_0 + i_1 + \cdots + i_m = k - m} ds_0 s_0^{i_0} ds_1 s_1^{i_1} \cdots ds_m s_m^{i_m} + o(\partial(\boldsymbol{s})) \\
&= \prod_{i=0}^{m} ds_i \left\{ \sum_{i=0}^{m} \rho_i s_i^k \right\} + o(\partial(\boldsymbol{s})).
\end{aligned}
$$

Consequently, on ignoring $o(\partial(\boldsymbol{s}))$ terms we have

$$
\begin{aligned}
P\{M = m, R_0 \in ds_0, \ldots, R_m \in ds_m\} &= \prod_{i=0}^{m} ds_i \left\{ \sum_{k=0}^{\infty} \sum_{i=0}^{m} \rho_i p_k s_i^k \right\} \\
&= \prod_{i=0}^{m} ds_i \left\{ \sum_{i=0}^{m} \rho_i \pi(s_i) \right\}.
\end{aligned}
$$

On integrating with respect to the s_i's we thus obtain,

$$
P(M = m) = \int_{s_0 < \cdots < s_m} \left\{ \sum_{i=0}^{m} \rho_i \pi(s_i) \right\} ds_0 \cdots ds_m.
$$

Since the integrand in the multiple integral above is a symmetric function of the s_i's, the following result holds:

Theorem 7.2.3 (Bunge and Nagaraja, 1991) *Let M denote the number of nontrivial records in the random record model. Then,*

$$
P(M = m) = \frac{1}{(m+1)!} \int_{0 < s_i < 1, 0 \le i \le m} \left\{ \sum_{i=0}^{m} \rho_i \pi(s_i) \right\} ds_0 \cdots ds_m,
$$

$$
(7.2.14)
$$

where $\pi(s)$ is the pgf of N and the ρ_i's are given by (7.2.7).

There is another representation for the pdf of M due to Westcott (1977). It connects the pdf of M to the pdf of N_n, the number of records among the first n observations in the classical model. Consider

$$
\begin{aligned}
P(M = m) &= \sum_{j=n}^{\infty} P(m+1 \text{ records among } X_1, \ldots, X_{j+1} \text{ and } N = j) \\
&= \sum_{j=n}^{\infty} P(m+1 \text{ records among } X_1, \ldots, X_{j+1}) p_j \\
&= \sum_{j=n}^{\infty} \frac{|S_{j+1}^{m+1}|}{(j+1)!} \, p_j,
\end{aligned}
\tag{7.2.15}
$$

where the last equality follows from (2.5.9). Sometimes, while there may not be a direct approach to evaluate the above infinite series, the multiple integral in (7.2.14) may lead to a compact expression for the pdf of M. Here is an example.

Example 7.2.2 (Geometric Model) Let $N + 1$ be a Geo(p) random variable. Then the pgf of N is given by $\pi(s) = p/(1 - qs)$, and the integrand on the R.H.S. of (7.2.14) simplifies as

$$
\begin{aligned}
\sum_{i=0}^{m} \rho_i \pi(s_i) &= p \sum_{i=0}^{m} \frac{\rho_i}{1 - qs_i} \\
&= pq^m \sum_{i=0}^{m} \frac{\rho_i^*}{1 - qs_i},
\end{aligned}
$$

where the ρ_i^*'s are the coefficients in (7.2.13). Consequently, we have

$$
\sum_{i=0}^{m} \rho_i \pi(s_i) = pq^m \prod_{i=0}^{m} \frac{1}{1 - qs_i}
$$

and hence, on using (7.2.14) we may write

$$
\begin{aligned}
P(M = m) &= \frac{1}{(m+1)!} pq^m \prod_{i=0}^{m} \int_{0 < s_i < 1} \frac{1}{1 - qs_i} \, ds_i \\
&= \frac{p}{q(m+1)!} \{ -\log(p) \}^{m+1}, \quad m \geq 0.
\end{aligned}
\tag{7.2.16}
$$

In other words, $M + 1$ has a zero-truncated Poisson distribution with parameter $-\log(p)$. ◯

Bunge and Nagaraja (1991) have obtained an explicit expression for the pdf of M when $N + r$ is a NBin(r, p) random variable. We omit the proof of their result, given below.

Theorem 7.2.4 *Let M denote the number of nontrivial records in the random record model where N has pgf $\pi(s) = (p/(1-qs))^r$. Then,*

$$P(M = m) = \frac{p^r}{q(m+1)!} \sum_{i_0 + \cdots + i_m = r-1} \prod_{j=0}^{m} D(i_j), \qquad (7.2.17)$$

where the i_j's are non-negative integers and $D(x) = (1 - p^x)/x p^x$, $x \geq 0$.

On noting that $D(0) = -\log(p)$, it is easy to see that when $r = 1$, (7.2.17) reduces to (7.2.16).

Example 7.2.3 When $N + r$ is NBin(r, p) with $r = 2$, we obtain

$$
\begin{aligned}
P(M = m) &= \frac{p^2}{q(m+1)!} (m+1)\{D(0)\}^m \{D(1)\} \\
&= \frac{e^{-\log(p)}\{-\log(p)\}^m}{m!}, \qquad m \geq 0.
\end{aligned}
$$

That is, M is a Poisson random variable with parameter $-\log(p)$. ○

Our discussion in this section assumed that the X_i's are i.i.d. random variables with common continuous cdf F. One may also consider the distribution theory for weak records assuming F to be discrete and for records from the Pfeifer model, introduced in Section 6.5. See Bunge and Nagaraja (1991) for some details. Note that expressions for $p_n(\cdot)$ appear more complex for those models. Recently, Hofmann (1997) has explored the properties of record statistics that arise from a random number of observations taken from an F^α model (discussed in Section 6.3). Exercise 7.3 considers some distributional results applicable to weak records and Exercise 7.4 presents the distribution of the number of observed records from two F^α models.

7.3 BASIC POINT PROCESS MODEL AND APPLICATIONS

Thus far, our tacit assumption was that the X_n's are observed at a random number of equally spaced time points. We now assume that the observations arrive at random time points determined by an independent *pacing* process $\boldsymbol{P}$. We assume $\boldsymbol{P}$, defined on $(0, \infty)$, is a simple point process. It can be described either by the sequence of interarrival times $\{V_n, n \geq 1\}$ or, equivalently, by the counting process $N(t) = \max\{n : \sum_{i=1}^{n} V_i \leq t\}$, which counts the number of observations that have arrived during $(0, t]$. We will use $\{N(t), t > 0\}$, $\boldsymbol{P}$ and $\{V_n, \ n \geq 1\}$ interchangeably in describing the pacing process. We assume X_1, the initial record value R_0, is observed at time point 0 and for $n \geq 2$, X_n is observed at time $\sum_{i=1}^{n-1} V_i$. We define T_n and R_n as in the classical model. Note that here T_n represents the nth record index and $W_n \equiv \sum_{i=1}^{T_n-1} V_i$, represents the actual arrival time of the nth record. This setup describes our basic *point process record model* (PPRM) having pacing process $\boldsymbol{P}$. Here we are interested in the counting process $M(t) = \max\{n : W_n \leq t\}$, representing the number of nontrivial record values encountered by time t and the associated sequence of the inter-record times $U_n \equiv \sum_{i=T_{n-1}}^{T_n-1} V_i$, $n \geq 1$. Note that U_n is the sum of Δ_n random variables.

Pickands (1971) was the first to consider a PPRM. He took $\boldsymbol{P}$ to be a homogeneous Poisson process with unit intensity. Gaver (1976) called PPRM a "random record model". Note that, for us, this represents a slightly different framework. Gaver, and Gaver and Jacobs (1978) obtained exact and asymptotic results for R_n, U_n, and W_n, when $\boldsymbol{P}$ is one of the common point processes. Westcott (1977) clarified the link between the classical model and the PPRM, and derived asymptotic properties of U_n, W_n and $M(t)$, under various assumptions about $\boldsymbol{P}$. Assuming $\boldsymbol{P}$ is a renewal process with interarrival time cdf F_V, Westcott (1977, 1979), Embrechts and Omey (1983), and Yakymiv (1986) studied the connection between the upper tails of the cdf's of W_n and U_n, and that of F_V. Meanwhile, Deheuvels (1982) obtained a simple distributional representation for W_n, when $\boldsymbol{P}$ is a homogeneous unit Poisson process. Through his extensive work on optimal rules in *secretary problems*, Bruss (e.g., 1988) has successfully found applications of record value theory to the best-choice problem. Recent work on records from PPRM's are reported in Bunge (1989), Bruss and Rogers (1991), Bunge and Nagaraja (1991, 1992a, 1992b), Nagaraja (1994b), Browne and Bunge (1995) and Hofmann (1997). We elaborate on some of these important developments in subsequent sections.

As we trek the trail of the point process models, let us recall that the stochastic structure of a counting process can be described in a variety of ways. One can describe the arrival rate of events, or examine the behavior of the interarrival times. Of course, one can also discuss the distribution of the count by certain time t. In that case, PPRM can be viewed as a random record model, described in Section 7.2. Consequently, the distribution theory developed for the record values, and the number of records there, is directly applicable here for the investigation of the record counting process when the pacing process P is observed over a finite interval. Note that N, the number of observations and M, the number of record values, correspondingly are represented by $N(t)$ and $M(t)$ here. However, for a comprehensive investigation of the record arrival process $M(t)$ as t evolves over $(0, \infty)$, we need to specify P. Assuming P is one of the common point processes such as a Poisson process, or a birth process, we will describe in the coming sections the probabilistic structure of the inter-record time sequence $\{U_n, n \geq 1\}$. But first, as motivation, we will describe some applications of the study of records where a PPRM provides a natural framework.

Climatology and High River Flows

In climatic and flood records, large observations tend to occur in clusters due to the high degree of short-range dependence among the series of observations. A common approach is to consider only the maximum value in the clusters that exceed a high threshold and treat them as the observations. This is known as the *peaks-over threshold* method or *partial duration series* method (Buishand, 1989). The arrival times of these peaks are modeled by a (possibly nonhomogeneous) Poisson process. Similar to this is the situation with high river flows (Leadbetter, Lindgren and Rootzén, 1983, pp. 281–282). Our discussion in Section 7.4 presents some exact distributional results for record statistics arising from such data.

System Strength and Shock Models

Consider a system that works without failure until a shock exceeding its strength arrives. Let us assume these shocks arrive according to a point process P. Let X_1 represent the magnitude of the system strength and let X_n, $n \geq 2$, represent the magnitudes of the arriving shocks. Gaver (1976) considered this model where he assumed that X_1 has cdf F_0 and the X_n's are i.i.d. with cdf F. Then, $W_1 (= U_1)$ represents the waiting time for the failure of the equipment. Gaver has given expressions for its cdf when P is one of the common point processes. See Exercise 7.7 for the Poisson process model.

Now, in the above setup, assume that upon failure, for safety reasons, modified components or system designs are used so that the system endures shocks up to the magnitude of the last shock. Then the successive failure times of the system are the record times and $M(t)$ counts the number of failures in the interval $(0, t]$. With $F_0 = F$, we can use the distribution theory discussed in this chapter to obtain exact distributions of these statistics.

Secretary Problems

Record counting statistics play aprominent role in the classical secretary problem, which dates back to Cayley (1875). The goal is to pick the maximum of a set of random variables that are observed in sequence with no recall possible and we have to devise a strategy that maximizes the probability of actually picking the best. We assume the no-information setup, where only the rank of the current observation relative to the previous ones is available. This brings us directly into the realm of record counting statistics from the classical model. The problem becomes more fascinating when the number of available observations is assumed to be random – a still more interesting situation is the case where these observations arrive at random time points. We will formally discuss these scenarios in Section 7.8. There are several variations of the classical secretary problem determined by the amount of the information, as well as the recall options that are permitted. While results developed here are not applicable to such situations, it is clear that, for all optimal rules, the final choice will be made at one of the record times.

7.4   RECORDS OVER POISSON PROCESSES

7.4.1   Homogeneous Poisson Pacing Process

In his pioneering work, Pickands (1971) introduced a point process model in which P is a homogeneous Poisson process. We now explore the distributional structure of the record arrival process for that model. We now assume that $N(t)$ is a Poisson process with constant intensity λ.

We might contemplate using Theorem 7.2.3 to obtain the pdf of $M(t)$. But we quickly discover that (7.2.14) does not appear to provide a compact form for $P(M(t) = m)$, when $\pi(s) = Es^{N(t)} = \exp\{\lambda t(s - 1)\}$. Nor does the representation in (7.2.15) provide a closed-form expression. However, there is an interesting representation for the inter-record time sequence $\{U_n, n \geq 1\}$ that characterizes the record arrival process. It was first derived by Deheuvels (1982) and will form Theorem 7.4.1, the main result of this section. Our proof is based on the work of Bunge and Nagaraja (1992a), while Deheuvels' equivalent representation was derived using lower records and standard uniform random variables.

It is convenient for our discussion to introduce the following two lemmas:

Lemma 7.4.1 *Let A_j, B_j, $1 \leq j \leq n$, be i.i.d. standard exponential random variables and let $C_j = \sum_{i=1}^{j} B_i$. Then, the joint characteristic function of $(C_1 + \log A_1, C_2 + \log A_2, \ldots, C_n + \log A_n)$ is given by*

$$
E\left(\exp\left\{ \sum_{j=1}^{n} z_j(C_j + \log A_j) \right\} \right) = \prod_{j=1}^{n} \Gamma(z_j + 1) \left(1 - \sum_{i=j}^{n} z_i \right)^{-1},
$$

(7.4.1)

where the z_j's are imaginary numbers and Γ is the gamma function.

Proof. The computation is straightforward and is omitted.   ○

Lemma 7.4.2 *Let x, y, and z be complex numbers such that real parts of $(y + 1)$, $(z - y)$, and $(z + x + 2)$ are all positive. Then*

$$
\eta(x, y, z) \equiv \sum_{k=1}^{\infty} \frac{\Gamma(x + k)\Gamma(y + k)}{\Gamma(z + x + 1 + k)\Gamma(k)} = \frac{\Gamma(x + 1)\Gamma(y + 1)\Gamma(z - y)}{\Gamma(z + 1)\Gamma(z + x - y + 1)}.
$$

Proof. See Abramowitz and Stegun (1965, p. 556).   ○

We are now ready to state the useful representation that describes the joint distribution of the inter-record times from the Poisson model.

Theorem 7.4.1 (Deheuvels, 1982) *Let $\boldsymbol{P}$ be a homogeneous Poisson process with unit intensity. Then, the inter-record times U_j satisfy the distributional identity*

$$
(\log U_1, \ldots, \log U_n) \overset{d}{=} (C_1 + \log A_1, C_2 + \log A_2, \ldots, C_n + \log A_n),
$$

(7.4.2)

where the A_j and C_j are as in Lemma 7.4.1.

Proof. (Bunge and Nagaraja, 1991) We show that the characteristic function of $(\log U_1, \ldots, \log U_n)$ is given by the R. H. S. expression in (7.4.1). Let $\boldsymbol{z} = (z_1, \ldots, z_n)$ and $\varphi(\boldsymbol{z}) = E\left(\exp\left\{ \sum_{j=1}^{n} z_j \log U_j \right\} \right)$ be the joint characteristic function of the $\log U_j$'s. We evaluate this expectation by conditioning on the values of the record indices, $T_1, \ldots, T_n$, and by exploiting the conditional independence of the U_i's given the T_i's. Thus,

$$\varphi(z) = E\left\{\prod_{j=1}^{n} U_j^{z_j}\right\}$$

$$= \sum_{1<t_1<t_2<\cdots<t_n} P(T_1 = t_1,\ldots,T_n = t_n)\prod_{j=1}^{n} E\left\{\left(\sum_{i=t_{j-1}}^{t_j-1} V_i\right)^{z_j}\right\}.$$

$$(7.4.3)$$

Note that $\sum_{i=t_{j-1}}^{t_j-1} V_i$ follows a Gamma$(k_j, 1)$ distribution where $k_j = t_j - t_{j-1}$, $1 \leq j \leq n$, and $t_0 = 1$. Hence the characteristic function of $\log(\sum_{i=t_{j-1}}^{t_j-1} V_i)$ is given by $\Gamma(k_j + z_j)/\Gamma(k_j)$. Further, from (2.5.3), we have $P(T_1 = t_1,\ldots,T_n = t_n) = \{(t_1 - 1)\cdots(t_n - 1)t_n\}^{-1}$. Thus (7.4.3) may be expressed as

$$\varphi(z) = \sum \frac{1}{(t_1 - 1)\cdots(t_n - 1)t_n}\prod_{j=1}^{n}\frac{\Gamma(k_j + z_j)}{\Gamma(k_j)}, \qquad (7.4.4)$$

where the summation is taken over the k_j's forming the set $\{(k_1,\ldots,k_n) : k_j \geq 1, \ 1 \leq j \leq n\}$. We now sum (7.4.4) over $k_n, k_{n-1},\ldots$ in order, and repeatedly use Lemma 7.4.2 and the relation $t_j = t_{j-1} + k_j$.

Note that $\{(t_n - 1)t_n\}^{-1}$ can be expressed as $\Gamma(t_{n-1}-1+k_n)/\Gamma(t_{n-1}+k_n+1)$. Thus, on summing over k_n, we obtain $\eta(t_{n-1}-1, z_n, 1)$ of Lemma 7.4.2. On using η's second form, the sum over k_n can be expressed as

$$\frac{\Gamma(z_n + 1)}{1 - z_n}\frac{\Gamma(2 - z_n)\Gamma(t_{n-1})}{\Gamma(t_{n-1}+1-z_n)}.$$

Next, summing over k_{n-1}, we get

$$\frac{\Gamma(z_n + 1)}{1 - z_n}\Gamma(2 - z_n)\eta(t_{n-2} - 1, z_{n-1}, 1 - z_n)$$

$$= \prod_{j=n-1}^{n}\frac{\Gamma(z_j + 1)}{(1 - \sum_{i=j}^{n} z_i)}\frac{\Gamma(2 - \sum_{i=n-1}^{n} z_i)\Gamma(t_{n-2})}{\Gamma(t_{n-2}+1-\sum_{i=n-1}^{n} z_i)}.$$

Continuing in this fashion, the final summation (over k_1) yields

$$\varphi(z) = \prod_{j=2}^{n}\frac{\Gamma(z_j + 1)}{(1 - \sum_{i=j}^{n} z_i)}\Gamma\left(2 - \sum_{i=2}^{n} z_i\right)\eta\left(t_0 - 1, z_1, 1 - \sum_{i=2}^{n} z_i\right)$$

$$= \prod_{j=1}^{n}\Gamma(z_j + 1)\left(1 - \sum_{i=j}^{n} z_i\right)^{-1},$$

the expression given in (7.4.1).   ◯

7.4.2   Comparison with the Classical Model

The representation for the inter-record times established in Theorem 7.4.1 is extremely useful in exploring the finite time, as well as asymptotic dependence structure of record times and inter-record times. For example, (7.4.2) implies that for every $n \geq 1$, $W_n \overset{d}{=} \sum_{j=1}^{n} A_j e^{C_j}$. In Chapter 2 we observed some distributional identities for the record counting statistics from the classical model (see Exercises 2.2, 2.5). Table 7.4.1, an extension of Table 7.1 in Bunge and Nagaraja (1992a), compares the properties of record counting statistics from these two models.

In Table 7.4.1, for the Poisson model, part (b) follows from part (a) after appropriate relabeling. While part (c) follows directly from (a), Gaver (1976) had earlier established this marginal distributional identity. From (c) it is easily verified that (d) and (e) hold. Part (f) follows from the fact that $U_{n+1}/U_n \overset{d}{=} (A_{n+1}/A_n)e^{B_{n+1}}$.

As noted in the table, similar results have been established for the classical model and the resemblances are striking. In regard to (c), note that as $n \to \infty$, $C_n \to \infty$ (almost surely), and consequently, an application of l'Hôpital's rule shows that $e^{C_n}/((-\log(1-e^{-C_n}))^{-1}) \to 1$ almost surely. Thus, the stochastic properties of Δ_n and $[U_n]$ are quite similar for large n. It is interesting to observe that the asymptotic results (d)–(f) for the classical model have their counterparts for the Poisson model that hold for all n! More can be said about part (f). Westcott (1977, Theorem 6.e) showed that the limit result of part (f) holds for any PPRM for which $\{V_j, j \geq 1\}$ is stationary and $\lim_{k \to \infty}(1/k)\sum_{j=1}^{k} V_j = 1$ almost surely.

Tata (1969) derived the limiting joint distribution of successive inter-record times for the classical model. Since $\Delta_n \equiv U_n$ in that situation, her result can be stated as

$$\lim_{n \to \infty} P\left\{\frac{\log U_n - n}{\sqrt{n}} > x, \; \frac{\log U_{n-1} - n}{\sqrt{n}} \leq y\right\}$$

$$= \begin{cases} 0 & x \geq y \\ \Phi(y) - \Phi(x) & x < y, \end{cases} \qquad (7.4.5)$$

where Φ is the standard normal cdf. It is now easy to show that (7.4.5) holds for the Poisson P model as well.

Theorem 7.4.2 *Let P be a homogeneous Poisson process with unit intensity. Then (7.4.5) holds.*

Table 7.4.1. Properties of Record Counting Statistics from Poisson Point Process and Classical Models

Unit Poisson $\boldsymbol{P}$ Model	Classical Model*
(a) $W_n \overset{d}{=} \sum_{j=1}^{n} A_j e^{C_j}$	$T_n \overset{d}{=} 1 + \sum_{j=1}^{n} \left(\left[\frac{A_j}{-\log(1-e^{-C_j})} \right] + 1 \right)$
(Theorem 7.4.1)	(Deheuvels, 1981)
(b) $W_{n+1} \overset{d}{=} e^{B_{n+1}}(W_n + A_{n+1})$	$T_{n+1} \overset{d}{=} [e^{B_n} T_n] + 1$
(From (a))	(Williams, 1973)
(c) $U_n \overset{d}{=} A_n e^{C_n}$	$\Delta_n \overset{d}{=} \left[\frac{A_n}{-\log(1-e^{-C_n})} \right] + 1$
(From (a); Gaver, 1976)	(Deheuvels, 1981)
(d) $E(\log U_n) = n - \gamma$	$E(\log \Delta_n) = n - \gamma + O\left(\frac{n}{2^n}\right)$
(From (c))	(Pfeifer, 1981)
(e) $\mathrm{Var}(\log U_n) = n + \frac{\pi^2}{6}$	$\mathrm{Var}(\log \Delta_n) = n + \frac{\pi^2}{6} + O\left(\frac{n^2}{2^n}\right)$
(From (c))	(Pfeifer, 1981)
(f) $P\left\{ \frac{U_{n+1}}{U_n} > w \right\}$ $= \frac{\log(1+w)}{w}, \; w > 0$	$\lim_{n\to\infty} P\left\{ \frac{\Delta_{n+1}}{\Delta_n} > w \right\}$ $= \frac{\log(1+w)}{w}, \; w > 0$
(From (c))	(Tata, 1969)

* [.] indicates the greatest integer function.

Proof. From Theorem 7.4.1, $(\log U_1, \ldots, \log U_n) \overset{d}{=} (C_1 + \log A_1, \ldots, C_n + \log A_n)$. Consequently,

$$P\left\{\frac{\log U_n - n}{\sqrt{n}} > x, \quad \frac{\log U_{n-1} - n}{\sqrt{n}} \le y\right\}$$

$$= P\left\{\frac{C_n + \log A_n - n}{\sqrt{n}} > x, \quad \frac{C_{n-1} + \log A_{n-1} - n}{\sqrt{n}} \le y\right\}$$

$$= P\left\{\frac{B_n + \log A_n}{\sqrt{n}} + \frac{C_{n-1} - n}{\sqrt{n}} > x, \quad \frac{\log A_{n-1}}{\sqrt{n}} + \frac{C_{n-1} - n}{\sqrt{n}} \le y\right\}.$$

Both $(B_n + \log A_n)/\sqrt{n}$ and $\log A_{n-1}/\sqrt{n}$ converge to 0 in probability as $n \to \infty$. Further, from the central limit theorem it follows that $\{(C_{n-1} - (n-1))/\sqrt{n-1}\} \overset{d}{\to} N(0,1)$. This also implies that $\{(C_{n-1} - n)/\sqrt{n}\} - \{(C_{n-1} - (n-1))/\sqrt{n-1}\} \overset{P}{\to} 0$. Therefore,

$$\lim_{n\to\infty} P\left\{\frac{\log U_n - n}{\sqrt{n}} > x, \quad \frac{\log U_{n-1} - n}{\sqrt{n}} \le y\right\}$$

$$= \lim_{n\to\infty}\left\{x < \frac{C_{n-1} - (n-1)}{\sqrt{n-1}} \le y\right\}$$

$$= \Phi(y) - \Phi(x),$$

whenever $x < y$ and is 0 otherwise. Thus, (7.4.5) holds. $\qquad \bigcirc$

7.4.3   Nonhomogeneous Poisson Pacing Process

Now suppose $\boldsymbol{P}$ is a Poisson process with nonconstant rate function $\lambda(t)$. Since such a Poisson process can be converted into a homogeneous Poisson process by transforming the time axis, Theorem 7.4.1 can be used to describe the structure of the inter-record times from such a pacing process.

Theorem 7.4.3 *Let $\boldsymbol{P}$ be a Poisson process with rate $\lambda(t)$. Let $\int_0^t \lambda(s)ds = \Lambda(t)$ denote the mean function, and let $\Lambda^{-1}(t) = \inf\{s : \Lambda(s) > t\}$, $t > 0$. Then*

$$(U_1, U_2, \ldots, U_n)$$
$$\overset{d}{=} \left(\Lambda^{-1}(A_1 e^{C_1}), \Lambda^{-1}(A_1 e^{C_1} + A_2 e^{C_2}) - \Lambda^{-1}(A_1 e^{C_1}), \ldots,\right.$$
$$\left.\Lambda^{-1}\left(\sum_{j=1}^n A_j e^{C_j}\right) - \Lambda^{-1}\left(\sum_{j=1}^{n-1} A_j e^{C_j}\right)\right),$$

where the A_j and C_j are as in Lemma 7.4.1.

Proof. Note that $\Lambda(V_1), \Lambda(V_1 + V_2)$, etc., are the arrival times of a homogeneous Poisson process. Consequently, $\Lambda(U_1), \Lambda(U_1 + U_2), \ldots, \Lambda(U_1 + \cdots + U_n)$ are the record times of a homogeneous Poisson process, and the assertion follows from Theorem 7.4.1. ○

Example 7.4.1 (Exponential intensity function) Consider a nonhomogeneous Poisson point process record model with $\lambda(t) = e^t$. We can use Theorem 7.4.3 to find the finite-time structure of its record arrival process. Here $\Lambda(t) = \int_0^t e^s \, ds = e^t - 1$ for $t \geq 0$, so that $\Lambda^{-1}(t) = \log(1 + t)$. Hence we have the following distributional identity:

$$
(U_1, U_2, \ldots, U_n)
$$
$$
\stackrel{d}{=} \left(\log(1 + A_1 e^{C_1}), \log\left(\frac{1 + A_1 e^{C_1} + A_2 e^{C_2}}{1 + A_1 e^{C_1}} \right), \ldots, \right.
$$
$$
\left. \log\left(\frac{1 + \sum_{j=1}^{n} A_j e^{C_j}}{1 + \sum_{j=1}^{n-1} A_j e^{C_j}} \right) \right). \tag{7.4.6}
$$

From (7.4.6) it can be shown that U_n converges in distribution to a standard exponential random variable as $n \to \infty$. ○

In the above example $\lambda(t)/\Lambda(t)$ converges to a constant as $t \to \infty$. In such cases the following general result holds:

Theorem 7.4.4 (Bunge and Nagaraja, 1992b) *In the nonhomogeneous Poisson process record model, suppose the rate function $\lambda(t)$ is such that*

$$
\lim_{t \to \infty} \frac{\lambda(t)}{\Lambda(t)} = c, \tag{7.4.7}
$$

where c is finite and positive. Then, as $n \to \infty$,

$$
(cU_{n+1}, \ldots, cU_{n+k}) \stackrel{d}{\to} (A_1, \ldots, A_k)
$$

for any finite k, where the A_j's are i.i.d. Exp(1) random variables.

We refer you to the original source for a proof. Theorem 7.4.4 implies that if the arrival rate is asymptotically exponential, the inter-record times are asymptotically i.i.d. exponential random variables. Later, in section 7.6.3, we present a pacing process P, for which the inter-record times are i.i.d. exponential random variables, even in finite time!

Gaver and Jacobs (1978) present a result similar to Theorem 7.4.4 for the inter-record times observed after time τ, where τ is large. Now let $U_1(\tau), \ldots, U_k(\tau)$ denote the first k inter-record times *after* time τ. From their work we may conclude that when $\lambda(t) = ce^{ct}$, $(cU_1(\tau), \ldots, cU_k(\tau)) \xrightarrow{d}$ $(A_1, \ldots, A_k)$ as $\tau \to \infty$ where, as above, the A_j's are i.i.d. Exp(1) random variables.

One can consider a PPRM in the context of an F^α model. Hofmann (1997) and Hofmann and Nagaraja (1997) present several results on inter-record times when observations from an F^α model arrive according to an independent Poisson process. Asymptotic properties of the inter-record times depend upon the properties of $S(\Lambda(t))$ where $S(t)$ is a smooth monotonic function defined for all $t > 0$ such that for a positive integer t, $S(t)$ is the partial sum of the $\alpha(i)$'s. See also Theorem 7.6.3.

7.5   RECORDS OVER A RENEWAL PROCESS

We will now assume that the pacing process $\boldsymbol{P}$ is a renewal process. In other words, the interarrival times of the observations, V_n, are i.i.d. random variables. Let F_V be the cdf of the interarrival distribution. As we remarked briefly earlier (in Section 7.3), the behavior of the cdf of the inter-record time U_n is closely related to that of F_V. Gaver (1976), Westcott (1977, 1979), and Embrechts and Omey (1983) have examined this connection. Below we follow the discussion in Westcott (1977).

Using a conditioning argument, the probability distribution of U_n can be expressed as

$$
\begin{aligned}
P(U_n > t) &= \sum_{i=1}^{\infty} P(T_n - T_{n-1} = i) P\left(\sum_{j=T_{n-1}}^{T_{n-1}+i-1} V_j > t \right) \\
&= \sum_{i=1}^{\infty} P(\Delta_n = i) P\left(\sum_{j=1}^{i} V_j > t \right),
\end{aligned} \tag{7.5.1}
$$

since the V_j are i.i.d. Recall that $P(\Delta_n = i)$ is the pdf of the inter-record time in the classical model (see (2.5.24) for an expression). From (7.5.1), it follows that the Laplace transform of U_n is related to that of the cdf F_V and the pgf of Δ_n. Let $L_V(s) = E(e^{-sV_1})$ and $L_n(s) = E(e^{-sU_n})$ be the two Laplace transforms. Then from (7.5.1) we have,

$$
\begin{aligned}
L_n(s) &= \sum_{i=1}^{\infty} P(\Delta_n = i)\{L_V(s)\}^i \\
&= E\{L_V(s)\}^{\Delta_n}.
\end{aligned} \tag{7.5.2}
$$

In view of (2.5.26), we may infer that (7.5.2) establishes an explicit relation between L_V and L_n.

When $n = 1$, the relation is rather simple to express. For, in that case, $E\{s^{\Delta_1}\} = 1 + s^{-1}(1 - s)\log(1 - s)$ and, consequently, the Laplace transform of the first inter-record time is given by

$$1 - L_1(s) = \frac{1 - L_V(s)}{L_V(s)} \log(1 - L_V(s)). \qquad (7.5.3)$$

We will now estimate the right tail of the distribution of U_1.

Theorem 7.5.1 (Westcott, 1977) *Suppose the pacing process is a renewal process whose interarrival times have mean μ and a finite variance. Then, as $t \to \infty$,*

$$P(U_1 > t) \approx \frac{\mu}{t}. \qquad (7.5.4)$$

Proof. Since $P(\Delta_1 = i) = 1/i(i+1)$, from (7.5.1) we have

$$P(U_1 > t) = \sum_{i=1}^{\infty} \frac{1}{i(i+1)} P(S_i > t), \qquad (7.5.5)$$

where $S_i = \sum_{j=1}^{i} V_j$. Following Westcott, we divide the range of i in the sum above into the following three regions:

$$(1, [t(\mu + \varepsilon)^{-1}]), \ \ ([t(\mu + \varepsilon)^{-1}] + 1, [t(\mu - \varepsilon)^{-1}]), \ \ ([t(\mu - \varepsilon)^{-1} + 1, \infty).$$

Let the parts of the sum on the R.H.S. of (7.5.5) corresponding to the above three regions be denoted by $\Sigma_1, \Sigma_2,$ and Σ_3, respectively. We show that the first two sums are negligible, while the last one is close to the approximation presented in (7.5.4).

For i in the range of Σ_1,

$$P(S_i > t) \leq P\left\{|S_i - i\mu| > \frac{\varepsilon t}{\mu + \varepsilon}\right\} \leq \frac{i(\mu + \varepsilon)^2 \mathrm{Var}(V_1)}{\varepsilon^2 t^2},$$

where the second inequality is just Chebyshev's. Hence,

$$\Sigma_1 \leq \frac{c}{t^2} \sum_{i=1}^{[t(\mu+\varepsilon)^{-1}]} (i+1)^{-1} \approx \frac{c \log t}{t^2} = o\left(\frac{1}{t}\right), \qquad (7.5.6)$$

as $t \to \infty$, where c is a generic constant.

Next, we note that

$$\Sigma_2 \leq \sum_{i=[t(\mu+\varepsilon)^{-1}]+1}^{[t(\mu-\varepsilon)^{-1}]} \left\{ \frac{1}{i} - \frac{1}{i+1} \right\}$$
$$\approx (\mu+\varepsilon)t^{-1} - (\mu-\varepsilon)t^{-1} = o(1/t), \tag{7.5.7}$$

as ε is arbitrary. Finally, let us consider Σ_3. For values of i in the third region, $i(\mu-\varepsilon) > t$. Hence $P(i^{-1}S_i > (\mu-\varepsilon)) \leq P(S_i > t)$. Thus we have the following inequalities:

$$\sum_{i=[t(\mu-\varepsilon)^{-1}]+1}^{\infty} \frac{1}{i(i+1)} P(i^{-1}S_i > \mu - \varepsilon)$$
$$\leq \Sigma_3 \leq \sum_{i=[t(\mu-\varepsilon)^{-1}]+1}^{\infty} \frac{1}{i(i+1)} < \frac{\mu-\varepsilon}{t}. \tag{7.5.8}$$

Since from weak law of large numbers we know that $i^{-1}S_i \xrightarrow{P} \mu$, as $i \to \infty$, (7.5.8) implies that as $t \to \infty$, $\Sigma_3 \approx \mu/t$. In view of (7.5.6) and (7.5.7) we conclude that (7.5.4) indeed holds. ◯

Theorem 7.5.1 implies that the right tail of U_1 is equivalent to that of a standard Pareto distribution. Further, even though $E(U_1)$ is infinite, $E(U_1^\delta)$ is finite for all $\delta < 1$. While we assumed the finiteness of the variance of F_V to establish (7.5.4), Westcott (1977) assumed just the finiteness of $(1+\delta)$th moment. Of course, we would then need stronger tools than the classical Chebyshev's inequality to carry through the arguments. Westcott also shows that (7.5.4) holds for a $\boldsymbol{P}$ for which the interarrival times are stationary and uniformly mixing with finite second moment. Further, he establishes upper and lower bounds for $P(U_1 > t)$ for finite t.

The concept of regular variation (see Resnick, 1987, for a formal exposition) has been used successfully to study the tail behavior of record times and inter-record times from a renewal process record model. For our purposes, just its definition will suffice here. We say that a non-negative function L defined on $(0, \infty)$ is *regularly varying at infinity with index α* and write $L \in RV_\alpha$ if $\lim_{t\to\infty}(L(tx)/L(t)) = x^\alpha$ for all $x > 0$. If α is 0, we say L is *slowly varying*. Assuming a regular variation property for $1 - F_V(x)$ or a slow variation property for $\int_0^t \{1 - F_V(x)\}dx$, Westcott (1977, 1979), and Embrechts and Omey (1983) obtain several interesting equivalent conditions on the tail behaviors of the cdf's of V_1, U_n and W_n. Embrechts and Omey observe that the cdf's of U_n and W_n are infinite mixtures of convolutions of F_V; in other words, these cdf's are *subordinated* to F_V (see Exercise 7.17 for a definition). They use general results on subordinated distributions to improve upon the results by Westcott. One of their results is the following:

Theorem 7.5.2 (Embrechts and Omey, 1983) *For* $-1 < \alpha \le 0$, *the following statements are equivalent.*

  (i) $P(V_1 > t) = 1 - F_V(t) \in RV_\alpha$,
  (ii) $P(U_n > t) \in RV_\alpha$,
  (iii) $P(W_n > t) \in RV_\alpha$.

Further, for large t,

$$P(U_n > t) \approx P(W_n > t) \approx \frac{1 - F_V(t)}{n!} \{-\log(1 - F_V(t))\}^n. \quad (7.5.9)$$

One can use the above relationships to approximate the tail probabilities of the nth record time or inter-record time when $1 - F_V$ is regularly varying. It is interesting to note that when F_V is continuous, the last expression in (7.5.9) is asymptotically equivalent to the survival function of the nth upper record value from F_V.

Yakymiv (1986) provides further results on the tail behavior of the record and inter-record times. The following result extends Theorem 7.5.1:

Theorem 7.5.3 (Yakymiv, 1986) *Suppose the pacing process is a renewal process whose interarrival time cdf F_V has finite mean μ and satisfies the condition $\int_0^t \{1 - F_V(x)\} dx = o((t \log t)^{-1})$ as $t \to \infty$. Then*

$$P(U_n > t) = \frac{\mu}{t} \{1 + o(1)\} \frac{\{\log(t)\}^{n-1}}{(n-1)!}.$$

7.6 RECORDS OVER BIRTH PROCESSES

7.6.1 General Birth Process

Consider a birth process $\boldsymbol{P}$ defined by the sequence of birth rates $\{\beta_j, j \ge 0\}$. We assume $\beta_j > 0$ for all $j \ge 0$ to ensure $P\{\lim_{t \to \infty} N(t) = \infty\} = 1$, and $\sum_{j=0}^{\infty} \beta_j^{-1} = \infty$ to guarantee $P\{N(t) < \infty\} = 1$ for all $t < \infty$ (see Feller, 1968, p. 452). The following result gives an expression for the record accrual rate when the observations are paced by such an independent birth process. We refer the reader to the original source for a proof.

Theorem 7.6.1 (Bunge and Nagaraja, 1992a) *Let $N(t)$ satisfy the condition* $P\{N(t+h) - N(t) = 1 \mid N(t) = j\} = \beta_j h + \varepsilon_j h$, $h > 0$, $t > 0$, $j \geq 0$, *where* $\varepsilon_j = \varepsilon_j(h) \to 0$ *as* $h \to 0$. *Assume* $\beta_j > 0$ *for all* $j \geq 0$, $\sum_{j=0}^{\infty} \beta_j^{-1}$ *diverges, and*

$$\sum_{j=n}^{\infty} \beta_j p_j(t) \frac{|S_{j+1}^{n+1}|}{(j+1)!} < \infty \qquad \textit{for all } n \geq 0, \qquad (7.6.1)$$

where $p_j(t) = P(N(t) = j)$. *Then, for* $t > 0$, $h > 0$, *and* $n \geq 0$,

$$P\{M(t+h) - M(t) = 1 \mid M(t) = n\}$$

$$= h \, \frac{\sum_{j=n}^{\infty} \beta_j p_j(t) \frac{|S_{j+1}^{n+1}|}{(j+2)!}}{P\{M(t) = n\}} + o(h), \qquad (7.6.2)$$

as $h \to 0$, *where* $M(t)$ *counts the number of records in* $(0, t]$.

If $\sum_{j=0}^{\infty} \beta_j p_j(t)$ is finite, then (7.6.1) holds. This condition is clearly satisfied if $\beta_j = \beta$ (this means $\boldsymbol{P}$ is Poisson). When $\beta_j = (j+m)\beta$, $N(t) + m$ is $\text{NBin}(m, p)$, where $p = \exp(-\beta t)$ (Karlin and Taylor, 1975, p. 123) and as $N(t)$ has a finite mean, $\sum \beta_j p_j(t)$ is again finite. Further, in this case, appealing to Theorem 7.2.4 we may conclude that

$$P(M(t) = n) = \frac{e^{-(m+1)\beta t}}{(1 - e^{-\beta t})(n+1)!} \sum_{\substack{i_0 + \cdots + i_n = m \\ i_j \geq 0, \, 0 \leq j \leq n}} \prod_{j=0}^{n} D(i_j, t), \qquad (7.6.3)$$

where $D(0, t) = \beta t$; $D(i, t) = (1 - e^{-i\beta t})/(i e^{-i\beta t})$, $i \geq 1$.

Using (7.6.2) and (7.6.3) we obtain the following corollary:

Corollary 7.6.1 *Suppose $\boldsymbol{P}$ is a birth process with birth rate* $\beta_j = \beta(m + 1 + j) > 0$, $j \geq 0$ *where* $m \geq 0$ *is a constant. Then, for* $t > 0$, $h > 0$, *and* $n \geq 0$,

$$P\{M(t+h) - M(t) = 1 \mid M(t) = n\}$$

$$= \beta h \Bigg[1 + (m-1)(n+1)! \bigg(\sum_{\substack{i_0 + \cdots + i_k = m \\ i_j \geq 0, \, 0 \leq j \leq k}} \prod_{j=0}^{k} D(i_j, t) \bigg)^{-1}$$

$$\times \sum_{k=n}^{\infty} \binom{m+k}{m} (1 - e^{-\beta t})^{k+1} \frac{|S_{k+1}^{n+1}|}{(k+2)!} \Bigg] + o(h), \qquad (7.6.4)$$

as $h \to 0$, *where* $D(i, t)$ *is as in (7.6.3).*

The above pacing process P can be interpreted as a pure birth process with rate parameter β and a single progenitor at time $t = 0$ and $m \geq 0$ immigration sources. Corollary 7.6.1 shows that the leading term on the R. H. S. of (7.6.2) can be simplified, to some extent, when the birth rate is linear. The simplification is rather substantial when m is small. For $m = 0$ and $m = 1$, we explore in detail the structure of $\{M(t), \ t \geq 0\}$ in the next two subsections.

7.6.2   Yule Process

When $m = 0$, $\beta_j = (j + 1)\beta$ and the birth process P is a Yule process with a single progenitor at 0. From (7.6.3) it follows immediately that

$$P\{M(t) = n\} = \frac{e^{-\beta t}(\beta t)^{n+1}}{(1 - e^{-\beta t})(n + 1)!}, \qquad n \geq 0. \tag{7.6.5}$$

This means $M(t)+1$ is a trucated Poisson random variable with parameter βt (being truncated at 0). The record arrival rate in (7.6.4) also simplifies after some tedious algebra (see Exercise 7.18 for some help). Bunge and Nagaraja (1992a) show that

$$P\{M(t + h) - M(t) = 1 \mid M(t) = n\}$$
$$= \ \beta h \left[1 - \frac{(n + 1)!}{(1 - e^{-\beta t})(\beta t)^{n+1}} \left(1 - e^{-\beta t} \sum_{j=0}^{n+1} \frac{(\beta t)^j}{j!} \right) \right] + o(h).$$
$$\tag{7.6.6}$$

In view of the relation $\{W_n > t\} \equiv \{M(t) < n\}$, (7.6.5) yields

$$P(W_n > t) = \frac{e^{-\beta t}}{(1 - e^{-\beta t})} \sum_{i=1}^{n} \frac{(\beta t)^i}{i!} = \sum_{i=1}^{n} \sum_{j=1}^{\infty} \frac{e^{-j\beta t}(\beta t)^i}{i! j!}. \tag{7.6.7}$$

This expression can be used to determine the first two moments of W_n. They involve the Reimann zeta function $\zeta(n) = \sum_{i=1}^{\infty} i^{-n}$ seen earlier in Theorem 3.7.1. On recalling $E(W_n^k) = k \int_0^{\infty} t^{k-1} P(W_n > t)dt$, we may conclude that

$$E(W_n) = \frac{1}{\beta} \sum_{j=1}^{n} \zeta(j + 1) \text{ and } E(W_n^2) = \frac{2}{\beta^2} \sum_{j=1}^{n} (j + 1)\zeta(j + 2). \tag{7.6.8}$$

From (7.6.8), $\text{Var}(W_n)$ can be computed. Note that $E(U_1) = E(W_1) = \pi^2/6\beta$. Further, (7.6.8) yields $E(U_n) = \zeta(n + 1)/\beta$, which indicates that the mean inter-record time monotonically decreases to $1/\beta$. Gaver and Jacobs (1978) show that in the limit as $t \to \infty$, the records indeed occur according to a homogeneous Poisson process with rate β.

7.6.3 Poisson Record Arrival Process

In Theorem 7.4.4 we noted that, when P is a nonhomogeneous Poisson process with exponentially increasing rate function, the interarrival times of the records are eventually i.i.d. exponential random variables. When P is a Yule process, from (7.6.5) we may infer that $M(t)$ is almost Poisson. Further, as noted above, the inter-record times are eventually determined by a Poisson process (Gaver and Jacobs, 1978). In other words, there are record models for which the records arrive eventually, according to a Poisson process. A natural question is whether this property can be achieved in a finite time.

We now display a birth process P, which produces an *exact* homogeneous Poisson record arrival process. First recall that when $m = 1$, so that the birth rate $\beta_j = (j+2)\beta$, (7.6.4) shows that $P\{M(t+h) - M(t) = 1 \mid M(t) = n\}$ is simply $\beta h + o(h)$. Further, when $m = 1$, from (7.6.3) (or from Example 7.2.1), we may note that $M(t)$ is Poisson with parameter βt. These suggest the possibility that the record arrival process is homogeneous Poisson; this indeed is the case, as was discovered independently by Bunge (1989) and Bruss and Rogers (1991). We follow the approach taken in Bunge and Nagaraja (1992a). First we need the following technical lemma:

Lemma 7.6.1 *Let $\xi_i, 1 \leq i \leq n$, be complex numbers with positive real parts. Let $t_0 = 1 < t_1 \cdots < t_n$ be positive integers. Then*

$$\Psi_n(\xi_1, \ldots, \xi_n) = \sum_{1 < t_1 < \cdots < t_n} \prod_{j=1}^{n} \frac{\Gamma(t_{j-1} + \xi_j)}{(t_j - 1)\Gamma(t_j + \xi_j)} \Gamma(t_n) \qquad (7.6.9)$$

can also be expressed as $\prod_{i=1}^{n} \xi_i^{-1}$.

Proof. The above claim holds if Ψ_n satisfies the recurrence relation $\Psi_n(\xi_1, \ldots, \xi_n) = \xi_n^{-1} \Psi_{n-1}(\xi_1, \ldots, \xi_{n-1})$ for $n \geq 2$, and $\Psi_1(\xi_1) = \xi_1^{-1}$. Let $k_n = t_n - t_{n-1}$ and recall the function η defined in Lemma 7.4.2. Now, for $n \geq 1$

$$\sum_{t_n = t_{n-1}+1}^{\infty} \frac{\Gamma(t_n)\Gamma(t_{n-1} + \xi_n)}{(t_n - 1)\Gamma(t_n + \xi_n)}$$

$$= \Gamma(t_{n-1} + \xi_n) \sum_{k_n=1}^{\infty} \frac{\Gamma(t_{n-1} + k_n)}{\Gamma(t_{n-1} + k_n + \xi_n)}$$

$$= \Gamma(t_{n-1} + \xi_n)\eta(t_{n-1} - 1, 0, \xi_n)$$

$$= \frac{\Gamma(t_{n-1})}{\xi_n}, \qquad (7.6.10)$$

from Lemma 7.4.2. Hence (7.6.9) implies that the recurrence relation holds for $n \geq 2$. When $n = 1$, (7.6.10) yields $\Psi_1(\xi_1) = \Gamma(t_0)\xi_1^{-1} = \xi_1^{-1}$. $\bigcirc$

Theorem 7.6.2 *Let P be a birth process with a single progenitor at $t = 0$ and be such that the interarrival times, the V's, are independent and V_i is $Exp(\{(i+1)\beta\}^{-1})$ for $i \geq 1$. Then the records arrive according to a homogeneous Poisson process with rate β; that is, $\{U_i, i \geq 1\}$ are independent $Exp(\beta^{-1})$ random variables.*

Proof. We consider the joint characteristic function of the first n inter-record times $U_1, \ldots, U_n$ given by

$$\varphi(\mathbf{z}) = E\left\{\exp\left\{\left(\sum_{j=1}^{n} z_j U_j\right)\right\}\right\},$$

where $\mathbf{z} = (z_1, \ldots, z_n)$, and the z_j's are imaginary numbers. As in the proof of Theorem 7.4.1, we evaluate this expectation by conditioning on the values of the record indices, $T_1, \ldots, T_n$ and exploit the conditional independence of the U_i's given the T_i's. This yields

$$\varphi(\mathbf{z}) = \sum_{1 < t_1 < t_2 < \cdots < t_n} \frac{1}{(t_1 - 1) \cdots (t_n - 1)t_n} \prod_{j=1}^{n} \prod_{i=t_{j-1}}^{t_j - 1} E(e^{z_j V_i}),$$

$$(7.6.11)$$

since the V_i's are independent. Further, as V_i is $Exp(\{(i+1)\beta\}^{-1})$,

$$E(e^{z_j V_i}) = \frac{(i+1)\beta}{(i+1)\beta - z_j} = \frac{i+1}{i+\xi_j},$$

where $\xi_j = 1 - (z_j/\beta)$. Hence (7.6.11) yields

$$\begin{aligned}
\varphi(\mathbf{z}) &= \sum_{1 < t_1 < t_2 < \cdots < t_n} \frac{1}{(t_1 - 1) \cdots (t_n - 1)t_n} \prod_{j=1}^{n} \prod_{i=t_{j-1}}^{t_j - 1} \left(\frac{i+1}{i+\xi_j}\right) \\
&= \sum_{1 < t_1 < t_2 < \cdots < t_n} \frac{1}{(t_1 - 1) \cdots (t_n - 1)t_n} \prod_{j=1}^{n} \frac{\Gamma(t_j + 1)}{\Gamma(t_{j-1} + 1)} \\
&\quad \times \frac{\Gamma(t_j + \xi_j)}{\Gamma(t_{j-1} + \xi_j)} \\
&= \sum_{1 < t_1 < t_2 < \cdots < t_n} \frac{1}{(t_1 - 1) \cdots (t_n - 1)} \Gamma(t_n) \prod_{j=1}^{n} \frac{\Gamma(t_{j-1} + \xi_j + 1)}{\Gamma(t_j + \xi_j + 1)},
\end{aligned}$$

due to the recursive property of the gamma function. The above series is nothing but $\Psi_n(\xi_1, \ldots, \xi_n)$ given by (7.6.9). On recalling Lemma 7.6.1 and substituting for ξ_j, we conclude that

$$\varphi(z_1, \ldots, z_n) = \frac{1}{\xi_1 \xi_2 \cdots \xi_n} = \frac{1}{(1 - (z_1/\beta))(1 - (z_2/\beta)) \cdots (1 - (z_n/\beta))} .$$

Thus, the U_i's are i.i.d. $\text{Exp}(\beta^{-1})$ random variables. ○

An alternate proof of the above result has been given by Bruss and Rogers (1991). They identify pure birth processes as mixed Poisson processes and study the record accrual processes from these models. We pursue some of those results in the next section.

One can also arrive at a homogeneous Poisson record arrival process from other point process models. An example is provided by an F^α model paced by a homogeneous Poisson process where the indices $\alpha(j)$'s are chosen such that the record rate remains a constant. This is easily accomplished by a Yang-type F^α model. More precisely, we have the following result due to Bunge and Nagaraja (1992a):

Theorem 7.6.3 *Let $\beta > 0$, $0 < p < 1$, $q = 1 - p$, and define $\alpha(1) = 1$ and $\alpha(j) = p/q^{j-1}$ for $j \geq 2$. Let F be a continuous cdf, and Y_j have cdf $F_j = F^{\alpha(j)}$, $j \geq 1$. Let Y_1 be observed at time 0 and for $j \geq 1$, Y_{j+1} be observed at the jth arrival time of a homogeneous Poisson process with rate β/p. Then the (nontrivial) records occur according to a homogeneous Poisson process with rate β.*

Proof. From Lemma 6.2.1, it follows that $P(Y_j \text{ is a record}) = \alpha(j)/S(j)$ (where $S(j) = \alpha(1) + \cdots + \alpha(j)$), and this simplifies to p for the choices of $\alpha(j)$ made in the theorem. This means $\{M(t), t > 0\}$ is a Poisson process with rate β (Karlin and Taylor, 1975, p. 153). ○

Theorems 7.6.2 and 7.6.3 indicate the possibility of having two distinct models involving independent observations that lead us to the same record arrival process. As this process is a *randomly thinned* version of the pacing process, we observe that such thinned processes do not identify the original model. This leads to the problem of nonidentifiability, in the sense that the randomly thinned process consisting of record arrival times fails to identify the observation arrival process. In Section 6.7.1 we witnessed nonidentifiability issues that arise when dependent observations are allowed. In that setup even the information about the bivariate sequence of record values and record times failed to identify the record generating phenomenon.

We conclude this section with a complete characterization of the record arrival process when P is a pure birth process with a single progenitor and m immigration sources where $m \geq 1$. The following result provides a fascinating probabilistic representation for the inter-record time sequence $\{U_n, n \geq 1\}$ in terms of a Markov renewal process (Karlin and Taylor, 1975, p. 207).

Theorem 7.6.4 (Browne and Bunge, 1995) *Let R_0 be observed at time 0. Suppose the pacing process P is a pure birth process characterized by the birth rate $\beta_j = (j + m)\beta$, for $j \geq 0$, where m is a positive integer. Then there exists a sequence of random variables θ_n such that $\{(U_n, \theta_n)\}$, $n \geq 0$, is a Markov renewal process possessing the following structure:*

(i) *θ_0 is degenerate at m and U_0 is degenerate at 0.*

(ii) *$\{\theta_n,\ n \geq 0\}$ is a Markov chain with state space $\{1, 2, \ldots, m\}$ and stationary transition probability $p_{ij} \equiv P(\theta_{n+1} = j \mid \theta_n = i) = 1/i$, $1 \leq j \leq i \leq m$.*

(iii) *Given the θ_n's, U_n's are (conditionally) independent and $P(U_n > t \mid \theta_n = j) = e^{-j\beta t}$, $t \geq 0$, $n \geq 1$.*

We refer you to the original source for a proof. It resembles the proof of Theorem 7.6.2 (which corresponds to the case $m = 1$), but is more involved. Note that while for $m = 1$ the U_n's are i.i.d. exponential random variables, for $m > 1$, as the Markov chain $\{\theta_n,\ n \geq 0\}$ gets ultimately absorbed in state 1, eventually these inter-record times become i.i.d. exponential random variables. In other words, in the long run, records are paced by a homogeneous Poisson process!

7.7   RECORDS OVER OTHER PACING PROCESSES

We now briefly discuss the structure of record arrival process for a few other PPRM's. It is now more convenient to begin observing the process at time 0 and assume that the first observation X_1 (the trivial record R_0) arrives at V_0, the first arrival time of the pacing process P_0. This is in contrast with the way the PPRM has been described thus far. Let $\{N_0(t),\ t \geq 0\}$ describe P_0 and let $\{M_0(t),\ t \geq 0\}$ describe the record counting process where we include the trivial record in our count of the records. Bruss and Rogers (1991) discuss the properties of M_0 when N_0 is a *Pascal process* and Browne and Bunge (1995) explore the record arrival process when N_0 is a *mixed Poisson process* or a *Mittag-Leffler renewal process*. These results generalize the findings described in Section 7.5 and 7.6.

Now let us introduce four counting processes observed over $[0, \infty)$. Let $N^*(t)$ be the homogeneous Poisson process with unit intensity, and $N_1(t)$ be a pure birth process with birth rate $\beta_j = (j+1)\beta$, $j \geq 0$. Let $N_2(t)$ be such that for all $t > 0$ and all $0 < t < s$, the conditional distribution of $N_2(s)$ given $N_2(u)$ for all $u \leq t$ is $\text{NBin}(N_2(t) + 1, p(t, s))$. Bruss and Rogers (1991) name $N_2(t)$ a *Pascal process*, and it is characterized by $p(t, s)$. When $p(t, s) = \exp\{-(s - t)\}$, $N_2(t)$ is called *standard Pascal process*. Finally, let $N_3(t) = N^*(W\Lambda(t))$, where W is a random variable independent of the process $N^*(t)$ and $\Lambda(t)$ is nondecreasing in t. The process $N_3(t)$ is known as a *mixed Poisson process*, where W is the mixing random variable. Then the following equivalence holds:

Lemma 7.7.1 (Bruss and Rogers, 1991) *Let $\{N_1(t), t \geq 0\}$ be a pure birth process with birth rate $\beta_j = (j+1)\beta$, $j \geq 0$ and $\{N_2(t), t \geq 0\}$ be a Pascal process with $p(t, s) = \exp\{-\beta(s - t)\}$, $t < s$. Further, suppose $\{N_3(t), t \geq 0\}$ is a mixed Poisson process where W is an $Exp(\beta^{-1})$ random variable and $\Lambda(t) = e^t - 1$. These three processes are identical.*

As the above result shows, the counting process $N_1(t)$ is a special case of both $N_2(t)$ and $N_3(t)$. When $N_0(t) = N_1(t)$, it follows from Theorem 7.6.2 that $M^*(t)$ is a homogeneous Poisson process. Thus, it is natural to ask what happens for the other two pacing processes in general. Bruss and Rogers focus on the Pascal process $N_2(t)$ and provide the following answer. We will omit the proof here.

Theorem 7.7.1 (Bruss and Rogers, 1991) *Let the pacing process $\{N^*(t), t \geq 0\}$ be a Pascal process with success parameter function $p(t, s)$. Then*

(i) *the record arrival process $M^*(t)$ is a nonhomogeneous Poisson process whose rate function $\lambda(t)$ satisfies the condition $\int_t^s \lambda(u)du = -\log\{p(t, s)\}$ for $t < s$;*

(ii) *for $k > 1$, the record arrival processes of the Type 1 j-records for $1 \leq j \leq k$, are i.i.d. processes beyond the time $V_1 + \cdots + V_{k-1}$, the arrival time of X_{k-1}. In other words, beyond that random time, they are all i.i.d. nonhomogeneous Poisson processes.*

Browne and Bunge (1995) focus on the model that uses a mixed Poisson process as the pacing process. Note that the record arrival process is a randomly thinned version of the pacing process as the observation associated with the nth arrival time is kept with probability $p_n = 1/n$. These authors pose the following question: Under what conditions is a randomly thinned mixed Poisson process also a (possibly nonhomogeneous) Poisson process? Suppose $\{N_3(t), \ t \geq 0\}$ is a mixed Poisson process with mixing random variable W. They show that its randomly thinned version is a Poisson process, either when W is degenerate and p_n is a constant, or when W is a Gamma random variable which shape parameter α and $p_n = c/(n - 1 + \alpha)$ for some $0 < c \leq \alpha$.

Browne and Bunge (1995) also consider a generalization of a homogeneous Poisson process known as *Mittag-Leffler renewal process* as a model for $\boldsymbol{P}$. It is a renewal process where the distribution of V_1 is best described by its Laplace transform $E(e^{-sV}) = 1/(1 + s^\alpha)$, where $0 \leq \alpha < 1$. For such a PPRM, their paper provides a representation for the inter-record times that resembles the one given in (7.4.2). We do not pursue the details here.

7.8   THE SECRETARY PROBLEM

In his interesting discussion paper, Ferguson (1989) defines the basic secretary problem as "a sequential observation and selection problem in which the payoff depends on the observations only through their relative ranks and not otherwise on their actual values." (See also Samuels, 1991.) There are several variations of this problem and there are as many variations in the name as well. The basic version of the secretary problem is the best-choice problem, where the goal is to maximize the probability of picking the maximum of n values observed sequentially and with no recall. Thus, the goal is to maximize the probability of choosing the last record value in the sequence. Hence the solution belongs to the domain of record counting statistics. Since the decisions are made at record epochs, the optimal policy is related to the record arrival process in other variations of the secretary problem as well. Gaver (1976) appears to be the first one to discuss this connection between the secretary problem and the record arrival process. In recent times, Bruss (e.g., 1988) has explored and exploited this link to describe the optimal strategy. Let us now elaborate on the basic secretary problem in the framework of the classical record model.

Suppose n candidates apply for a position and are interviewed sequentially. On interviewing the ith candidate, the employer has only the information about rank of that candidate relative to the previous ones. Now he/she will have to choose candidate i or decide to interview the next candidate. Any candidate not chosen at his/her turn is unavailable later (no recall). The goal is to maximize the probability of picking the best candidate. This is the *classical secretary problem* or the *no-information best-choice problem*. What is the optimal strategy here? To get at it, first let us denote the score associated with the ith candidate by X_i. Then our goal is to maximize the chance of identifying the index of the last record value in the sequence $X_1, \ldots, X_n$, where we will now assume that the X_i's are i.i.d. continuous random variables.

Now consider the strategy that lets the first $i-1$ go by and selects the first relatively best one (if any) from the ith observation onwards. This strategy picks the last record value in the sequence iff there is exactly one record value among $X_i, \ldots, X_n$. Hence, the probability of best choice for this strategy is given by

$$p_n(i) = P\left(\sum_{j=i}^{n} I_j = 1\right),$$

where the I_j's are the record indicator variables. It is well known (by now) that the I_j's are independent and I_j is a $\mathrm{Ber}(1/j)$ random variable. Consequently,

$$
\begin{aligned}
p_n(i) &= \sum_{j=i}^{n} P(I_j = 1) \prod_{k=i\neq j}^{n} P(I_k = 0) \\
&= \frac{1}{n} + \frac{i-1}{n} \sum_{j=i}^{n-1} \frac{1}{j}, \quad i \geq 1.
\end{aligned}
\tag{7.8.1}
$$

From (7.8.1) we observe that

$$p_n(i) - p_n(i+1) = \frac{1}{n}\left\{\sum_{j=i}^{n-1}\frac{1}{j} - 1\right\}.$$

The above equation implies that $p_n(i)$ increases as i increases to $i_0(n)$ and then it decreases, where

$$i_0(n) = 1 + \max\left\{i : \sum_{j=i}^{n-1} j^{-1} \leq 1\right\}.
\tag{7.8.2}$$

Thus the optimal strategy (among such policies) is to wait until i_0 candidates are seen and select the first relatively best one, if any, from the i_0th candidate onwards. Let us call this policy $D(i_0)$. This is the *best one-step-ahead* strategy. From (7.8.2) we observe that as $n \to \infty$, $i_0(n)/n \to 1/e$, and from (7.8.1) we may then conclude that $p_n(i_0) \to 1/e$ as well. Thus, the optimal policy waits until about n/e observations are examined and then picks the next record if it occurs.

In fact, $D(i_0)$ is the best among all strategies that use only the sequential relative ranks of the observations. To establish this claim, we argue as follows (see also Bruss, 1988): Let W be the random waiting time at or after which we will accept the first record, if observed. Suppose that if W is related to the X_i's, it depends on them only through the sequential ranks. Then, the probability that this policy chooses the best is given by $p_n(W) = \sum_{i=1}^{n} P(W = i)p_n(i)$, since for $j \geq i$, the I_j's are independent of the relative ranks of $X_1, \ldots, X_{i-1}$. Obviously this probability cannot exceed $p_n(i_0)$.

Now let us see what interesting twists appear if we consider the classical secretary problem, where N, the number of available observations, or options as they are commonly called, is random. This is quite natural in the Bayesian context, as the probability distribution of N represents its prior distribution. We see this as precisely the random record model introduced in Section 7.2. The optimal one-step-ahead policy in this situation is a little more involved than the one available for the classical problem (see Exercise 7.24). It need not be the overall optimal policy. In the case of PPRM, it becomes more difficult, in general, to identify the best strategy. Let us now consider some specific PPRM's and discuss the optimal rules.

Suppose $\boldsymbol{P}_0$, observed over $(0, T]$, represents the pacing process. (As was done in Section 7.7, it is now more convenient to assume that X_1 or R_0 arrives after a random time rather than to reset the clock to time 0 at the time X_1 arrives.) Then the arrival times of the observations contain vital information about the conditional distribution of the number of future arrivals in the remaining time over $(0, T]$. Let us now assume that $\boldsymbol{P}_0$ is a homogeneous Poisson process with known rate λ. Cowan and Zabczyk (1978) provide the optimal best choice rule for this model. It is quite involved. (See Exercise 7.25 for a formal description.) Basically, for $m \geq 0$, one chooses the mth record value if its arrival time W_m $(=\sum_{j=0}^{m} U_j)$ is in the interval $(T - x_m/\lambda, T]$, where x_m is the unique solution to the following equation:

$$\sum_{n=0}^{\infty} \frac{x^n}{n!(m+n+1)} = \sum_{n=1}^{\infty} \frac{x^n}{n!(m+n+1)} \left\{ \sum_{i=1}^{n} \frac{1}{i+m} \right\}. \qquad (7.8.3)$$

Since a nonhomogeneous Poisson process can be transformed into a homogeneous one by a deterministic time change, a parallel result can be stated for the case where P_0 is a nonhomogeneous Poisson process with known intensity function.

What is our optimal strategy in the Poisson process model if λ is unknown? When λ is assumed to have a noninformative prior distribution over $(0, \infty)$, it turns out to be a stationary one, in the sense that the threshold value for W_m is free of m! To be precise, the following result holds:

Theorem 7.8.1 (Bruss, 1987) *Let P_0 be a homogeneous Poisson process with unknown random rate λ that has uniform (improper) prior density over $(0, \infty)$. If the PPRM with P_0 as its pacing process is observed over $(0, T]$, the optimal solution to the best-choice problem is to pick the first record value after time T/e.*

Bruss shows that if the prior distribution of λ is exponential, the optimal policy is again stationary. See Exercise 7.26 for a description of that strategy.

From Theorem 7.7.1 we observe that in the PPRM setup when P_0 is a Pascal process, the record arrival process is a (possibly nonhomogeneous) Poisson process. When this occurs the optimal strategy for the basic secretary problem turns out to be nice and simple. Simplest is the situation where the record arrival process is a homogeneous Poisson process.

Theorem 7.8.2 *Let the record arrival process $\{M_0(t),\ t \geq 0\}$ be a Poisson process with intensity β. Suppose the goal is to choose the rule that maximizes the probability of selecting the last record value in $(0, T]$. Then,*

    (i) *the optimal policy observes without taking action until the fixed time t_0 and then chooses the next record, if any, in $(t_0, T]$ where $t_0 = \max(T - 1/\beta, 0)$;*

    (ii) *for the optimal policy, the probability of selecting the best is $1/e$ if $\beta T \geq 1$ and it is $\beta T\ e^{-\beta T}$, otherwise.*

Proof. Let $D(t)$ denote the decision rule which waits until time t without any action and then chooses the next record, if any, in $(t, T]$. Then the probability that our choice is the best is nothing but $P(M_0(T) - M_0(t) = 1) = \beta(T - t) \exp\{-\beta(T - t)\}$. This will attain its maximum as a function of t, when $\beta(T - t) = 1$ if $T \geq 1/\beta$, or when $t = 0$ if $T < 1/\beta$. Thus $D(t_0)$ is the optimal policy where t_0 is given in part (i) and the probability that it selects the best is as claimed in part (ii) of the theorem. To claim that $D(t_0)$ is indeed the overall optimal rule, we can argue just as we did in the classical secretary problem. The independent increment property of the $M_0(t)$ process plays a crucial role in the proof.

$\bigcirc$

If the observations arrive according to a pure birth process with birth rate $\beta_j = (j+1)\beta$, $j \geq 0$, from Theorem 7.6.2, then it follows that the records $\{R_n, n \geq 0\}$ arrive according to a Poisson process with intensity β. Consequently, the optimal rule for the secretary problem turns out to be quite simple and stationary. If $T < 1/\beta$, pick the first arrival; otherwise choose the first record (if any) that arrives after $T - 1/\beta$.

Bruss (1988) has considered some modifications to the classical problem that allow for several choices or employ more general objective functions. He has provided optimal rules that turn out to be rather simple and even stationary, in some cases. See also Bruss (1998) and references therein.

EXERCISES

1. In the random record model, suppose F is absolutely continuous and $N + 1$ is a Geo(p) random variable. Find the conditional joint pdf of $R_0, \ldots, R_n$, given that there are at least n nontrivial records.

2. In the random record model, find the pdf of M, when $N + 3$ is a NBin($3, p$) random variable.

3. Consider the function $g(s)$ given by (7.2.3), where $s = (s_0, \ldots, s_m)$.

   (a) If $0 \leq s_0 = s_1 = \cdots = s_m = s < 1$, show that

   $$g(s) = (1-s)^{-(m+1)} \left(1 - \sum_{j=0}^{m} \frac{\pi^{(j)}(s)(1-s)^j}{j!} \right),$$

   where $\pi^{(j)}(s)$ is the jth derivative of the pgf of N.

   (b) Consider the sequence of weak records from a random record model where F is a discrete uniform cdf over $[0, n]$, and pgf of N is $\pi(s) = p/(1 - qs)$. Give an expression for $P(R_0 = R_1 = R_2 = x)$.

   (c) Obtain a closed-form expression for $g(s)$ when $s_0, \ldots, s_{r-1}$ are equal to, say, s and the rest of the s_i's are all distinct and distinct from s for $r \leq i \leq m$.

This is helpful in finding expressions for $p_n(\cdot)$ when F is discrete and we are dealing with weak records. For example, $P(R_0 = R_1 = R_2 = x, R_3 = y)$ for $x < y$ can be evaluated using such an expression. [Bunge and Nagaraja, 1991]

4. Consider a record model where F is a continuous cdf, X_i has cdf $F^{\alpha(i)}$, $i \geq 1$, and the number of available observations $(N + 1)$ is a random variable independent of the X_i's. This is the F^α version of the random record model discussed in Section 7.2. Suppose $N + 1$ is a Geo(p) random variable. Recall that M is the number of nontrivial records.

   (a) Let $\alpha(1) = c$ and $\alpha(i) = (c-1)c^{i-1}$ for $i > 1$, where $c > 1$; that is, the $\alpha(i)$'s are geometrically increasing. Show that $M + 1$ is a Geo(p_0) random variable where $p_0 = pc/(c - 1 + p)$.

   (b) Let $\alpha(i) = i$, $i \geq 1$. Show that, for $m \geq 0$,

   $$P(M = m) = 2^{m+1} \frac{p}{q^2} \frac{1}{(m+1)!} \int_0^{-log(p)} e^{-y} y^{m+1} dy.$$

   [Hofmann, 1997; Hofmann and Nagaraja, 1997]

5. Suppose we observe a PPRM having a homogeneous Poisson pacing process with rate λ over $(0, t]$. Show that the pdf of the number of nontrivial records can be expressed as

$$P\{M(t) = n\} = \frac{1}{(n+1)!} \int_0^1 \cdots \int_0^1 \sum_{j=0}^n \frac{e^{\lambda t(s_j - 1)}}{\prod_{\substack{i=0 \\ i \neq j}}^n (s_j - s_i)} ds_n \cdots ds_0.$$

Evaluate the above probability explicitly when $n = 0$ and $n = 1$. (*Hint*: Use Theorem 7.2.3.)

6. (a) Assume $\boldsymbol{P}$ is a homogeneous Poisson process with unit intensity. Using Theorem 7.4.1 or Exercise 7.5 above, obtain an expression for $P(U_1 > t)$, $t \geq 0$.

   (b) Now assume $\boldsymbol{P}$ is a Poisson process with mean function $\Lambda(t)$.

   (i) Prove that

   $$P(U_1 > t) = \begin{cases} (1 - e^{-\Lambda(t)})/\Lambda(t), & \Lambda(t) > 0 \\ 1, & \text{otherwise.} \end{cases}$$

(ii) Show that $E(U_1)$ is finite iff $\int_x^\infty \frac{dt}{\Lambda(t)}$ is finite for large x.

[Gaver, 1976]

7. In a PPRM, let X_0 represent the strength of a component subjected to shocks that arrive according to a Poisson process with mean function $\Lambda(t)$. For $n \geq 1$, let X_n represent the magnitude of the nth shock to arrive. Suppose the component fails at the arrival time of a shock whose magnitude exceeds its strength. In this context U_1 represents the failure time of the component. Now assume the X_n's are i.i.d. continuous random variables with cdf F for $n \geq 1$ and X_0 is continuous with cdf F_0.

 (a) Show that $P(U_1 > t \mid X_0 = x) = \exp(-\Lambda(t)\{1 - F(x)\})$.

 (b) Determine $P(U_1 > t)$ when $F_0(x) = \{F(x)\}^k$ for some $k \geq 1$.

[Gaver, 1976]

8. Let $U_1(\tau)$ denote the waiting time for the first record after time τ. Assume $\boldsymbol{P}$ is a Poisson process with mean function $\Lambda(t)$. Determine $P(U_1(\tau) > t)$. (*Hint:* Let $Y_0 = \max(X_0, \ldots, X_{N(t)})$. Find its cdf F_0 first and follow the conditioning approach used in Exercise 7.7. Gaver considers a slightly different setup.) [Gaver, 1976]

9. When records are observed over a Poisson paced intervals with unit intensity, show that the inter-record times have the following moments:

$$E\{\log(U_{n+1}/U_n)\} = 1 \text{ and } \text{Var}\{\log(U_{n+1}/U_n)\} = 1 + (\pi^2/3).$$

10. Let the A_j's and the C_j's be defined as in Lemma 7.4.1. Show that

$$\log\left(\frac{1 + \sum_{j=1}^n A_j e^{C_j}}{1 + \sum_{j=1}^{n-1} A_j e^{C_j}}\right) \xrightarrow{d} A \text{ as } n \to \infty,$$

where A is a standard exponential random variable. Hence, establish that the inter-record times from a Poisson paced record process with exponential intensity function eventually become exponentially distributed.

11. Consider an F^α model where $\alpha(j) = \alpha j$, $j \geq 1$, for some $\alpha > 0$. Suppose the observations arrive at time points that form a homogeneous

Poisson process with unit intensity. Show that the inter-record times U_j satisfy the distributional identity

$$(\log U_1, \ldots, \log U_n) \overset{d}{=} \left(\frac{1}{2}C_1 + \log A_1, \frac{1}{2}C_2 + \log A_2, \ldots, \frac{1}{2}C_n + \log A_n\right),$$

where the A_j and C_j are as in Lemma 7.4.1. Compare this with Theorem 7.4.1. [Hofmann, 1997; Hofmann and Nagaraja, 1997]

12. Consider an F^α model paced by a Poisson process $\boldsymbol{P}$ with rate function $\lambda(t)$. Assume that the α's are geometrically increasing such that $\alpha(1) = c$, $\alpha(j) = (c-1)c^{j-1}$, $j \geq 2$, where $c > 1$. If either one of the following holds, then show that $\lambda(t)$ is necessarily a constant.

   (a) U_1, U_2 are independent.

   (b) U_1, U_2 are identically distributed.

   From Theorem 7.6.3 it follows that when $\boldsymbol{P}$ is homogeneous Poisson, the inter-record times are i.i.d. exponential random variables. Thus the above provide characterizations of homogeneneous Poisson arrival process of the observations based on the properties of first two record arrival times. (*Hint:* Remember CFE and ICFE!)

[Hofmann, 1997; Hofmann and Nagaraja, 1997]

13. In the renewal process record model, find an explicit expression for the Laplace transform of U_n in terms of that of V_1.

[Westcott, 1977]

14. Assume the pacing process is a renewal process and consider Theorem 7.5.1.

   (a) Show that (7.5.4) holds if $E(V_1^{1+\delta})$ exists for some $\delta > 0$. (This is an improvement over Theorem 7.5.1.)

   (b) Show that the condition $E\{V_1 \log(\max(1, V_1))\}$ is finite is sufficient for (7.5.4) to hold. (This is an improvement over (a).)

   (c) Assuming $E(V_1)(= \mu)$ is finite show that (7.5.4) holds iff $t \log(t)\{1 - F_V(t)\} \to 0$ as $t \to \infty$.

[Westcott, 1977; 1979]

15. Show that (7.5.4) holds for any pacing process for which the inter-arrival times are stationary and uniformly mixing with finite second moment.   [Westcott, 1977]

16. Suppose the pacing process is a renewal process where the interarrival time random variable has a stable distribution with exponent α, $0 < \alpha < 1$. (Its Laplace transform is given by $E(e^{-sV}) = \exp(-s^{\alpha})$.)

    (a) Find the Laplace transform of U_1.

    (b) Show that

    $$P(U_1 > t) \approx \alpha t^{-\alpha} \log(t)/\Gamma(1 - \alpha),$$

    for large t.   [Gaver, 1976]

17. Let $X_1, X_2, \ldots$ be i.i.d. non-negative random variables with common cdf F_1, and let N be an independent integer valued random variable with $P(N = n) = p_n$. Then the random sum S_N (where $S_0 = 0$) has the cdf given by $F_2(x) = \sum_{n=0}^{\infty} p_n F_1^{(n)}(x)$, where $F_1^{(n)}$ is the n-fold convolution of F_1 and $F_1^{(0)}$ is degenerate at 0. If F_1 and F_2 are related by this relation, we say F_2 is *subordinated* to F_1 with subordinator $\{p_n, n \geq 0\}$. Show that in a renewal process record model, both the cdf's of U_n and W_n are subordinated to F_V, the cdf representing the interarrival time of the observations. Identify the sequence $\{p_n\}$ in each of these cases.   [Embrechts and Omey, 1983]

18. (a) For the Yule Process model with rate β, show that (7.6.4) simplifies to

    $$P\{M(t + h) - M(t) = 1 \mid M(t) = n\}$$
    $$= \beta h \left(1 - \frac{q(n+1)!}{(-\log p)^{n+1}} \sum_{k=n}^{\infty} q^k \frac{|S_{k+1}^{n+1}|}{(k+2)!}\right) + o(h),$$

    where $p = e^{-\beta t}$ and $q = 1 - p$.

    (b) For $n \geq 1$ and $0 < q < 1$, show that

    (i) $\sum_{k=n}^{\infty} q^k \frac{|S_{k+1}^{n+1}|}{(k+2)!}$
    $$= q^n \int_{0 < s_0 < \cdots < s_n < 1} (1 - s_n) \prod_{i=0}^{n} (1 - q s_i)^{-1} ds_0 \cdots ds_n$$

    and

    (ii) the R.H.S. in (i) can be expressed as

    $$q^{-(n+2)} \left(1 - p \sum_{j=0}^{n+1} \frac{(-\log p)^j}{j!}\right).$$

    Note that these steps lead to the claim made in (7.6.6).

                                              [Bunge and Nagaraja, 1992a]

19. For the Yule Process record model with rate parameter β, show that $E(W_n) \approx (n+1)/\beta$ as $n \to \infty$.

20. (a) Prove Theorem 7.6.4.

    (b) Using the theorem or otherwise find an expression for the joint pdf of $(U_1, \ldots, U_n)$ when $m = 2$.

    [Browne and Bunge, 1995; Bunge and Nagaraja, 1992a]

21. Prove Lemma 7.7.1.

22. Consider the classical secretary problem introduced in Section 7.8. Let $D(i_0)$ be the optimal one-step ahead policy.

    (a) What happens to $p_n(i)$ as $i/n \to x (0 \le x \le 1)$ where $p_n(i)$ is given by (7.8.1)?

    (b) Show that $i_0(n)/n \to 1/e$ as $n \to \infty$, where $i_0(n)$ is given by (7.8.2). What is the limiting value of $p_n(i_0)$?

23. Consider the classical secretary problem in the F^α model setting. In other words, let X_i have cdf $F^{\alpha(i)}$.

    (a) Determine $p_n(i)$, the probability that the strategy which picks the first record at or after X_i is observed actually picks the best.

    (b) Describe the optimal one-step-ahead policy.

    (c) Is it an overall optimal policy?

    [Pfeifer, 1989]

24. In the secretary problem, let the number of available candidates be a random variable, say N, independent of the X_i's.

    (a) Show that the optimal one-step-ahead policy is to accept the first record whose arrival number j satisfies

    $$E\left(\frac{j}{N} \mid N \ge j\right) \ge \sup_{k > j} \sum_{n=k}^{\infty} p_n(k) P(N = n \mid N \ge j),$$

    where $p_n(k)$ is as defined in the original secretary problem (in (7.8.1)). (The N here is like the $N+1$ in the random record model of Section 7.2.)

(b) Describe the optimal rule explicitly when N is a Geo(p) random variable.

[Bruss, 1988]

25. In the secretary problem, let us assume the X_i's arrive according to a homogeneous Poisson process with unit intensity. Suppose we observe the Poisson pacing process over $(0, T]$.

   (a) Show that the overall optimal policy can be described as follows:

   For $m \geq 0$, choose the mth record value if its arrival time t_m satisfies the condition $t_m > T - x_m$, where x_m is the unique solution of (7.8.3). Otherwise, wait for the $(m + 1)$st record value and see whether its arrival time satisfies the equation obtained by replacing m by $m + 1$ in (7.8.3).

   (b) Evaluate x_m for $0 \leq m \leq 5$ and describe the optimal policy explicitly when $T = 10$.

[Cowan and Zabczyk, 1978]

26. (a) Prove Theorem 7.8.1.

   (b) Under the setup described in Theorem 7.8.1, now assume that the prior distribution of λ is standard exponential. Show that the optimal policy is to accept the first record value if any after time t^* where $t^* = \max(0, (T + 1)/e - 1)$.

[Bruss, 1987]

27. Generalize the result in Theorem 7.8.2 to the case where $\{M_0(t),\ t \geq 0\}$ is a nonhomogeneous Poisson process. The optimal rule thus obtained is applicable to the PPRM, where $\{N_0(t),\ t \geq 0\}$ is a Pascal process.

[Bruss, 1987]

CHAPTER 8

HIGHER DIMENSIONAL PROBLEMS

8.1 HOW SHOULD WE DEFINE MULTIVARIATE RECORDS?

The concept of a record value in a sequence of i.i.d. univariate random variables is relatively straightforward. Whenever we see an X that is bigger than any observation that preceded it, we recognize a record. If the observations, the X's, are multidimensional, life is more complicated. Vectors in m dimensional space are only partially ordered by coordinatewise inequality. So observations will be encountered that are not smaller than nor larger than the current record (however it is defined). The complexity of the problem increases slightly as the dimension of the X's increases, but the radical step is going from one to two dimensions. Our discussion will focus on the bivariate case, with only occasional mention of higher dimensional extensions.

Our scenario is then one of observing a sequence $(X_1, Y_1), (X_2, Y_2), \ldots$ of i.i.d. bivariate random variables with common joint cdf $F(x, y)$, assumed continuous, to eliminate tie problems. There will then be no loss of generality and some potential advantage to assuming that the distribution $F(x, y)$ has standard exponential marginals. In such a setting, we renew our asterisk convention from Chapter 2, thus (X_i^*, Y_i^*) will be possibly dependent standard exponential random variables. We can extend the asterisk notation and use $F^*(x, y)$ to denote their corresponding joint cdf. As before, an asterisk signifies an exponential distribution (though only marginally in the current context). There will be one special situation when the bivariate record problem will be quite tractable. That will correspond to the independent case, that is, when $F(x,y)$ factors into the form $F_1(x)F_2(y)$ or, equivalently, after a marginal transformation, when $F^*(x,y) = (1 - e^{-x})(1 - e^{-y})$, $x, y > 0$.

Now let us return to our incoming stream of observations (X_1, Y_1), $(X_2, Y_2), \ldots$ arriving at "times" $1, 2, \ldots$. As in the univariate case, we declare the first observation, (X_1, Y_1), to be a record, the zeroth record. The zeroth record occurs at time 1 and will be denoted by $(R_0(X), R_0(Y))$. Subsequent records, occurring at times $T_1, T_2, \ldots$ will be denoted by $(R_n(X), R_n(Y))$, $n \geq 1$. There are many possible criteria that might be used to designate record times in this context. We will only describe four possibilities.

Definition 8.1.1 *A new record of the first kind occurs at time k when either X_k exceeds all preceding X_i's or Y_k exceeds all preceding Y_i's, or both. Denoting the corresponding sequence of record times by $\{_1T_n, n \geq 0\}$, the corresponding sequence of record values is given by*

$$(_1R_n(X), \ _1R_n(Y)) = \max_{i \leq _1T_n}(X_i, Y_i). \tag{8.1.1}$$

Definition 8.1.2 *A new record of the second kind occurs at time k when either X_k exceeds the current record X value or Y_k exceeds the current record Y value, or both. Denoting the corresponding sequence of record times by $\{_2T_n, n \geq 0\}$, the corresponding sequence of record values is given by*

$$(_2R_n(X), \ _2R_n(Y)) = (X_{_2T_n}, Y_{_2T_n}). \tag{8.1.2}$$

Definition 8.1.3 *A new record of the third kind occurs at time k if X_k exceeds the current X record **and** Y_k exceeds the current Y record. Denoting the corresponding sequence of record times by $\{_3T_n, n \geq 0\}$, the corresponding sequence of record values is given by*

$$(_3R_n(X), \ _3R_n(Y)) = (X_{_3T_n}, Y_{_3T_n}). \tag{8.1.3}$$

Definition 8.1.4 *A new record of the fourth kind occurs at time k when X_k exceeds all preceding X_i's **and** Y_k exceeds all preceding Y_i's. Denoting the corresponding sequence of record times by $\{_4T_n, n \geq 0\}$, the corresponding sequence of record values is given by*

$$(_4R_n(X), \ _4R_n(Y)) = \max_{i \leq _4T_n}(X_i, Y_i).$$

In all these cases, the triple $(_jT_n, \; _jR_n(X), \; _jR_n(Y))$ forms a Markov chain. The rules governing transitions depend on which of the four definitions is chosen. Other definitions are possible (see, e.g., Kinoshita and Resnick, 1989) but four provide more than adequate potential for confusion. To illustrate the differences between the four definitions, here is an example.

Example 8.1.1 (Bivariate Records) Consider the following sequence of 10 observations on (X, Y)

$$\binom{1.2}{1.3}, \binom{0.7}{1.9}, \binom{4.1}{3.6}, \binom{4.2}{0.5}, \binom{9.1}{0.6}, \binom{1.2}{3.7}, \binom{1.9}{4.0}, \binom{4.3}{3.7}, \binom{11.4}{6.5}, \binom{10.4}{7.1}.$$

Using Definition 8.1.1, records occur at times 1, 2, 3, 4, 5, 6, 7, 9 and 10 and the corresponding record sequence begins

$$\binom{1.2}{1.3}, \binom{1.2}{1.9}, \binom{4.1}{3.6}, \binom{4.2}{3.6}, \binom{9.1}{3.6}, \binom{9.1}{3.7}, \binom{9.1}{4.0}, \binom{11.4}{6.5}, \binom{11.4}{7.1}.$$

Using Definition 8.1.2, records occur at times 1, 2, 3, 4, 5, 6, 7, 8, 9 and 10 and the corresponding record sequence begins

$$\binom{1.2}{1.3}, \binom{0.7}{1.9}, \binom{4.1}{3.6}, \binom{4.2}{0.5}, \binom{9.1}{0.6}, \binom{1.2}{3.7}, \binom{1.9}{4.0}, \binom{4.3}{3.7}, \binom{11.4}{6.5}, \binom{10.4}{7.1}.$$

Using Definition 8.1.3, records occur at times 1, 3, 8 and 9 and the corresponding record sequence begins

$$\binom{1.2}{1.3}, \binom{4.1}{3.6}, \binom{4.3}{3.7}, \binom{11.4}{6.5}.$$

Using Definition 8.1.4 records occur at times 1, 3 and 9 and the corresponding record sequence begins

$$\binom{1.2}{1.3}, \binom{4.1}{3.6}, \binom{11.4}{6.5}.$$

8.2 BIVARIATE RECORDS WITH INDEPENDENT COORDINATES

Let the common joint distribution of the (X_i, Y_i)'s be quite arbitrary. To get a feeling for the nature of the kinds of results we might expect, it is convenient to consider simple special cases. A trivial special case occurs when the joint distribution is singular, that is, when $Y_i = aX_i + b$, almost surely. In this case, all four definitions of bivariate records will coincide and $(R_n(X), R_n(Y)) = (R_n, aR_n + b)$, where R_n is the univariate record sequence corresponding to the X_i's. All the distributional results from Chapter 2 then apply to R_n and trivially extend to $(R_n(X), R_n(Y))$: easy, but not interesting.

A second case in which simplicity (but not triviality) is encountered is the case of independent marginals. We assume that $F(x, y)$ factors into the form $F_1(x)F_2(y)$ and, without loss of generality, assume a marginal transformation to exponentiality has been made. Thus the common distribution of the (X_i, Y_i)'s is

$$F(x, y) = (1 - e^{-x})(1 - e^{-y}), \qquad x > 0, \ y > 0.$$

The lack of memory property of the exponential distribution will be crucial.

First we remark on the record time distributions in the various definitions. There are two quite simple cases. On careful scrutiny, it is clear that in Definition 8.1.1, a record occurs at time k iff either a univariate X record occurs at time k, or a univariate Y record occurs at time k or both. If we let $_1I_k = 1$ iff k is a record time according to this definition, then, assuming independence of the X's and Y's and recalling material from Section 2.5 on univariate record times, we conclude that the $_1I_k$'s are independent Bernoulli random variables with

$$P(_1I_k = 1) = (2k - 1)/k^2. \qquad (8.2.1)$$

So, letting $_1N_n = \sum_{k=1}^{n} {_1I_k}$, the number of records among the first n observed pairs $(X_1, Y_1), \ldots, (X_n, Y_n)$, we find

$$\begin{aligned} E(_1N_n) &= \sum_{k=1}^{n} \frac{2k-1}{k^2} \\ &\approx 2\log n + 2\gamma - \frac{\pi^2}{6} \end{aligned} \qquad (8.2.2)$$

and

$$\begin{aligned} \mathrm{Var}(_1N_n) &= \sum_{k=1}^{n} \frac{2k-1}{k^2} \cdot \left(\frac{k-1}{k}\right)^2 \\ &\approx 2\log n \end{aligned} \qquad (8.2.3)$$

and applying the Liapounov condition we find

$$({}_1N_n - 2\log n)/\sqrt{2\log n} \xrightarrow{d} N(0,1). \tag{8.2.4}$$

If we turn to Definition 8.1.4, we find that, in this setting, a record occurs at time k iff both a univariate X record occurs at time k *and* a univariate Y record occurs at time k. Now let ${}_4I_k = 1$ iff k is a record time then, assuming X's and Y's are independent and referring again to Section 2.5, we conclude that the ${}_4I_k$'s are independent Bernoulli random variables with

$$P({}_4I_k = 1) = \frac{1}{k^2} . \tag{8.2.5}$$

Now something perhaps surprising occurs. If we count the number of records among the first n observed pairs, i.e. ${}_4N_n = \sum_{k=1}^n {}_4I_k$, we find

$$
\begin{aligned}
E({}_4N_n) &= \sum_{k=1}^n \frac{1}{k^2} \\
&\approx \pi^2/6
\end{aligned}
\tag{8.2.6}
$$

and

$$
\begin{aligned}
\mathrm{Var}({}_4N_n) &= \sum_{k=1}^n \frac{1}{k^2}\left(1 - \frac{1}{k^2}\right) \\
&\approx \pi^2/6.
\end{aligned}
\tag{8.2.7}
$$

The expected number of records is about 1.64 and we know the first observed pair is always a record! So we expect only 0.64 non-trivial records. Note that there are no unbreakable records here, but there is positive probability that the first record stands for all time (since the infinite product $\prod_{k=2}^\infty (1 - k^{-2})$ is convergent). One consequence of this sparsity of records is that ${}_4N_n$ will not be asymptotically normal. In fact, ${}_4N_n$ converges in distribution (without any normalization) to a random variable whose pgf is

$$\prod_{j=1}^\infty \left(1 - \frac{(1-s)}{j^2}\right) = \sin(\pi\sqrt{1-s})/(\pi\sqrt{1-s}). \tag{8.2.8}$$

If we relax the condition of independence of the X's and Y's, it is still possible to determine whether the number of records (using Definition 8.1.4) is almost surely finite or infinite. The relevant computations were provided by Goldie and Resnick (1989) (see also Goldie and Resnick, 1995). Whether $\sum_{k=1}^\infty {}_4I_k$ is finite or infinite with probability one depends on whether (X, Y) are asymptotically independent or not.

We say that $F_{X,Y}(x,y)$ is in the domain of (maximal) attraction of a bivariate extreme value cdf $G(x,y)$ if

$$\left(\max_{i\leq n} X_i, \max_{i\leq n} Y_i\right) ,$$

suitably normalized converges in distribution to G. If $F_{X,Y}(x,y)$ is in the domain of attraction of an extreme value cdf G which is a product measure, then $\sum_{k=1}^{\infty} {}_4I_k$ is finite a.s. If the extremal distribution is not a product measure then $\sum_{k=1}^{\infty} {}_4I_k = \infty$ a.s. For example, if (X,Y) has a bivariate normal distribution, then, provided its correlation is not equal to 1, it is attracted to a bivariate extreme value distribution which *is* a product measure (Sibuya, 1960). Consequently, as Goldie and Resnick point out, only a finite number of Definition 8.1.4 records can be encountered in a sequence of observations from such a bivariate normal population.

If we use Definition 8.1.2, it is difficult to determine the probability that a record will occur at time k. It is evident that, using (8.1.2) to define records, we will encounter records quite frequently. If, say, ${}_2R_n(X)$ is large, then the next record is likely to occur because of a Y observation larger than ${}_2R_n(Y)$. The X value corresponding to this outstanding Y value is not likely to be large (recall X_i's and Y_i's are independent), and this not too big X value will, using (8.1.2), be ${}_2R_{n+1}(X)$ replacing ${}_2R_n(X)$. Using this logic, $({}_2R_n(X), {}_2R_n(Y))$ will not grow, and so records will continually be relatively frequent. This is in contrast to all three other definitions where records become increasingly sparse.

Definition 8.1.3 is also difficult to deal with. Using ${}_3I_k$ as an indicator random variable for the occurrence of a Definition 8.1.3 record at time k, we can easily verify that, for any positive integer m,

$$P({}_3I_1 = 1, \ldots, {}_3I_m = 1) = \frac{\binom{2m}{m}}{(2m)!}$$

$$= \frac{1}{(m!)^2} .$$

However enumeration of cases indicates that

$$P({}_3I_1 = 1, \ {}_3I_2 = 0, \ {}_3I_3 = 0) = \frac{103}{180} .$$

It is evident that the ${}_3I_k$'s are not independent random variables in this case. The precise nature of the joint distribution of the ${}_3I_k$'s presents an interesting combinatorial problem.

Now we turn to discuss the distribution of the nth bivariate record $(R_n(X), R_n(Y))$ under the various definitions. One case is particularly easy. If we use Definition 8.1.3 then, since a new record is registered when both observed X and Y values exceed the current record, and if X and Y are assumed to both be exponential and hence memoryless, and since X and Y are assumed to be independent, it follows that

$$[(_3R_n(X),\ _3R_n(Y)] \stackrel{d}{=} \left(\sum_{i=0}^{n} X_i^*, \sum_{i=0}^{n} Y_i^* \right), \qquad (8.2.9)$$

where, as usual, the X_i^*'s and Y_i^*'s are independent Exp(1) random variables.

If, more generally, the X_i's are i.i.d. with cdf F_X and the Y_i's are i.i.d. with cdf F_Y (and the X_i's and Y_i's are independent), then

$$[(_3R_n(X),\ _3R_n(Y)] \stackrel{d}{=} \left[\psi_{F_X} \left(\sum_{i=0}^{n} X_i^* \right), \psi_{F_Y} \left(\sum_{i=0}^{n} Y_i^* \right) \right], \qquad (8.2.10)$$

where ψ_F is the inverse hazard function defined in (2.3.14). The distribution theory of nth records and the associated asymptotic distribution may then be directly obtained from Chapter 2 results applied to each coordinate of $(R_n(X), R_n(Y))$ (conveniently $R_n(X)$ and $R_n(Y)$ are independent in this scenario).

Turning to the other definitions, life becomes more complicated. We remarked earlier on the fact that with Definition 8.1.2, the process $(R_n(X), R_n(Y))$ will not automatically increase as n increases. In fact, it is evidently a Markov process with a non-trivial stationary distribution. The integral equation that defines the stationary distribution is quite complicated, and a solution is not available. Efforts to determine the distribution of $(R_n(X), R_n(Y))$, using Definitions 8.1.1 or 8.1.4, have also been unsuccessful. The study of bivariate records, even with independent X_i's and Y_i's, seems to be more difficult than it should be.

8.3   CONCOMITANTS OF RECORDS

The study of record concomitants was initiated by Houchens (1984). The scenario is a precise parallel to that associated with concomitants of order statistics (see David and Nagaraja, 1998). In this setting, we consider i.i.d. bivariate observations $(X_1, Y_1), (X_2, Y_2), \ldots$ with common joint cdf $F(x, y)$. For convenience, we will assume that F is absolutely continuous with joint pdf $f(x, y)$ and we will denote the family of conditional densities of Y given $X = x$ by $f_{Y|X}(y|x)$. The pdf, cdf of X and its inverse will be denoted by $f_X(x), F_X(x)$ and $F_X^{-1}(u)$. As usual, we let R_n denote the nth record value in the sequence of X's. The corresponding random variable Y, i.e. the Y value paired with or, if you wish, observed with the X value which qualified as the nth record will be called the nth record concomitant and will be denoted by $R_{[n]}$. Since the (X_i, Y_i) pairs are independent, it follows readily that

$$f_{R_{[n]}|R_n}(y|x) = f_{Y|X}(y|x). \qquad (8.3.1)$$

Referring to (2.3.7), we can immediately write the density of nth record concomitant in the form

$$f_{R_{[n]}}(y) = \int_{-\infty}^{\infty} f_{Y|X}(y|x) f_X(x) \, \frac{[-\log(1 - F_X(x))]^n}{n!} \, dx. \qquad (8.3.2)$$

A change of variable suggested by representation (2.3.3) enables us to write this in the alternative form

$$f_{R_{[n]}}(y) = \int_{0}^{\infty} f_{Y|X}\left[y|F_X^{-1}(1 - e^{-z})\right] \frac{z^n e^{-z}}{n!} \, dz. \qquad (8.3.3)$$

Expressions for moments of $R_{[n]}$ can be computed in principle from the conditional moments of Y, given X, using the following result:

$$E(R_{[n]}^k) = \int_{-\infty}^{\infty} E(Y^k|X = x) f_X(x) \, \frac{\{-\log[1 - F_X(x)]\}^n}{n!} \, dx.$$

In some cases, a more explicit representation of the nth record concomitant is possible. Here are three examples.

Example 8.3.1 (Bivariate Normal Distribution) Let (X_i, Y_i) be bivariate normal having mean vector (μ_X, μ_Y), variances σ_X^2 and σ_Y^2 and correlation ρ. Then the conditional distribution of Y given $X = x$ is described by

$$Y|X = x \sim N\left(\mu_Y + \rho \, \frac{\sigma_Y}{\sigma_X}(x - \mu_X), \sigma_Y^2(1 - \rho^2)\right).$$

It follows that the concomitant of the nth record, $R_{[n]}$, will satisfy

$$\frac{R_{[n]} - \mu_Y}{\sigma_Y} \overset{d}{=} \rho R_n^0 + \sqrt{1 - \rho^2} Z, \qquad (8.3.4)$$

where R_n^0 is the nth record for a sequence of standard normal random variables and Z is a standard normal random variable independent of R_n^0. The asymptotic normality of $R_{[n]}$ may be deduced from (8.3.4). Recall (2.3.27) to note that

$$R_n^0 - \Phi^{-1}(1 - e^{-n}) \overset{d}{\longrightarrow} N\left(0, \frac{1}{2}\right).$$

Consequently, (8.3.4) implies that

$$\frac{R_{[n]} - \mu_Y}{\sigma_Y} - \rho\Phi^{-1}(1 - e^{-n}) \xrightarrow{d} N\left(0, 1 - \rho^2/2\right). \qquad (8.3.5)$$

A second example in which the analysis is tractable is provided by the following.

Example 8.3.2 (Mardia's Bivariate Pareto Distribution) Let the joint pdf of (X, Y) be given by

$$f_{X,Y}(x, y) = \frac{\alpha(\alpha + 1)}{\sigma_X \sigma_Y} \left(1 + \frac{x}{\sigma_X} + \frac{y}{\sigma_Y}\right)^{-(\alpha+2)}, \qquad x > 0, \quad y > 0, \quad (8.3.6)$$

where $\alpha, \sigma_X, \sigma_Y > 0$. The corresponding marginal pdf of X and conditional density of Y given $X = x$ are then of the form

$$f_X(x) = \frac{\alpha}{\sigma_X}\left(1 + \frac{x}{\sigma_X}\right)^{-(\alpha+1)}, \qquad x > 0 \qquad (8.3.7)$$

and, for $x > 0$,

$$f_{Y|X}(y|x) = \frac{(\alpha + 1)}{c(x)}\left(1 + \frac{y}{c(x)}\right)^{-(\alpha+2)}, \qquad y > 0, \qquad (8.3.8)$$

where

$$c(x) = \sigma_Y\left(1 + \frac{x}{\sigma_X}\right). \qquad (8.3.9)$$

Thus $c(X)$ is Pareto(α, σ_Y), and conditioned on $X = x$, $Y + c(x)$ is a Pareto$((\alpha + 1), c(x))$ random variable. It follows that the nth record concomitant admits the representation

$$R_{[n]} \stackrel{d}{=} c(R_n)(V - 1), \qquad (8.3.10)$$

where V is a Pareto$((\alpha+1), 1)$ random variable and $c(R_n)$ behaves like the nth record from a Pareto(α, σ_Y) distribution. Further, V is independent of R_n. We have obtained in (2.4.9) a representation for the nth record from a Pareto$(\alpha, 1)$ distribution that is based on i.i.d. standard uniform random variables. Hence we have the following representation for the concomitant of R_n from Mardia's bivariate Pareto distribution:

$$R_{[n]} \stackrel{d}{=} \sigma_Y\left(\prod_{j=0}^{n} U_j\right)^{-1/\alpha}(V - 1), \qquad (8.3.11)$$

where the U_j's are i.i.d. Uniform$(0, 1)$ and the random variable V is Pareto$((\alpha + 1), 1)$ and is independent of them.

Example 8.3.3 (Farlie-Gumbel-Morgenstern Distribution) For (X, Y) with marginals f_X and f_Y, let the joint density be given by

$$f_{X,Y}(x, y) = f_X(x)f_Y(y)[1 + \theta(2F_X(x) - 1)(2F_Y(y) - 1)], \qquad (8.3.12)$$

where $\theta \in (-1, 1)$. This is the family of *Farlie-Gumbel-Morgenstern distributions*. Referring to (8.3.2), we can immediately write the following expression for the density of the nth record concomitant:

$$\begin{aligned} f_{R_{[n]}}(y) &= \int_{-\infty}^{\infty} f_X(x)f_Y(y)[1 + \theta(2F_X(x) - 1)(2F_Y(y) - 1)] \\ &\times \frac{(-\log[1 - F_X(x)])^n}{n!} \, dx. \end{aligned}$$

If we make the substitution $z = -\log(1 - F_X(x))$, the integration is readily accomplished, yielding

$$f_{R_{[n]}}(y) = f_Y(y)[1 + \theta(1 - 2^{-n})(2F_Y(y) - 1)]. \qquad (8.3.13)$$

The asymptotic distribution of $R_{[n]}$ is evident from (8.3.13). Without any normalization, we have

$$\lim_{n \to \infty} f_{R_{[n]}}(y) = f_Y(y)[1 + \theta(2F_Y(y) - 1)].$$

Further discussion of this example, including expressions for moments of $R_{[n]}$, may be found in Houchens (1984, pp. 125–127).   ◯

8.4   LOWER AND UPPER RECORDS AND THE RECORD RANGE

It is quite conceivable that interest will focus on lower and upper records simultaneously. Houchens (1984) surveyed results in this setting. The following material is abstracted from his presentation.

Suppose that $X_1, X_2, \ldots$ are i.i.d. with common cdf F. Following notation introduced earlier, we will denote the nth upper record from the sequence by R_n and the mth lower record (from the same sequence) by R'_m. The zeroth upper and lower records coincide, both being $R_0 = X_1$. Assuming F is absolutely continuous with corresponding pdf f, the joint density of

$$\left(R'_m, R'_{m-1}, \ldots, R'_1, R_0, R_1, R_2, \ldots, R_n \right)$$

is given by

$$\frac{f(r_0)f(r_m')f(r_n)}{F(r_0)(1-F(r_0))} \prod_{i=1}^{m-1} \frac{f(r_i')}{F(r_i')} \prod_{j=1}^{n-1} \frac{f(r_j)}{[1-F(r_j)]} \tag{8.4.1}$$

for $r_m' < r_{m-1}' < \ldots < r_1' < r_0 < r_1 < \ldots < r_n$. It is evident that, for any fixed $m \geq 1$ and $n \geq 1$, R_m' and R_n are conditionally independent, given R_0. This observation allows us to readily obtain the joint pdf of R_m' and R_n in the form

$$f_{R_m', R_n}(r_m', r_n) = \frac{f(r_m')f(r_n)}{(m-1)!(n-1)!}$$
$$\times \int_{r_m'}^{r_n} \left[\log \frac{F(r_0)}{F(r_m')}\right]^{m-1} \left[\log \frac{1-F(r_0)}{1-F(r_n)}\right]^{n-1} \frac{f(r_0)dr_0}{F(r_0)(1-F(r_0))}.$$
$$\tag{8.4.2}$$

In the case $m = n = 1$, we get

$$f_{R_1', R_1}(r', r) = f(r')f(r) \log\left\{\frac{F(r)(1-F(r'))}{F(r')(1-F(r))}\right\}, \quad r' < r. \tag{8.4.3}$$

If the upper *and* the lower tail of the distribution F behave appropriately, it is possible to determine the asymptotic joint distribution of (R_m', R_n) as $m, n \to \infty$, using results from Chapter 2.

In the above development, one is not keeping track of the smallest X observed so far and the largest X observed so far. For example, R_1' may not be observed before R_6, if we happen to get an increasing run of large observations. Instead, we may choose to keep track of the largest X value observed and the smallest X value observed at the times when a new record of either kind occurs. In this scenario, a new "record" occurs at time j if either $X_j > \max\{X_1, \ldots, X_{j-1}\}$ or if $X_j < \min\{X_1, \ldots, X_{j-1}\}$; that is, the initial rank [defined in (2.10.2)] of X_j is either 1 or j. Let S_n' denote the value of the smallest observation after observing n such "records" and let S_n denote the value of the largest observation after observing n records. We define $S_0' = S_0 = X_1$. The interval (S_n', S_n) could be called the *record coverage* and the difference $S_n - S_n'$, the *record range*.

Denoting the common cdf and pdf of the X_i's by F and f, respectively, Houchens (1984) uses an inductive argument to verify that the joint density of S_n', S_n is

$$f_{S_n', S_n}(s_1, s_2) = 2^n f(s_1)f(s_2)\{-\log[1 - F(s_2) + F(s_1)]\}^{n-1}/(n-1)!$$
$$-\infty < s_1 < s_2 < \infty \tag{8.4.4}$$

and the corresponding marginal densities are

$$f_{S_n'}(s_1) = 2^n f(s_1) \left[1 - F(s_1) \sum_{j=0}^{n-1} \frac{(-\log F(s_1)^j}{j!} \right]$$

and

$$f_{S_n}(s_2) = 2^n f(s_2) \left[1 - (1 - F(s_2)) \sum_{j=0}^{n-1} \frac{\{-\log(1 - F(s_2))\}^j}{j!} \right]. \qquad (8.4.5)$$

It may be observed that if the X_i's are i.i.d. Exp(1) random variables, then $S_n \overset{d}{=} X_1^* + \frac{1}{2}\sum_{j=2}^{n+1} X_j^*$, where, as usual, the asterisks on the X_i's imply they are i.i.d. Exp(1) random variables. The results on the asymptotic distribution of R_n from Chapter 2 will thus require trivial modification to obtain the limiting distribution of S_n, when it exists, for a more general cdf F.

Next we turn to the record range, i.e. $W_n = S_n - S_n'$. Directly from (8.4.4) we get

$$f_{W_n}(w)$$
$$= \frac{2^n}{(n-1)!} \int_{-\infty}^{\infty} f(w+u)f(u)[-\log(1 - F(w+u) + F(u))]^{n-1} \, du,$$
$$0 < u < \infty. \qquad (8.4.6)$$

The only case in which this can be further simplified is when the X_i's are i.i.d. uniform random variables.

Example 8.4.1 (Uniform Distribution) Let $\{\tilde{X}_i\}$ denote Uniform$(0,1)$ random variables and denote the corresponding record range by $\tilde{W}_n$. Then the integration in (8.4.6) is readily performed to yield

$$f_{\tilde{W}_n}(w) = \frac{2^n}{(n-1)!}(1-w)[-\log(1-w)]^{n-1}, \quad 0 < w < 1. \qquad (8.4.7)$$

This nth record range has the same distribution as the $(n-1)$st Type 2 2-record value $(R_{(n-1)(2)})$ from the standard uniform distribution.   ◯

There are other ways of viewing the record range. We could for each k, denote by $Y_k = X_{k:k} - X_{1:k}$, the sample range of the observations $X_1, \ldots, X_k$. The record range W_n is then the ordinary $(n-1)$th record value in the sequence $Y_1, Y_2, \ldots$ (a non-independent sequence). With a view to higher dimensional extensions, we could instead describe Y_k as the measure of the convex hull of the k points $X_1, X_2, \ldots, X_k$. This definition makes sense for higher dimensional X_i's. We of course would not expect simple results, unless the X_i's were uniform random variables. We know the distribution of $\tilde{W}_n$ when the X_i's are univariate standard uniform random variables. The next trivial generalization would be to go to two dimensions. We would then consider $\mathbf{X_1}, \mathbf{X_2}, \ldots$ to be i.i.d. random variables each uniform on the unit square. For each n we let Y_k denote the Lebesgue measure (area) of the convex hull of the n points $\mathbf{X_1}, \ldots, \mathbf{X_k}$. The determination of the distribution of the upper record values from the Y_k sequence is a challenging but tantalizing problem.

8.5   RECORDS IN PARTIALLY ORDERED SETS

The material on multivariate records can be viewed as a special case of records in a partially ordered set. In this paradigm, our random variables $X_1, X_2, \ldots$ assume values in an arbitrary partially ordered set S. We say that X_n is a record if $X_n > X_i$, $\forall\, i < n$. Details are provided in Goldie and Resnick (1989). They show, under minimal measurability conditions, that the record sequence $\{R_n\}$ is a Markov chain with stationary transition distributions. The actual distribution of R_n, for $n > 1$, is generally not obtainable. Expressions (albeit rather complicated ones) can be given for the distribution of number of records which fall in any given set $B \subset S$.

EXERCISES

1. Consider bivariate observations (X_i, Y_i) from the Gumbel type 2 bivariate exponential density

$$f_{X,Y}(x, y) = e^{-x-y}[1 + \theta(2e^{-x} - 1)(2e^{-y} - 1)]\, I(x > 0, y > 0),$$

   where $|\theta| < 1$. Let R_n denote the nth record in the sequence of observed X's and let $R_{[n]}$ denote the concomitant of R_n. Evaluate

   (a) $E(R_n), E(R_{[n]})$

   (b) $\mathrm{Var}(R_n),\ \mathrm{Var}(R_{[n]})$

   (c) $\mathrm{Cov}(R_n, R_{[n]}),\ \mathrm{Cov}(R_{[n]}, R_{[n+k]})$.

2. Verify the expression given in (8.4.4) for the joint density of (S'_n, S_n).

3. The following table (from Klüppelberg and Schwere, 1995) lists the average July and August temperatures at Neuenberg, Switzerland, during 1864–1993. Compute the observed values $(_jR_n(X), _jR_n(Y))$ and $_jT_n$ $(j = 1, 2, 3, 4)$, for the four different definitions of bivariate records discussed in the chapter. Is there any evidence for global warming in the bivariate record data?

July	19.0	20.1	18.4	17.4	19.7	21.0	21.4	19.2	19.9	20.4
August	17.3	16.7	15.8	19.0	18.7	17.5	16.1	19.1	17.1	19.7
July	20.9	17.2	20.2	17.8	18.1	15.6	19.4	21.7	16.2	16.4
August	16.6	18.7	19.3	18.7	17.6	19.2	17.0	18.5	16.4	17.5
July	19.0	20.6	19.0	20.7	15.8	17.7	16.8	17.1	18.1	18.4
August	18.6	18.1	17.9	17.9	16.5	16.6	16.7	16.9	18.9	20.0
July	18.7	18.7	18.4	19.2	18.0	18.7	20.7	19.4	19.2	17.4
August	17.1	17.9	15.1	17.8	20.1	20.0	17.4	17.2	16.8	17.3
July	22.0	21.4	19.3	16.8	18.2	16.2	15.9	22.1	17.5	15.3
August	19.5	17.8	19.9	18.8	16.6	17.6	17.1	21.7	14.1	16.7
July	16.5	17.4	17.0	18.3	18.3	15.3	18.2	21.5	17.0	21.6
August	17.4	16.5	17.5	16.3	18.0	20.5	16.3	18.7	17.4	19.4
July	18.2	18.1	17.6	18.2	22.6	19.9	17.1	17.2	17.3	19.4
August	14.5	17.3	18.3	16.8	20.1	17.9	17.4	15.6	20.3	19.8
July	20.1	20.1	17.0	19.4	17.5	16.8	17.0	19.9	18.2	19.2
August	17.0	16.9	17.8	18.6	17.6	17.9	17.0	16.6	18.7	20.3
July	18.5	20.8	19.5	21.1	15.8	21.3	21.2	18.8	22.3	18.6
August	21.9	17.7	18.0	22.0	17.4	19.7	18.4	17.3	19.3	18.7
July	16.8	18.2	17.2	18.4	18.7	21.1	16.3	17.4	18.0	19.5
August	16.4	17.7	15.3	17.1	18.5	18.3	16.8	17.7	20.5	16.6
July	21.2	16.8	17.4	20.7	18.4	19.8	18.7	20.5	18.3	18.2
August	18.2	17.0	16.6	18.2	16.1	17.8	18.4	19.3	17.5	19.8
July	18.2	19.2	20.2	18.2	17.4	19.2	16.3	17.4	20.3	23.4
August	19.2	18.7	17.5	16.8	16.8	16.5	18.9	19.3	17.8	19.6
July	19.2	20.2	19.3	19.0	18.8	20.3	19.7	20.7	19.6	18.1
August	18.6	18.4	18.1	18.6	19.2	19.1	20.2	21.1	21.5	19.0

BIBLIOGRAPHY

ABRAMOWITZ, M. and STEGUN, I. A. (Eds.) (1965). *Handbook of Mathematical Functions with Formulas, Graphs, and Mathematical Tables*, Dover, New York.

ADKE, S. R. (1993). Records generated by Markov sequences, *Statistics & Probability Letters*, **18**, 257–263.

AHSANULLAH, M. (1978). Record values and the exponential distribution, *Annals of the Institute of Statistical Mathematics*, **30**, 429–433.

AHSANULLAH, M. (1979). Characterization of the exponential distribution by record values, *Sankhyā, Series B*, **41**, 116–121.

AHSANULLAH, M. (1980). Linear prediction of record values for the two parameter exponential distribution, *Annals of the Institute of Statistical Mathematics*, **32**, 363–368.

AHSANULLAH, M. (1981a). On a characterization of the exponential distribution by weak homoscedasticity of record values, *Biometrical Journal*, **23**, 715–717.

AHSANULLAH, M. (1981b). Record values of exponentially distributed random variables, *Statistische Hefte*, **22**, 121–127.

AHSANULLAH, M. (1982). Characterizations of the exponential distribution by some properties of the record values, *Statistische Hefte*, **23**, 326–332.

AHSANULLAH, M. (1986). Estimation of the parameters of a rectangular distribution by record values, *Computational Statistics Quarterly*, **2**, 119–125.

AHSANULLAH, M. (1987a). Record statistics and the exponential distribution, *Pakistan Journal of Statistics*, **3A**, 17–40.

AHSANULLAH, M. (1987b). Two characterizations of the exponential distribution, *Communications in Statistics—Theory and Methods*, **16**, 375–381.

AHSANULLAH, M. (1988). *Introduction to Record Statistics*, Ginn Press, Needham Heights, Massachusetts.

AHSANULLAH, M. (1990). Estimation of the parameters of the Gumbel distribution based on m record values, *Computational Statistics Quarterly*, **6**, 231–239.

AHSANULLAH, M. (1995). *Record Statistics*, Nova Science Publishers, Commack, New York.

AHSANULLAH, M. and HOLLAND, B. (1984). Record values and the geometric distribution, *Statistische Hefte*, **25**, 319–327.

AHSANULLAH, M. and HOLLAND, B. (1987). Distributional properties of record values from the geometric distribution, *Statistica Neerlandica*, **41**, 129–137.

AHSANULLAH, M. and KIRMANI, S. N. U. A. (1991). Characterizations of the exponential distribution through a lower record, *Communications in Statistics—Theory and Methods*, **20**, 1293–1299.

AITCHISON, J. and DUNSMORE, I. R. (1975). *Statistical Prediction Analysis*, Cambridge University Press, Cambridge, England.

ARNOLD, B. C. (1983). *The Pareto Distributions*, International Cooperative Publishing House, Fairland, Maryland.

ARNOLD, B. C. and BALAKRISHNAN, N. (1989). *Relations, Bounds and Approximations for Order Statistics*, Lecture Notes in Statistics **53**, Springer-Verlag, New York.

ARNOLD, B. C., BALAKRISHNAN, N., and NAGARAJA, H. N. (1992). *A First Course in Order Statistics*, John Wiley & Sons, New York.

ARNOLD, B. C. and VILLASEÑOR, J. A. (1997). Gumbel records and related characterizations, In *Advances in the Theory and Practice of Statistics – A Volume in Honor of Samuel Kotz* (Eds., N. L. Johnson and N. Balakrishnan), pp. 441–453, John Wiley & Sons, New York.

AZLAROV, T. and VOLODIN, N. A. (1986). *Characterization Problems Associated With the Exponential Distribution*, Springer-Verlag, New York.

BALABEKYAN, V. A. and NEVZOROV, V. B. (1986). Numbers of records in a sequence of series of nonidentically distributed random variables, In *Rings and Modules – Limit Theorems of Probability Theory*, Vol. 1, pp. 147–153, Leningrad University.

BALAKRISHNAN, N. (1990). Improving the Hartley-David-Gumbel bound for the mean of extreme order statistics, *Statistics & Probability Letters*, **9**, 291–294.

BALAKRISHNAN, N. (Ed.) (1992). *Handbook of the Logistic Distribution*, Marcel Dekker, New York.

BALAKRISHNAN, N. (1993). A simple application of binomial-negative binomial relationship in the derivation of sharp bounds for moments of order statistics based on greatest convex minorants, *Statistics & Probability Letters*, **18**, 301–305.

BALAKRISHNAN, N. and AHSANULLAH, M. (1994a). Recurrence relations for single and product moments of record values from generalized Pareto distribution, *Communications in Statistics—Theory and Methods*, **23**, 2841–2852.

BALAKRISHNAN, N. and AHSANULLAH, M. (1994b). Relations for single and product moments of record values from Lomax distribution, *Sankhyā, Series B*, **56**, 140–146.

BALAKRISHNAN, N. and AHSANULLAH, M. (1995). Relations for single and product moments of record values from exponential distribution, *Journal of Applied Statistical Science*, **2**, 73–87.

BALAKRISHNAN, N., AHSANULLAH, M., and CHAN, P. S. (1992). Relations for single and product moments of record values from Gumbel distribution, *Statistics & Probability Letters*, **15**, 223–227.

BALAKRISHNAN, N., AHSANULLAH, M., and CHAN, P. S. (1995). On the logistic record values and associated inference, *Journal of Applied Statistical Science*, **2**, 233–248.

BALAKRISHNAN, N. and BALASUBRAMANIAN, K. (1995). A characterization of geometric distribution based on record values, *Journal of Applied Statistical Science*, **2**, 277–282.

BALAKRISHNAN, N., BALASUBRAMANIAN, K., and PANCHAPAKE-
SAN, S. (1997). δ-exceedance records, *Journal of Applied Statistical
Science*, **4**, 123–132.

BALAKRISHNAN, N. and BENDRE, S. M. (1993). Improved bounds
for expectations of linear functions of order statistics, *Statistics*, **24**,
161–165.

BALAKRISHNAN, N. and CHAN, P. S. (1994). Record values from
Rayleigh and Weibull distributions and associated inference, *NIST
Special Publication 866, Proceedings of the Conference on Extreme
Value Theory and Applications*, Vol. **3** (Eds., J. Galambos, J. Lech-
ner and E. Simiu), pp. 41–51.

BALAKRISHNAN, N. and CHAN, P. S. (1995). On the normal record
values and associated inference, *Technical Report*, McMaster University, Hamilton, Ontario, Canada.

BALAKRISHNAN, N. and CHAN, P. S. (1998). On the normal record
values and associated inference, *Statistics & Probability Letters* (to
appear).

BALAKRISHNAN, N., CHAN, P. S., and AHSANULLAH, M. (1993).
Recurrence relations for moments of record values from generalized
extreme value distribution, *Communications in Statistics—Theory
and Methods*, **22**, 1471–1482.

BALAKRISHNAN, N. and COHEN, A. C. (1991). *Order Statistics and
Inference: Estimation Methods*, Academic Press, San Diego.

BALAKRISHNAN, N. and NEVZOROV, V. B. (1997). Stirling numbers and records, In *Advances in Combinatorial Methods and Applications to Probability and Statistics* (Ed., N. Balakrishnan), pp.
189–200, Birkhäuser, Boston.

BALAKRISHNAN, N. and RAO, C. R. (1997). A note on the best
linear unbiased estimation based on order statistics, *The American
Statistician*, **51**, 181–185.

BALAKRISHNAN, N. and RAO, C. R. (1998). Some efficiency properties of best linear unbiased estimators, *Journal of Statistical Planning and Inference* (to appear).

BALLERINI, R. (1987). Another characterization of the type I extreme value distribution, *Statistics & Probability Letters*, **5**, 83–85.

BALLERINI, R. (1994). A dependent F^α–scheme, *Statistics and Probability Letters*, **21**, 21–25.

BALLERINI, R. and RESNICK, S. I. (1985). Records from improving populations, *Journal of Applied Probability*, **22**, 487–502.

BALLERINI, R. and RESNICK, S. I. (1987a). Records in the presence of a linear trend, *Advances in Applied Probability*, **19**, 801–828.

BALLERINI, R. and RESNICK, S. I. (1987b). Embedding sequences of successive maxima in extremal processes, *Journal of Applied Probability*, **24**, 827–837.

BARLOW, R. E. and PROSCHAN, F. (1981). *Statistical Theory of Reliability and Life Testing: Probability Models*, Second edition, To Begin With, Silver Spring, Maryland.

BASAK, P. (1996). Lower record values and characterizations of exponential distribution, *Calcutta Statistical Association Bulletin*, **46**, 1–7.

BERNARDO, J. M. (1976). Psi (digamma) function. Algorithm AS 103, *Applied Statistics*, **25**, 315–317.

BIONDINI, R. and SIDDIQUI, M. M. (1975). Record values in Markov sequences, In *Statistical Inference and Related Topics*, Vol. 2 (Ed., M. L. Puri), pp. 291–352, Academic Press, New York.

BROWNE, S. and BUNGE, J. (1995). Random record process and state dependent thinning, *Stochastic Processes and Their Applications*, **55**, 131–148.

BRUSS, F. (1987). On an optimal selection problem of Cowan and Zabczyk, *Journal of Applied Probability*, **24**, 918–928.

BRUSS, F. (1988). Invariant record processes and applications to best choice modelling, *Stochastic Processes and Their Applications*, **30**, 303–316.

BRUSS, F. (1998). Quick solutions for general best choice problems in continuous case, *Stochastic Models*, **14** (to appear).

BRUSS, F. and ROGERS, B. (1991). Pascal processes and their characterization, *Stochastic Processes and Their Applications*, **37**, 331–338.

BUISHAND, T. (1989). Statistics of extremes in climatology, *Statistica Neerlandica*, **43**, 1–30.

BUNGE, J. (1989). Distribution Theory for Record Statistics from Random Record Models, *Ph.D. Dissertation*, Ohio State University, Columbus, Ohio.

BUNGE, J. and NAGARAJA, H. N. (1991). The distribution of certain record statistics from a random number of observations, *Stochastic Processes and Their Applications*, **38**, 167–183.

BUNGE, J. and NAGARAJA, H. N. (1992a). Exact distribution theory for some point process record models, *Advances in Applied Probability*, **29**, 587–596.

BUNGE, J. and NAGARAJA, H. N. (1992b). Dependence structure of Poisson-paced records, *Journal of Applied Probability*, **24**, 20–44.

BURDEN, R. L. and FAIRES, J. D. (1985). *Numerical Analysis*, Third edition, Prindle, Weber & Schmidt, Boston.

CARLIN, B. P. and GELFAND, A. E. (1993). Parametric likelihood inference for record breaking problems, *Biometrika*, **80**, 507–515.

CAYLEY, A. (1875). Mathematical questions with their solutions, *The Educational Times*, **23**, 18–19. See *The Collected Mathematical Papers of Arthur Cayley*, **10**, 587–588 (1896), Cambridge University Press, Cambridge, England.

CHAN, P. S. (1998). Interval estimation of parameters of life based on record values, *Statistics & Probability Letters* (to appear).

CHANDLER, K. N. (1952). The distribution and frequency of record values, *Journal of the Royal Statististical Society, Series B*, **14**, 220–228.

COWAN, R. and ZABCZYK, J. (1978). An optimal selection problem associated with the Poisson process, *Theory of Probability and Its Applications*, **23**, 584–592.

CRAMÉR, H. (1970). *Random Variables and Probability Distributions*, Third edition, Cambridge University Press, Cambridge, England.

DALEY, D. J. and VERE-JONES, D. (1988). *An Introduction to the Theory of Point Processes*, Springer-Verlag, New York.

DALLAS, A. C. (1981a). Record values and the exponential distribution, *Journal of Applied Probability*, **18**, 949–951.

DALLAS, A. C. (1981b). A characterization using conditional variance, *Metrika*, **28**, 151–153.

DALLAS, A. C. (1982). Some results on record values from the exponential and Weibull law, *Acta Math. Acad. Sci. Hungary*, **40**, 307–311.

DALLAS, A. C. (1989). Some properties of record values coming from the geometric distribution, *Annals of the Institute of Statistical Mathematics*, **41**, 661–669.

DAVID, H. A. (1981). *Order Statistics*, Second edition, John Wiley & Sons, New York.

DAVID, H. A. and NAGARAJA, H. N. (1998). Concomitants of order statistics, In *Handbook of Statistics – 16: Order Statistics: Theory and Methods* (Eds., N. Balakrishnan and C.R. Rao), pp. 487–513, North-Holland, Amsterdam, The Netherlands.

DAVIS, H. T. (1935). *Tables of the Higher Mathematical Functions*, Vols. 1 and 2, Principia Press, Bloomington, Indiana.

DE HAAN, L. (1970). On Regular Variation and Its Application to the Weak Convergence of Sample Extremes, *Mathematical Centre Tract* **32**, Mathematisch Centrum, Amsterdam, The Netherlands.

DE HAAN, L. and RESNICK, S. I. (1973). Almost sure limit points of record values, *Journal of Applied Probability*, **10**, 528–542.

DE HAAN, L. and VERKADE, E. (1987). On extreme value theory in the presence of a trend, *Journal of Applied Probability*, **24**, 62–76.

DEHEUVELS, P. (1981). The strong approximation of extremal processes, *Zeitschrift fuer Wahrscheinlichkeitstheorie und Verwandte Gebiete*, **58**, 1–6.

DEHEUVELS, P. (1982). Spacings, record times and extremal processes, In *Exchangeability in Probability and Statistics* (Eds., G. Koch and F. Spizzichino), pp. 233–243, North-Holland, Amsterdam, The Netherlands.

DEHEUVELS, P. (1983). The strong approximation of extremal processes (II), *Zeitschrift fuer Wahrscheinlichkeitstheorie und Verwandte Gebiete*, **62**, 7–15.

DEHEUVELS, P. (1984). The characterization of distributions by order statistics and record values – a unified approach, *Journal of Applied Probability*, **21**, 326–334. (Correction **22**, p. 997.)

DEKEN, J. G. (1978). Record values, scheduled maxima sequences, *Journal of Applied Probability*, **15**, 543–551.

DENY, J. (1961). Sur l'équation de convolution $\mu * \sigma = \mu$, *Semin. Theor. Potent. M. Brelot.*, Fac. Sci. Paris, 4 e ann.

DIERSEN, J. and TRENKLER, G. (1996). Records tests for trend in location, *Statistics*, **28**, 1–12.

DOGANAKSOY, N. and BALAKRISHNAN, N. (1997). A useful property of best linear unbiased predictors with applications to life-testing, *The American Statistician*, **51**, 22–28.

DUNSMORE, I. R. (1983). The future occurrence of records, *Annals of the Institute of Statistical Mathematics*, **35**, 267–277.

DWASS, M. (1960). Some k-sample rank order tests, In *Contributions to Probability and Statistics*, pp. 198–202, Stanford University Press, Stanford, California.

DWASS, M. (1964). Extremal processes, *Annals of Mathematical Statistics*, **36**, 1718–1725.

DZIUBDZIELA, W. and KOPOCIŃSKI, B. (1976). Limiting properties of the kth record values, *Zastosowania Matematyki*, **15**, 187–190.

EMBRECHTS, P. and OMEY, E. (1983). On subordinated distributions and random record processes, *Mathematical Proceedings of the Cambridge Philosophical Society*, **93**, 339–353.

ENGELEN, R., TOMASSEN, P., and VERVAAT, W. (1987). Ignatov's theorem: A new and short proof, In *A Celebration of Applied Probability*, Special Volume of *Journal of Applied Probability*, **25A**, 229–236.

FELLER, W. (1965). *An Introduction to Probability Theory and Its Applications*, Vol. 2, John Wiley & Sons, New York.

FELLER, W. (1968). *An Introduction to Probability Theory and Its Applications*, Vol. 1, Third Edition, John Wiley & Sons, New York.

FERGUSON, T. S. (1967). On characterizing distributions by properties of order statistics, *Sankhyā, Series A*, **29**, 265–278.

FERGUSON, T. S. (1989). Who solved the secretary problem?, *Statistical Science*, **4**, 282–296.

FEUERVERGER, A. and HALL, P. (1996). On distribution-free inference for record-value data with trend, *Annals of Statistics*, **24**, 2655–2678.

FISHER, R. A. (1934). Two new properties of the mathematical likelihood, *Proceedings of the Royal Society, Series A*, **144**, 285–307.

FOSTER, F. G. and STUART, A. (1954). Distribution-free tests in time-series based on the breaking of records (with discussions), *Journal of the Royal Statistical Society, Series B*, **16**, 2–22.

FOSTER, F. G. and TEICHROEW, D. (1955). A sampling experiment on the powers of the records tests for trend in a time series, *Journal of the Royal Statistical Society, Series B*, **17**, 115–121.

GALAMBOS, J. (1987). *The Asymptotic Theory of Extreme Order Statistics*, Second edition, Krieger, Malabar, Florida.

GALAMBOS, J. and KOTZ, S. (1978). *Characterizations of Probability Distributions*, Lecture Notes in Mathematics **675**, Springer-Verlag, Berlin.

GALAMBOS, J. and SENETA, E. (1975). Record times, *Proceedings of the American Mathematical Society*, **50**, 383–387.

GAVER, D. (1976). Random record models, *Journal of Applied Probability*, **13**, 538–547.

GAVER, D. and JACOBS, P. (1978). Nonhomogeneously paced random records and associated extremal processes, *Journal of Applied Probability*, **15**, 543–551.

GLICK, N. (1978). Breaking records and breaking boards, *American Mathematical Monthly*, **85**, 2–26.

GNEDENKO, B. V. (1943). Sur la distribution limite du terme maximum d'une serie aleatoire, *Annals of Mathematics*, **44**, 423–453.

GOLDBERGER, A. S. (1962). Best linear unbiased predictors in the generalized regression model, *Journal of the American Statistical Association*, **57**, 369–375.

GOLDIE, C. M. (1983). On Records and Related Topics in Probability Theory, *Ph.D. Thesis*, The University of Sussex, School of Mathematical and Physical Sciences, Sussex, England.

GOLDIE, C. M. and RESNICK, S. I. (1989). Records in a partially ordered set, *Annals of Probability*, **17**, 678–689.

GOLDIE, C. M. and RESNICK, S. I. (1995). Many multivariate records, *Stochastic Processes and Their Applications*, **59**, 185–216.

GOLDIE, C. M. and ROGERS, L. C. G. (1984). The k-record processes are i.i.d., *Zeitschrift fuer Wahrscheinlichkeitstheorie und Verwandte Gebiete*, **67**, 197–211.

GOVINDARAJULU, Z. (1963). On moments of order statistics and quasi-ranges from normal populations, *Annals of Mathematical Statistics*, **34**, 633–651.

GRAYBILL, F. A. (1983). *Matrices with Applications in Statistics*, Second edition, Wadsworth, Belmont, California.

GRUDZIEŃ, Z. and SZYNAL, D. (1985). On the expected values of kth record values and associated characterizations of distributions, In *Probability and Statistical Decision Theory, Vol. A* (Eds., F. Konecny, J. Mogyoródi and W. Wertz), pp. 119–127, Reidel, Dordrecht, The Netherlands.

GULATI, S. and PADGETT, W. J. (1992). Kernel density estimation from record-breaking data, In *Survival Analysis: State of the Art* (Eds., J. P. Klein and P. K. Goel), pp. 197–210, Kluwer Academic Publishers, Amsterdam, The Netherlands.

GULATI, S. and PADGETT, W. J. (1994a). Nonparametric quantile estimation from record-breaking data, *Australian Journal of Statistics*, **36**, 211–223.

GULATI, S. and PADGETT, W. J. (1994b). Estimation of nonlinear
statistical functions from record-breaking data: A review, *Nonlinear
Times and Digest*, **1**, 97–112.

GULATI, S. and PADGETT, W. J. (1994c). Smooth nonparametric
estimation of the distribution and density functions from record-
breaking data, *Communications in Statistics–Theory and Methods*,
23, 1259–1274.

GULATI, S. and PADGETT, W. J. (1994d). Smooth nonparametric
estimation of the hazard and the hazard rate functions from record-
breaking data, *Journal of Statistical Planning and Inference*, **42**,
331–341.

GULATI, S. and PADGETT, W. J. (1997). Nonparametric function
estimation from inversely sampled record-breaking data, *Canadian
Journal of Statistics* (to appear).

GUPTA, R. C. (1984). Relationships between order statistics and record
values and some characterization results, *Journal of Applied Probability*, **21**, 425–430.

GUPTA, R. C. and KIRMANI, S. N. U. A. (1988). Closure and monotonicity properties of nonhomogeneous Poisson processes and record
values, *Probability in the Engineering and Informational Sciences*,
2, 475–484.

GUTHRIE, G. L. and HOLMES, P. T. (1975). On record and inter-
record times for a sequence of random variables defined on a Markov
chain, *Advances in Applied Probability*, **7**, 195–214.

HAIMAN, G. and NEVZOROV, V. B. (1996). Stochastic ordering of the
number of records, In *Statistical Theory and Applications—Papers
in Honor of Herbert A. David* (Eds., H. N. Nagaraja, P. K. Sen and
D. F. Morrison), pp. 105–116, Springer-Verlag, New York.

HAMMING, R. W. (1973). *Numerical Methods for Scientists and Engineers*, McGraw-Hill, New York.

HOFMANN, G. (1997). A Family of General Record Models, *Ph.D.
Dissertation*, Ohio State University, Columbus, Ohio.

HOFMANN, G. and NAGARAJA, H. N. (1997). Random and point
process record models in the F^α setup, *Submitted for publication*.

HOINKES, L. A. and PADGETT, W. J. (1994). Maximum likelihood estimation from record-breaking data for the Weibull distribution, *Quality and Reliability Engineering International*, **10**, 5–13.

HOUCHENS, R. L. (1984). Record Value Theory and Inference, *Ph.D. Dissertation*, University of California, Riverside, California.

HUANG, W-J. and LI, S-H. (1993). Characterization results based on record values, *Statistica Sinica*, **3**, 583–599.

HUANG, W-J. and SU, J-C. (1994). On certain problems involving order statistics – A unified approach through order statistics property of point processes, (Abstract in *IMS Bulletin* (1994), **23**, pp. 400–401).

IGNATOV, Z. (1977). Ein von der Variationsreihe ergeugter Poissonscher Punktprozess, *Annuaire de l'Université de Sofia, Fac. Math. Méch*, **71**, 79–94 (published 1986).

IGNATOV, Z. (1978). Point processes generated by order statistics and their applications, *Colloquia Mathematica Societatis János Bolyai*, Keszthely, Hungary, **24**, *Point Processes and Queueing Problems* (Eds., P. Bártfai and J. Tomkó), pp. 109–116, North-Holland, Amsterdam, The Netherlands.

IIYAMA, Y., NISHIMURA, K., and SIBUYA, M. (1995). Power of record-breaking test, *Japanese Journal of Applied Statistics*, **24**, 13–26 (in Japanese).

JOHNSON, N. L., KOTZ, S., and BALAKRISHNAN, N. (1994). *Continuous Univariate Distributions — Vol. 1*, Second edition, John Wiley & Sons, New York.

JOHNSON, N. L., KOTZ, S., and BALAKRISHNAN, N. (1995). *Continuous Univariate Distributions — Vol. 2*, Second edition, John Wiley & Sons, New York.

KAGAN, A. M., LINNIK, Yu. V., and RAO, C. R. (1973). *Characterization Problems in Mathematical Statistics*, John Wiley & Sons, New York.

KAMPS, U. (1992). Identities for the difference of moments of successive order statistics and record values, *Metron*, **50**, 1179–1187.

KAMPS, U. (1994). Reliability properties of record values from non-identically distributed random variables, *Communications in Statistics—Theory and Methods*, **23**, 2101–2112.

KAMPS, U. (1995a). *A Concept of Generalized Order Statistics*, Teubner, Stuttgart, Germany.

KAMPS, U. (1995b). A concept of generalized order statistics, *Journal of Statistical Planning and Inference*, **48**, 1–23.

KARLIN, S. and TAYLOR, H. M. (1975). *A First Course in Stochastic Processes*, Academic Press, New York.

KINOSHITA, K. and RESNICK, S. I. (1989). Multivariate records and shape, In *Extreme Value Theory*, Lecture Notes in Statistics **51** (Eds., J. Hüsler and R.-D. Reiss), pp. 222–233, Springer-Verlag, New York.

KIRMANI, S. N. U. A. and BEG, M. I. (1984). On characterization of distributions by expected records, *Sankhyā, Series A*, **46**, 463–465.

KLÜPPELBERG, C. and SCHWERE, P. (1995). Records in time series: An investigation to global warming, *Berichte zur Stochastik und Verwandten Gebieten*, Johannes Gutenberg-Universitt, Mainz, Tech. Rep. 95-4, June 1995.

KNOPP, K. (1956). *Infinite Sequences and Series*, Dover, New York (Translated by F. Bagemihl).

KOCHAR, S. C. (1990). Some partial ordering results on record values, *Communications in Statistics—Theory and Methods*, **19**, 299–306.

KOCHAR, S. C. (1996). A note on dispersive ordering of record values, *Calcutta Statistical Association Bulletin*, **46**, 63–67.

KORWAR, R. M. (1984). On characterizing distributions for which the second record value has a linear regression on the first, *Sankhyā, Series B*, **46**, 108–109.

KOTLARSKI, I. I. (1972). On a characterization of some probability distributions by conditional moments, *Sankhyā, Series A*, **34**, 461–466.

LAU, K-S. and RAO, C. R. (1982). Integrated Cauchy functional equations and characterizations of the exponential law, *Sankhyā, Series A*, **44**, 72–90. (Corrections **44**, p. 452.)

LAU, K-S. and PRAKASA RAO, B. L. S. (1990). Characterization of the exponential distribution by the relevation transform, *Journal of Applied Probability*, **27**, 726–729.

LAWLESS, J. F. (1982). *Statistics Models & Methods for Lifetime Data*, John Wiley & Sons, New York.

LEADBETTER, M., LINDGREN, G., and ROOTZÉN, H. (1983). *Extremes and Related Properties of Random Sequences and Processes*, Springer-Verlag, New York.

LIN, G. D. (1987). On characterizations of distributions via moments of record values, *Probability Theory & Related Fields*, **74**, 479–483.

LIN, G. D. (1988a). Characterizations of distributions via relationships between two moments of order statistics, *Journal of Statistical Planning and Inference*, **19**, 73–80.

LIN, G. D. (1988b). Characterizations of uniform distributions and of exponential distributions, *Sankhyā, Series A*, **50**, 64–69.

LIN, G. D. and HUANG, J. S. (1987). A note on the sequence of expectations of maxima and of record values, *Sankhyā, Series A*, **49**, 272–273.

LOMAX, K. S. (1954). Business failures. Another example of the analysis of failure data, *Journal of the American Statistical Association*, **49**, 847–852.

MALOV, S. V. and NEVZOROV, V. B. (1997). Characterizations using ranks and order statistics, In *Advances in the Theory and Practice of Statistics — A Volume in Honor of Samuel Kotz* (Eds., N. L. Johnson and N. Balakrishnan), pp. 479–489, John Wiley & Sons, New York.

MANN, N. R. (1969). Optimum estimators for linear functions of location and scale parameters, *Annals of Mathematical Statistics*, **40**, 2149–2155.

MOHAN, N. R. and NAYAK, S. S. (1982). A characterization based on the equidistribution of the first two spacings of record values, *Zeitschrift fuer Wahrscheinlichkeitstheorie und Verwandte Gebiete*, **60**, 219–221.

MORENO REBOLLO, J. L., BARRANCO CHAMORRO, I., LÓPEZ BLÁZQUEZ, F., and GÓMEZ GÓMEZ, T. (1996). On the estimation of the unknown sample size from the number of records, *Statistics & Probability Letters*, **31**, 7–12.

MORIGUTI, S. (1953). A modification of Schwarz's inequality with applications to distributions, *Annals of Mathematical Statistics*, **24**, 107–113.

NAGARAJA, H. N. (1977). On a characterization based on record values, *Australian Journal of Statistics*, **19**, 70–73.

NAGARAJA, H. N. (1978). On the expected values of record values, *Australian Journal of Statistics*, **20**, 176–182.

NAGARAJA, H. N. (1982). Record values and extreme value distributions, *Journal of Applied Probability*, **19**, 233–239.

NAGARAJA, H. N. (1984). Asymptotic linear prediction of extreme order statistics, *Annals of the Institute of Statistical Mathematics*, **36**, 289–299.

NAGARAJA, H. N. (1988a). Record values and related statistics – A review, *Communications in Statistics—Theory and Methods*, **17**, 2223–2238.

NAGARAJA, H. N. (1988b). Some characterizations of continuous distributions based on regressions of adjacent order statistics and record values, *Sankhyā, Series A*, **50**, 70–73.

NAGARAJA, H. N. (1994a). Record occurrence in the presence of a linear trend, *Technical Report No. 546*, Department of Statistics, The Ohio State University, September 1994.

NAGARAJA, H. N. (1994b). Record statistics from point process models, In *Extreme Value Theory and Applications* (Eds., J. Galambos, J. Lechner and E. Simiu), pp. 355–370, Kluwer, Dordrecht, The Netherlands.

NAGARAJA, H. N. and NEVZOROV, V. B. (1996). Correlations between functions of records can be negative, *Statistics & Probability Letters*, **29**, 95–100.

NAGARAJA, H. N. and NEVZOROV, V. B. (1997). On characterizations based on record values and order statistics, *Journal of Statistical Planning and Inference*, **63**, 271–284.

NAGARAJA, H. N., SEN, P., and SRIVASTAVA, R. C. (1989). Some characterizations of geometric tail distributions based on record values, *Statistische Hefte*, **30**, 147–155.

NAYAK, S. S. (1981). Characterizations based on record values, *Journal of the Indian Statistical Association*, **19**, 123–127.

NAYAK, S. S. (1985). Record values for and partial maxima of a dependent sequence, *Journal of the Indian Statistical Association*, **23**, 109–125.

NAYAK, S. S. and INGINSHETTY, S. (1995). On record values, *Journal of the Indian Society for Probability and Statistics*, **2**, 43–55.

NEUTS, M. (1967). Waiting times between record observations, *Journal of Applied Probability*, **4**, 206–208.

NEVZOROV, V. B. (1985). Record and interrecord times for sequences of nonidentically distributed random variables, Notes of *Sci. Semin. LOMI*, **142**, 109–118. (English translation in *Journal of Soviet Mathematics*, **36**, 1987, 510–516).

NEVZOROV, V. B. (1986). Two characterizations using records, In *Stability Problems for Stochastic Models* (Eds., V. V. Kalashnikov, B. Penkov and V. M. Zolotarev), Lecture Notes in Mathematics **1233**, pp. 79–85, Springer-Verlag, Berlin.

NEVZOROV, V. B. (1987). Records, *Theory of Probability and Applications*, **32**(2), 201–228 (English translation).

NEVZOROV, V. B. (1990). Records for nonidentically distributed random variables, In *Proceedings of the 5th Vilnius Conference*, Volume 2 (Eds., B. Grigelionis, Yu. V. Prohorov, V. V. Sazano and V. Statulevicius), pp. 227–233, VSP, Mokslas.

NEVZOROV, V. B. (1992). A characterization of exponential distributions by correlations between records, *Mathematical Methods of Statistics*, **1**, 49–54.

NEVZOROV, V. B. (1995). Asymptotic distributions of records in non-stationary schemes, *Journal of Statistical Planning and Inference*, **45**, 261–273.

NEVZOROV, V. B. and BALAKRISHNAN, N. (1998). A record of records, In *Handbook of Statistics — 16: Order Statistics: Theory and Methods* (Eds., N. Balakrishnan and C. R. Rao), pp. 515–570, North-Holland, Amsterdam, The Netherlands.

NEVZOROVA, L. N., NEVZOROV, V. B., and BALAKRISHNAN, N. (1997). Characterizations of distributions by extremes and records in Archimedean copula processes, In *Advances in the Theory and Practice of Statistics — A Volume in Honor of Samuel Kotz* (Eds., N. L. Johnson and N. Balakrishnan), pp. 469–478, John Wiley & Sons, New York.

PFEIFER, D. (1981). Asymptotic expansions for the mean and variance of logarithmic inter-record times, *Methods in Operations Research*, **39**, 113–121.

PFEIFER, D. (1982). Characterizations of exponential distributions by independent non-stationary record increments, *Journal of Applied Probability*, **19**, 127–135. (Correction **19**, p. 906.)

PFEIFER, D. (1984). Limit laws for inter-record times from nonhomogeneous record values, *Journal of Organizational Behavior and Statistics*, **1**, 69–74.

PFEIFER, D. (1989). Extremal processes, secretary problems and the $1/e$ law, *Journal of Applied Probability*, **27**, 823–834.

PICKANDS, J. (1971). The two dimensional Poisson process and extremal processes, *Journal of Applied Probability*, **8**, 745–756.

PIRAZZOLI, P. (1982). Maree estreme a Venezia (periodo 1872–1981), *Acqua Aria*, **10**, 1023–1039.

PRAKASA RAO, B. L. S. (1992). *Identifiability in Stochastic Models – Characterization of Probability Distributions*, Academic Press, Boston.

RAMACHANDRAN, B. and LAU, K-S. (1991). *Functional Equations in Probability Theory*, Academic Press, Boston.

RAO, C. R. and SHANBHAG, D. N. (1986). Recent results on characterization of probability distributions: A unified approach through extensions of Deny's theorem, *Advances in Applied Probability*, **18**, 660–678.

RAO, C. R. and SHANBHAG, D. N. (1994). *Choquet-Deny Type Functional Equations with Applications to Stochastic Models*, John Wiley & Sons, Chichester, England.

RAO, C. R. and SHANBHAG, D. N. (1998). Recent approaches to characterizations based on order statistics and record values, In *Handbook of Statistics – 16: Order Statistics: Theory and Methods* (Eds., N. Balakrishnan and C. R. Rao), pp. 231–256, North-Holland, Amsterdam, The Netherlands.

RAQAB, M. Z. (1997). Bounds based on greatest convex minorants for moments of record values, *Statistics & Probability Letters*, **36**, 35–41.

RAQAB, M. Z. and AMIN, W. A. (1997). A note on reliability properties of k-record statistics, *Metrika*, **46**, 245–251.

RÉNYI, A. (1962). Theorie des elements saillants d'une suite d'observations, *Colloquium on Combinatorial Methods in Probability Theory*, Nathematisk Institut, Aarhus University, Aarhus, Denmark. English translation in *Selected Papers of Alfred Rényi*, Volume 2, Academic Press, New York.

RESNICK, S. I. (1973a). Limit laws for record values, *Stochastic Processes and Their Applications*, **1**, 67–82.

RESNICK, S. I. (1973b). Record values and maxima, *Annals of Probability*, **4**, 650–662.

RESNICK, S. I. (1973c). Extremal processes and record value times, *Journal of Applied Probability*, **10**, 864–868.

RESNICK, S. I. (1974). Inverses of extremal processes, *Advances in Applied Probability*, **6**, 392–406.

RESNICK, S. (1987). *Extreme Values, Regular Variation, and Point Processes*, Springer-Verlag, New York.

ROBERTS, E. M. (1979). Review of statistics of extreme values with application to air quality data. Part II. Applications, *Journal of Air Pollution Control Association*, **29**, 733–740.

ROGERS, L. C. G. (1989). Ignatov's Theorem: An abbreviation of the proof of Engelen, Tommassen and Vervaat, *Advances in Applied Probability*, **21**, 933–934.

ROHATGI, V. K. and SZÉKELY, G. J. (1992). On the background of some correlation inequalities, *Journal of Statistical Computation and Simulation*, **40**, 220–226.

ROSS, S. (1983). *Stochastic Processes*, John Wiley & Sons, New York.

ROY, D. (1990). Characterization through record values, *Journal of the Indian Statistical Association*, **28**, 99–103.

SAMANIEGO, F. J. and KAISER, L. D. (1978). Estimating value in a uniform auction, *Naval Research Logistics Quarterly*, **25**, 621–632.

SAMANIEGO, F. J. and WHITAKER, L. R. (1986). On estimating population characteristics from record-breaking observations. I. Parametric results, *Naval Research Logistics Quarterly*, **33**, 531–543.

SAMANIEGO, F. J. and WHITAKER, L. R. (1988). On estimating population characteristics from record-breaking observations II. Nonparametric results, *Naval Research Logistics*, **35**, 221–236.

SAMUELS, S. M. (1991). Secretary problems, In *Handbook of Sequential Analysis* (Eds., B. K. Ghosh and P. K. Sen), pp. 381–406, Marcel Dekker, New York.

SAMUELS, S. M. (1992). An all-at-once proof of Ignatov's Theorem, *Contemporary Mathematics*, **125**, 231–237.

SARMANOV, O. V. (1958). Maximum correlation coefficient, *Dokl. Akad. Nauk SSSR*, **120**, 715–718 and **121**, 52–55.

SCHNEIDER, B. E. (1978). Trigamma function. Algorithm AS121, *Applied Statistics*, **27**, 97–99.

SHANBHAG, D. N. (1977). An extension of the Rao-Rubin characterization of the Poisson distribution, *Journal of Applied Probability*, **14**, 640–646.

SHOHAT, J. A. and TAMARKIN, J. D. (1943). *The Problem of Moments*, Mathematical Surveys No. 1, American Mathematical Society, New York.

SHORROCK, R. W. (1972a). A limit theorem for inter-record times, *Journal of Applied Probability*, **9**, 219–223. (Correction **9**, p. 877.)

SHORROCK, R. W. (1972b). On record values and record times, *Journal of Applied Probability*, **9**, 316–326.

SHORROCK, R. W. (1974). On discrete time extremal processes, *Advances in Applied Probability*, **6**, 580–592.

SIBUYA, M. (1960). Bivariate extreme statistics, *Annals of the Institute of Statistical Mathematics*, **11**, 195–210.

SIBUYA, M. and NISHIMURA, K. (1997). Prediction of record-breakings, *Statistica Sinica*, **7**, 893–906.

SMITH, R. L. (1986). Extreme value theory based on the r largest annual events, *Journal of Hydrology*, **86**, 27–43.

SMITH, R. L. (1988). Forecasting records by maximum likelihood, *Journal of the American Statistical Association*, **83**, 331–338.

SMITH, R. L. and MILLER, J. E. (1986). A non-Gaussian state space model and application to the prediction of records, *Journal of the Royal Statistical Society, Series B*, **48**, 79–88.

SRIVASTAVA, R. C. (1979). Two characterizations of the geometric distribution by record values, *Sankhyā, Series B*, **40**, 276–278.

SRIVASTAVA, R. C. (1981a). On some characterizations of the geometric distribution, In *Statistical Distributions in Scientific Work*, Vol. 4 (Eds., C. Taillie, G. P. Patil and B. A. Baldessari), pp. 349–356, Reidel, Dordrecht, The Netherlands.

SRIVASTAVA, R. C. (1981b). Some characterizations of the exponential distribution based on record values, In *Statistical Distributions in Scientific Work*, Vol. 4 (Eds., C. Taillie, G. P. Patil and B. A. Baldessari), pp. 411–416, Reidel, Dordrecht, The Netherlands.

STAM, A. J. (1985). Independent Poisson processes generated by record values and inter-record times, *Stochastic Processes and Their Applications*, **19**, 315–325.

STEPANOV, A. V. (1990). Characterizations of a geometric class of distributions, *Theory of Probability and Mathematical Statistics*, **41**, 133–136 (English translation).

STEPANOV, A. V. (1992). Limit theorems for weak records, *Theory of Probability and Its Applications*, **37**, 570–574 (English translation).

STEPANOV, A. V. (1993). A characterization theorem for weak records, *Theory of Probability and Its Appplications*, **38**, 762–764 (English translation).

STRAWDERMAN, W. and HOLMES, P. T. (1970). On the law of iterated logarithm for inter-record times, *Journal of Applied Probability*, **7**, 432–439.

STUART, A. (1954). Asymptotic relative efficiencies of distribution-free tests of randomness against normal alternatives, *Journal of the American Statistical Association*, **49**, 147–157.

STUART, A. (1956). The efficiencies of tests of randomness against normal regression, *Journal of the American Statistical Association*, **51**, 285–287.

STUART, A. (1957). The efficiency of the records test for trend in normal regression, *Journal of the Royal Statistical Society, Series B*, **19**, 149–153.

SUGIURA, N. (1962). On the orthogonal inverse expansion with an application to the moments of order statistics, *Osaka Mathematical Journal*, **14**, 253–263.

SULTAN, K. S. and BALAKRISHNAN, N. (1997a). Higher order moments of record values from Rayleigh distribution and Edgeworth approximate inference, *Submitted for publication*.

SULTAN, K. S. and BALAKRISHNAN, N. (1997b). Higher order moments of record values from Weibull distribution and Edgeworth approximate inference, *Submitted for publication*.

SULTAN, K. S. and BALAKRISHNAN, N. (1997c). Higher order moments of record values from Gumbel distribution with applications to inference, *Submitted for publication*.

SZÉKELY, G. J. and MÓRI, T. F. (1985). An extremal property of rectangular distributions, *Statistics & Probability Letters*, **3**, 107–109.

TAILLIE, C. (1981). A note on Srivastava's characterization of the exponential distribution based on record values, In *Statistical Distributions in Scientific Work*, Vol. 4 (C. Taillie, G. P. Patil, B. A. Baldessari, Eds.), pp. 417–418, Reidel, Dordrecht, The Netherlands.

TATA, M. N. (1969). On outstanding values in a sequence of random variables, *Zeitschrift fuer Wahrscheinlichkeitstheorie und Verwandte Gebiete*, **12**, 9–20.

TRYFOS, P. and BLACKMORE, R. (1985). Forecasting records, *Journal of the American Statistical Association*, **80**, 46–50.

VAN ZWET, W. R. (1964). *Convex Transformations of Random Variables*, Mathematical Centre Tracts 7, Mathematisch Centrum, Amsterdam, The Netherlands.

VERVAAT, W. (1973). Limit theorems for records from discrete distributions, *Stochastic Processes and Their Applications*, **1**, 317–334.

WEISSMAN, I. (1978). Estimation of parameters and large quantiles based on the k largest observations, *Journal of the American Statistical Association*, **73**, 812–815.

WEISSMAN, I. (1995). Records from a power model of independent observations, *Journal of Applied Probability*, **32**, 982–990.

WESTCOTT, M. (1977). The random record model, *Proceedings of the Royal Society of London, Series A*, **356**, 529–547.

WESTCOTT, M. (1979). On the tail behavior of record time distributions in a random record process, *Annals of Probability*, **7**, 868–873.

WILLIAMS, D. (1973). On Rényi's 'record' problem and Engel's series, *Bulletin of the London Mathematical Society*, **5**, 235–237.

WITTE, H-J. (1988). Some characterizations of distributions based on the integrated Cauchy functional equation, *Sankhyā, Series A*, **50**, 59–63.

YANG, M. C. K. (1975). On the distribution of the inter-record times in an increasing population, *Journal of Applied Probability*, **12**, 148–154.

YAKYMIV, A. L. (1986). Asymptotic properties of the times the states change in a random record process, *Theory of Probability and Its Applications*, **31**, No. 3, 508–512.

Author Index

Abramowitz, M., 59, 66–68, 237, 279

Adke, S. R., 209, 210, 220, 279

Ahsanullah, M., 2, 27, 52–54, 56, 57, 60–62, 66, 67, 90, 91, 102, 103, 105–108, 112, 117, 118, 138, 150, 153, 157, 166, 168, 176, 177, 219, 279, 280–282

Aitchison, J., 161, 280

Amin, W. A., 117, 296

Arnold, B. C., 3, 16, 51, 61, 64, 76, 80, 94, 100, 109, 114, 117, 128, 154, 180, 201, 221, 280

Azlarov, T., 93, 103, 218

Balabekyan, V. A., 217, 281

Balakrishnan, N., 2, 3, 16, 51–57, 60–64, 66, 67, 72, 74, 76, 80, 82, 90, 91, 94, 100, 109, 112, 114, 118, 127–129, 131–133, 135, 136, 138, 146, 151, 154, 157, 162, 164–166, 168–170, 176, 177, 180, 201, 213, 221, 222, 280–282, 285, 286, 290, 292, 294–296, 299

Balasurbramanian, K., 112, 118, 222, 281, 282

Baldessari, B. A., 298, 299

Ballerini, R., 184, 188, 191–193, 195–197, 206, 209, 213, 215, 216, 221, 283

Barlow, R. E., 47, 80, 96, 97, 283

Barranco Chamorro, I., 181, 292

Bartfai, P., 290

Barton, D. E., 2

Basak, P., 108, 118, 283

Beg, M. I., 95, 291

Bendre, S. M., 74, 282

Bernardo, J. M., 59, 283

Biondini, R., 209, 210, 283

Blackmore, R., 300

Browne, S., 234, 252–254, 263, 283

Bruss, F., 234, 249, 251–254, 256–258, 264, 283, 284

Buishand, T., 235, 284

Bunge, J., 224, 228, 231–234, 236, 237, 239, 242, 247–249, 251–254, 259, 262, 263, 283, 284

Burden, R. L., 149, 284

Carlin, B. P., 198, 284

Cayley, A., 236, 284

Chan, P. S., 55–57, 60, 63, 64, 66, 67, 131, 132, 138, 146, 147, 149, 157, 160, 164–167, 176, 177, 179, 281, 282, 284

Chandler, K. N., 1, 2, 7, 170, 284

Cohen, A. C., 51, 127, 133, 169, 170, 282

Cowan, R., 256, 264, 284

Cramér, H., 93, 284

Daley, D. J., 41, 285
Dallas, A. C., 102, 285
David, F. N., 2
David, H. A., 51, 80, 133, 271, 285
Davis, H. T., 59, 285
de Haan, L., 2, 18, 19, 197, 217, 285
Deheuvels, P., 2, 43, 45, 102, 112, 114, 236, 237, 240, 285, 286
Deken, J. G., 209, 286
Deny, J., 104, 286
Diersen, J., 173, 286
Doganaksoy, N., 151, 286
Dunsmore, I. R., 161–164, 280, 286
Dwass, M., 2, 41, 286
Dziubdziela, W., 43, 286

Embrechts, P., 234, 243–246, 262, 286
Engelen, R., 43, 286

Faires, J. F., 149, 284
Feller, W., 19, 48, 213, 226, 246, 286, 287
Ferguson, T. S., 100, 254, 287
Feuerverger, A., 198, 287
Fisher, R. A., 146, 287
Foster, F. G., 170, 171, 173, 179, 180, 287

Gómez Gómez, T., 181, 292
Galambos, J., 2, 44, 93, 103, 108, 212, 282, 287, 293
Gaver, D., 235, 239, 242, 248, 249, 254, 260, 262, 287
Gelfand, A. E., 198, 294
Ghosh, B. K., 297
Glick, N., 2, 26, 287
Gnedenko, B. V., 16, 288

Goel, P. K., 288
Goldberger, A. S., 150, 288
Goldie, C. M., 2, 5, 41–43, 269, 270, 277, 288
Govindarajulu, Z., 64, 288
Graybill, F. A., 128, 175, 288
Grigelionis, B., 294
Grudzień, Z., 82, 87, 97, 102, 117, 288
Gulati, S., 170, 288, 289
Gupta, R. C., 47, 96, 102, 103, 114, 116, 289
Guthrie, G. L., 43, 209, 289

Hüsler, J., 291
Haiman, G., 222, 289
Hall, P., 198, 287
Hamming, R. W., 25, 289
Hofmann, G., 184, 188, 214, 215, 224, 233, 234, 243, 259, 261, 289
Hoinkes, L. A., 179, 289
Holland, B., 112, 219, 280
Holmes, P. T., 29, 43, 209, 289, 299
Houchens, R. L., 22, 32, 33, 63, 271, 274, 275, 290
Huang, J. S., 96, 292
Huang, W-J., 106, 112, 115, 117, 119, 290

Ignatov, Z., 41, 43, 290
Iiyama, Y., 173, 290
Inginshetty, S., 19, 212, 284

Jacobs, P., 242, 248, 249, 287
Johnson, N. L., 146, 162, 280, 290, 292, 295

Kagan, A. M., 93, 290
Kaiser, L. D., 131, 297
Kalashnikov, V. V., 294

Kamps, U., 46, 89, 115, 202, 218, 290, 291
Karlin, S., 247, 251, 252, 291
Kinoshita, K., 267, 291
Kirmani, S. N. U. A., 47, 95, 107, 108, 118, 280, 289, 291
Klüppelberg, C., 49, 278, 291
Klein, J. P., 288
Knopp, K., 189, 291
Koch, G., 285
Kochar, S. C., 47, 48, 98, 99, 291
Konecny, F., 288
Kopociński, B., 43, 286
Korwar, R. M., 114, 119, 291
Kotlarski, I. I., 97, 291
Kotz, S., 93, 103, 108, 146, 162, 287, 290

Lau, K-S., 102–106, 291, 295
Lawless, J. F., 146, 292
Leadbetter, M., 235, 292
Lechner, J., 282, 293
Li, S-H., 106, 112, 117, 119, 290
Lin, G. D., 95, 96, 116–118, 292
Lindgren, G., 235, 292
Linnik, Yu. V., 93, 290
Lomax, K. S., 61, 292
López Blázquez, F., 181, 292

Móri, T. F., 102, 115, 299
Mallows, C. L., 2
Malov, S. V., 292
Mann, N. R., 143, 153, 292
Miller, J. E., 170, 298
Mogyoródi, J., 288
Mohan, N. R., 112, 118, 292
Moreno Rebollo, J. L., 181, 292
Moriguti, S., 82, 293
Morrison, D. F., 289

Nagaraja, H. N., 2, 16, 29, 30, 47, 51, 69, 70, 72, 74, 77, 80, 94, 97, 99–102, 109–114, 116, 118, 119, 128, 154, 176, 180, 195, 201, 206, 219, 224, 228, 231, 232–234, 236, 237, 239, 242, 243, 247–249, 251, 259, 261–263, 271, 280, 284, 285, 289, 293
Nayak, S. S., 10, 102, 112, 118, 212, 221, 292, 294
Neuts, M., 2, 26, 294
Nevzorov, V. B., 2, 45, 47, 101, 110, 111, 116–118, 184, 185, 187, 188, 190–193, 204, 205, 213, 215–217, 219, 221, 222, 281, 282, 289, 292–295
Nevzorova, L. N., 213, 221, 295
Nishimura, K., 173, 220, 290, 298

Omey, E., 234, 243, 245, 246, 262, 286

Pickands, J., 41, 43, 234, 236, 295
Proschan, F., 47, 80, 96, 97, 283
Prakasa Rao, B. L. S., 94, 106, 219, 295
Pfeifer, D., 102, 185, 198, 199, 201, 207, 218, 219, 240, 263, 295
Padgett, W. J., 170, 179, 288, 289
Panchapakesan, S., 222, 282
Puri, M. L., 283
Penkov, B., 294
Prohorov, Yu. V. 294
Pirazzoli, P., 295
Patil, G. P., 298, 299

Ramachandran, B., 103–105, 295
Rao, C. R., 93, 102–106, 129, 135, 282, 285, 290, 291, 294–296
Raqab, M. Z., 82, 85, 89, 117, 296

Reiss, R-D., 291
Rényi, A., 24, 26, 28, 42, 296
Resnick, S. I., 2, 5, 12, 15–19, 41,
  42, 46, 48, 49, 184, 188,
  191–193, 195–197, 209, 215,
  216, 221, 245, 267, 269,
  270, 277, 283, 285, 288,
  291, 296
Roberts, E. M., 165, 168, 296
Rogers, B., 234, 249, 251–253, 284
Rogers, L. C. G., 42, 43, 288, 296
Rohatagi, V. K., 101, 296
Rootzén, H., 235, 292
Ross, S., 97, 297
Roy, D., 103, 297

Samaniego, F. J., 122, 131, 169,
  170, 297
Samuels, S. M., 43, 254, 297
Sarmanov, O. V., 101, 297
Sazano, V. V., 294
Schneider, B. E., 59, 297
Schwere, P., 49, 278, 291
Sen, P., 112, 113, 119, 219, 293
Sen, P. K., 289, 297
Seneta, E., 44, 287
Shanbhag, D. N., 102–104, 106,
  112, 295–297
Shohat, J. A., 95, 297
Shorrock, R. W., 2, 28, 41, 42,
  199, 218, 297
Sibuya, M., 173, 220, 270, 290,
  298
Siddiqui, M. M., 209, 210, 283
Simiu, E., 282, 293
Smith, R. L., 165, 170, 198, 298
Spizzichino, F., 285
Srivastava, R. C., 102, 112, 113,
  119, 219, 293, 298
Stam, A. J., 43, 298
Statulevicius, V., 294

Stegun, I. A., 59, 66–68, 237, 279
Stepanov, A. V., 34, 37, 38, 112,
  119, 298
Strawderman, W., 29, 299
Stuart, A., 2 170, 171, 173, 179,
  180, 287, 299
Su, J-C., 115, 290
Sugiura, N., 74, 299
Sultan, K. S., 136, 146, 168, 176,
  299
Székely, G. J., 101, 102, 115, 296,
  299
Szynal, D., 82, 87, 97, 102, 117,
  288

Taillie, C., 102, 198, 299
Tamarkin, J. D., 95, 297
Tata, M. M., 2, 14, 26, 45, 46, 48,
  101, 102, 105, 207, 239,
  240, 299
Taylor, H. M., 247, 251, 252, 291
Teichroew, D., 173, 187
Tomassen, P., 43, 286
Tomko, J., 290
Trenkler, G., 173, 286
Tryfos, P., 300

van Zwet, W. R., 77, 80, 300
Vere-Jone, D., 41, 285
Verkade, E., 197, 217, 285
Vervaat, W., 19, 34, 35, 37, 43,
  286, 300
Villaseñor, J. A., 109, 117, 280
Volodin, N. A., 94, 103, 281

Weissman, I., 217, 300
Wertz, W., 288
Westcott, M., 224, 232, 234, 243–
  245, 261, 300
Whitaker, L. R., 122, 169, 170,
  297
Williams, D., 44, 240, 300

Witte, H-J., 102, 103, 105, 106, 300

Yakymiv, A. L., 234, 246, 300
Yang, M. C. K., 4, 183, 184, 187, 191, 300

Zabczyk, J., 256, 264, 284
Zolotarev, V. M., 294

Subject Index

Applications
    air quality control, 165, 167, 168
    best choice problem, *see* secretary problems
    climatology, 94, 50, 165, 235, 278
    rainfall study, 173, 174, 180
    sea-level study, 165, 166
    secretary problems, 224, 234, 236, 254, 255, 263, 264
    shock models, 199, 235
    system strength, 235

Bivariate records, 266
    first kind, 266, 268
    fourth kind, 266, 269
    second kind, 266
    third kind, 266, 270, 271
    with independent coordinates, 267
Borel-Cantelli lemma, 26, 186, 189, 196

Cauchy functional equation, 103, 109
Cauchy-Schwarz inequality, 69, 70, 72, 75, 76, 82, 83, 85, 87
Characterizations, 93, 202
    based on lower records, 107, 109
    based on moments, 95, 116
    based on regression, 96, 99
    based on reliability properties, 97
    Cauchy functional equation, 103, 109
    exponential, 101, 105–107, 115, 117, 118, 204, 207, 219
    extreme value, 109
    geometric, 106, 118
    geometric-tail distribution, 111–114, 118, 119, 208, 219
    Gumbel, 109, 206
    in F^α models, 203, 204
    integrated Cauchy functional equation, 102, 104–106
    of F^α models, 202, 219
    Pareto, 100
    power function, 100
    uniform, 115
    Weibull, 118
Conditional confidence interval, 167
Correlation, 33, 101, 115
Covariance, 33, 40, 47, 65, 67, 68

Dependent record model, 208
    Archimedean copula, 213, 211
    exchangeable sequence, 211
    Markov sequence, 209, 210
    stationary Gaussian sequence, 212
Digamma function, 59
Distributions
    bivariate normal, 272
    bivariate Pareto, 273

DFR, 48, 80, 97–99, 218

DMRL, 98

exp-gamma, 163

exponential, 9–11, 33, 48, 49, 52, 69, 88, 100, 101, 105–108, 115, 117, 118, 122, 123, 128, 134, 143, 145, 150, 153, 156, 161–164, 204, 207, 219, 242, 250, 268

extreme value, 13, 22, 32, 34, 109, 136, 148, 152, 154, 160, 167, 179

Farlie-Gumbel-Morgenstern, 274

Fréchet, 15, 154

gamma, 9-11, 238

generalized extreme value, 50, 60, 178

generalized Pareto, 90, 91, 179

geometric, 39, 46, 106, 111, 118, 187, 191, 232

geometric-tail, 111–114, 118, 119, 208, 219

Gumbel, 16, 56, 109, 126, 168, 195, 201, 206

IFR, 47, 80, 97, 99, 117, 218

IFRA, 47

location-scale, 125, 133, 146, 159

log-gamma, 136

logistic, 13, 65, 68, 126, 138, 141, 158, 159, 166

lognormal, 16

Lomax, 61

NBU, 47, 97, 106

NBUE, 96

negative binomial, 40, 228, 233, 247

negative lognormal, 16

normal, 16, 32, 34, 63–65, 73, 78, 125, 138, 139, 157, 158, 165

NWU, 97, 106

NWUE, 96

Pareto, 14, 21, 22, 31, 33, 49, 88, 100, 108, 178, 273

Poisson, 233, 248

power function, 21, 31, 100, 108, 178

Rayleigh, 55, 131, 137, 164

uniform, 11, 23, 27, 33, 47, 49, 50, 115, 123, 124, 129, 136, 144, 152, 175, 276

Weibull, 13, 15, 20, 30, 54, 69, 88, 108, 118, 132, 154–165, 167

Domain of attraction

maximal, 17

record, 17, 18

Duality theorem, 17

Efficiency

determinant, 135, 176

relative, 130

trace, 134, 176

Equivariant estimators, 147

Estimation, 121

best linear invariant, 143

best linear unbiased, 127, 133

interval, 145

maximum likelihood, 122

minimum variance unbiased, 122

F^α record model, 184, 187, 251, 259–261, 263

asymptotic results, 190

characterizations in, 203, 204

characterizations of, 202, 219

dependent, 213

finite-sample results, 188

inter-record times, 191

model 1, 188, 194, 216
model 2, 188, 193, 194
record times, 189
record values, 193
Failure rate function, 98

Generalized order statistics, 46, 115, 202
Greatest convex minorant, 83, 85, 87

Hazard function, 9, 12, 35, 98, 100

Initial ranks, 42
Integrated Cauchy functional equation, 102–106, 112

k-record, 81, 86, 89, 92, 97, 119
    lower, 118
    times, 81
    Type 1, 43, 97
    Type 2, 43, 50, 81, 97, 101, 114, 117, 276
    values, 41, 43, 44
Lau-Rao theorem, 104
Law of the iterated logarithm, 19, 25, 38
Legendre polynomials, 75, 76
Linear draft record model, 184, 194
    characterizations, 206
    Gumbel, 195, 206
    number of records, 196

Markov chain, 28, 29, 45
Martingale, 45, 189, 190, 215
Moments, 29, 51, 52, 116
    approximations, 51, 68, 78
    bounds, 51, 68, 78, 86, 89
    c-ordering, 79
    exponential, 52

extreme value, 32
generalized extreme value, 59, 60
generalized Pareto, 90, 91
Gumbel, 56
logistic, 65, 68
Lomax, 61
normal, 32, 63, 65
Pareto, 31
power function, 31
recurrence relations, 51, 55, 56, 58, 61, 88, 90, 91
s-ordering, 79
Weibull, 30, 54

Order statistics, 64, 100, 104, 109, 114, 115, 154
Orthonormal system, 74, 75

Pfeifer record model, 198, 218
    characterizations, 207, 208, 219
    inter-record times, 199, 201
    record values, 199, 201
Point process record model, 223, 233, 234
    basic model, 234
    inter-record times, 236, 237, 240–243, 246, 250
    number of records, 247, 248, 259
    paced by
        birth process, 246, 250, 252, 258
        Mittag-Leffler renewal process, 253, 254
        mixed Poisson process, 253
        Pascal process, 253
        point process, 41
        Poisson process, 236
        renewal process, 243

Yule process, 248, 262, 263
Poisson process, 11, 41, 236, 242,
    248–251, 256, 257, 264
    nonhomogeneous, 241, 253, 256,
        257, 264
Prediction, 150
    asymptotic linear, 154
    best linear invariant, 153
    best linear unbiased, 150
    tolerance region, 161
Prediction intervals, 156
    based on BLUE's, 156
    Bayesian, 162
    conditional, 159, 167

Random power record model, 214
Random record model, 223, 224
    dependence structure, 229
    inter-record times, 230, 231
    number of records, 229, 231,
        259
    record values, 224, 229
Record (upper)
    arrival process, 249, 250, 251
    concomitants of, 271
    counting process, 41
    coverage, 275
    δ-exceedance, 222
    discrete
        regular, 34
        weak, 34, 37, 258
    in partially ordered sets, 277
    increment, 8, 22
    indicators, 23, 185, 213, 255,
        268, 269
    inter-, 8, 27, 169, 185, 199,
        201, 211, 239, 240, 243,
        244
    k-, see k-records
    lower, 47, 58, 60, 65, 107, 109,
        126, 169, 201, 274

models
    classical, 9, 226, 239, 243
    dependent, 208, 213
    F^α, 184, 187, 188, 251, 259–
        261, 263
    linear drift, 184, 194
    Pfeifer, 198, 218
    point process, 223, 233, 234
    random, 223, 224
    random power, 214
    Yang, 184
    moments, see moments
    multivariate, 265
        types, 266
    number of, 8, 24, 181, 196,
        217, 220, 268, 269
    range, 274, 275
    spacings, 96, 161–163
    times, 8, 22, 24, 27, 210, 240
    values, 8, 28, 114, 199, 201,
        210–212, 228, 229, 274
Regularly varying function, 18,
    216, 245
Riemann zeta function, 27, 66, 67,
    248

Secretary problem, 224, 234, 254,
    255, 263, 264
Shanbhag's lemma, 104
Shorrock process, 42, 43
Stirling number of first kind, 25,
    232, 247
Strong law of large numbers, 19,
    25, 38
Subordinated distribution, 246, 262

Tests of hypotheses, 145
    conditional, 146–149
    distribution-free, 170, 171
    unconditional, 145, 146

WILEY SERIES IN PROBABILITY AND STATISTICS
ESTABLISHED BY WALTER A. SHEWHART AND SAMUEL S. WILKS

Editors
Vic Barnett, Ralph A. Bradley, Noel A. C. Cressie, Nicholas I. Fisher,
Iain M. Johnstone, J. B. Kadane, David G. Kendall, David W. Scott,
Bernard W. Silverman, Adrian F. M. Smith, Jozef L. Teugels;
J. Stuart Hunter, Emeritus

Probability and Statistics Section

*ANDERSON · The Statistical Analysis of Time Series
ARNOLD, BALAKRISHNAN, and NAGARAJA · A First Course in Order Statistics
ARNOLD, BALAKRISHNAN, and NAGARAJA · Records
BACCELLI, COHEN, OLSDER, and QUADRAT · Synchronization and Linearity:
   An Algebra for Discrete Event Systems
BASILEVSKY · Statistical Factor Analysis and Related Methods: Theory and
   Applications
BERNARDO and SMITH · Bayesian Statistical Concepts and Theory
BILLINGSLEY · Convergence of Probability Measures
BOROVKOV · Asymptotic Methods in Queuing Theory
BRANDT, FRANKEN, and LISEK · Stationary Stochastic Models
CAINES · Linear Stochastic Systems
CAIROLI and DALANG · Sequential Stochastic Optimization
CONSTANTINE · Combinatorial Theory and Statistical Design
COVER and THOMAS · Elements of Information Theory
CSÖRGŐ and HORVÁTH · Weighted Approximations in Probability Statistics
CSÖRGŐ and HORVÁTH · Limit Theorems in Change Point Analysis
DETTE and STUDDEN · The Theory of Canonical Moments with Applications in
   Statistics, Probability, and Analysis
*DOOB · Stochastic Processes
DRYDEN and MARDIA · Statistical Analysis of Shape
DUPUIS and ELLIS · A Weak Convergence Approach to the Theory of Large Deviations
ETHIER and KURTZ · Markov Processes: Characterization and Convergence
FELLER · An Introduction to Probability Theory and Its Applications, Volume 1,
   Third Edition, Revised; Volume II, *Second Edition*
FULLER · Introduction to Statistical Time Series, *Second Edition*
FULLER · Measurement Error Models
GELFAND and SMITH · Bayesian Computation
GHOSH, MUKHOPADHYAY, and SEN · Sequential Estimation
GIFI · Nonlinear Multivariate Analysis
GUTTORP · Statistical Inference for Branching Processes
HALL · Introduction to the Theory of Coverage Processes
HAMPEL · Robust Statistics: The Approach Based on Influence Functions
HANNAN and DEISTLER · The Statistical Theory of Linear Systems
HUBER · Robust Statistics
IMAN and CONOVER · A Modern Approach to Statistics
JUREK and MASON · Operator-Limit Distributions in Probability Theory
KASS and VOS · Geometrical Foundations of Asymptotic Inference

*Now available in a lower priced paperback edition in the Wiley Classics Library.

Probability and Statistics (Continued)

KAUFMAN and ROUSSEEUW · Finding Groups in Data: An Introduction to Cluster Analysis

KELLY · Probability, Statistics, and Optimization

LINDVALL · Lectures on the Coupling Method

McFADDEN · Management of Data in Clinical Trials

MANTON, WOODBURY, and TOLLEY · Statistical Applications Using Fuzzy Sets

MARDIA and JUPP · Statistics of Directional Data, *Second Edition*

MORGENTHALER and TUKEY · Configural Polysampling: A Route to Practical Robustness

MUIRHEAD · Aspects of Multivariate Statistical Theory

OLIVER and SMITH · Influence Diagrams, Belief Nets and Decision Analysis

*PARZEN · Modern Probability Theory and Its Applications

PRESS · Bayesian Statistics: Principles, Models, and Applications

PUKELSHEIM · Optimal Experimental Design

RAO · Asymptotic Theory of Statistical Inference

RAO · Linear Statistical Inference and Its Applications, *Second Edition*

*RAO and SHANBHAG · Choquet-Deny Type Functional Equations with Applications to Stochastic Models

ROBERTSON, WRIGHT, and DYKSTRA · Order Restricted Statistical Inference

ROGERS and WILLIAMS · Diffusions, Markov Processes, and Martingales, Volume I: Foundations, *Second Edition;* Volume II: Îto Calculus

RUBINSTEIN and SHAPIRO · Discrete Event Systems: Sensitivity Analysis and Stochastic Optimization by the Score Function Method

RUZSA and SZEKELY · Algebraic Probability Theory

SCHEFFE · The Analysis of Variance

SEBER · Linear Regression Analysis

SEBER · Multivariate Observations

SEBER and WILD · Nonlinear Regression

SERFLING · Approximation Theorems of Mathematical Statistics

SHORACK and WELLNER · Empirical Processes with Applications to Statistics

SMALL and McLEISH · Hilbert Space Methods in Probability and Statistical Inference

STAPLETON · Linear Statistical Models

STAUDTE and SHEATHER · Robust Estimation and Testing

STOYANOV · Counterexamples in Probability

TANAKA · Time Series Analysis: Nonstationary and Noninvertible Distribution Theory

THOMPSON and SEBER · Adaptive Sampling

WELSH · Aspects of Statistical Inference

WHITTAKER · Graphical Models in Applied Multivariate Statistics

YANG · The Construction Theory of Denumerable Markov Processes

Applied Probability and Statistics Section

ABRAHAM and LEDOLTER · Statistical Methods for Forecasting

AGRESTI · Analysis of Ordinal Categorical Data

AGRESTI · Categorical Data Analysis

ANDERSON, AUQUIER, HAUCK, OAKES, VANDAELE, and WEISBERG · Statistical Methods for Comparative Studies

ARMITAGE and DAVID (editors) · Advances in Biometry

*ARTHANARI and DODGE · Mathematical Programming in Statistics

ASMUSSEN · Applied Probability and Queues

*BAILEY · The Elements of Stochastic Processes with Applications to the Natural Sciences

*Now available in a lower priced paperback edition in the Wiley Classics Library.

Applied Probability and Statistics (Continued)

BARNETT and LEWIS · Outliers in Statistical Data, *Third Edition*

BARTHOLOMEW, FORBES, and McLEAN · Statistical Techniques for Manpower Planning, *Second Edition*

BATES and WATTS · Nonlinear Regression Analysis and Its Applications

BECHHOFER, SANTNER, and GOLDSMAN · Design and Analysis of Experiments for Statistical Selection, Screening, and Multiple Comparisons

BELSLEY · Conditioning Diagnostics: Collinearity and Weak Data in Regression

BELSLEY, KUH, and WELSCH · Regression Diagnostics: Identifying Influential Data and Sources of Collinearity

BHAT · Elements of Applied Stochastic Processes, *Second Edition*

BHATTACHARYA and WAYMIRE · Stochastic Processes with Applications

BIRKES and DODGE · Alternative Methods of Regression

BLOOMFIELD · Fourier Analysis of Time Series: An Introduction

BOLLEN · Structural Equations with Latent Variables

BOULEAU · Numerical Methods for Stochastic Processes

BOX · Bayesian Inference in Statistical Analysis

BOX and DRAPER · Empirical Model-Building and Response Surfaces

BOX and DRAPER · Evolutionary Operation: A Statistical Method for Process Improvement

BUCKLEW · Large Deviation Techniques in Decision, Simulation, and Estimation

BUNKE and BUNKE · Nonlinear Regression, Functional Relations and Robust Methods: Statistical Methods of Model Building

CHATTERJEE and HADI · Sensitivity Analysis in Linear Regression

CHOW and LIU · Design and Analysis of Clinical Trials

CLARKE and DISNEY · Probability and Random Processes: A First Course with Applications, *Second Edition*

*COCHRAN and COX · Experimental Designs, *Second Edition*

CONOVER · Practical Nonparametric Statistics, *Second Edition*

CORNELL · Experiments with Mixtures, Designs, Models, and the Analysis of Mixture Data, *Second Edition*

*COX · Planning of Experiments

CRESSIE · Statistics for Spatial Data, *Revised Edition*

DANIEL · Applications of Statistics to Industrial Experimentation

DANIEL · Biostatistics: A Foundation for Analysis in the Health Sciences, *Sixth Edition*

DAVID · Order Statistics, *Second Edition*

*DEGROOT, FIENBERG, and KADANE · Statistics and the Law

DODGE · Alternative Methods of Regression

DOWDY and WEARDEN · Statistics for Research, *Second Edition*

DUNN and CLARK · Applied Statistics: Analysis of Variance and Regression, *Second Edition*

ELANDT-JOHNSON and JOHNSON · Survival Models and Data Analysis

EVANS, PEACOCK, and HASTINGS · Statistical Distributions, *Second Edition*

FLEISS · The Design and Analysis of Clinical Experiments

FLEISS · Statistical Methods for Rates and Proportions, *Second Edition*

FLEMING and HARRINGTON · Counting Processes and Survival Analysis

GALLANT · Nonlinear Statistical Models

GLASSERMAN and YAO · Monotone Structure in Discrete-Event Systems

GNANADESIKAN · Methods for Statistical Data Analysis of Multivariate Observations, *Second Edition*

GOLDSTEIN and LEWIS · Assessment: Problems, Development, and Statistical Issues

GREENWOOD and NIKULIN · A Guide to Chi-Squared Testing

*HAHN · Statistical Models in Engineering

HAHN and MEEKER · Statistical Intervals: A Guide for Practitioners

*Now available in a lower priced paperback edition in the Wiley Classics Library.

Applied Probability and Statistics (Continued)

HAND · Construction and Assessment of Classification Rules

HAND · Discrimination and Classification

HEIBERGER · Computation for the Analysis of Designed Experiments

HINKELMAN and KEMPTHORNE: · Design and Analysis of Experiments, Volume 1: Introduction to Experimental Design

HOAGLIN, MOSTELLER, and TUKEY · Exploratory Approach to Analysis of Variance

HOAGLIN, MOSTELLER, and TUKEY · Exploring Data Tables, Trends and Shapes

HOAGLIN, MOSTELLER, and TUKEY · Understanding Robust and Exploratory Data Analysis

HOCHBERG and TAMHANE · Multiple Comparison Procedures

HOCKING · Methods and Applications of Linear Models: Regression and the Analysis of Variables

HOGG and KLUGMAN · Loss Distributions

HOLLANDER and WOLFE · Nonparametric Statistical Methods

HOSMER and LEMESHOW · Applied Logistic Regression

HØYLAND and RAUSAND · System Reliability Theory: Models and Statistical Methods

HUBERTY · Applied Discriminant Analysis

JACKSON · A User's Guide to Principle Components

JOHN · Statistical Methods in Engineering and Quality Assurance

JOHNSON · Multivariate Statistical Simulation

JOHNSON and KOTZ · Distributions in Statistics
Continuous Multivariate Distributions

JOHNSON, KOTZ, and BALAKRISHNAN · Continuous Univariate Distributions, Volume 1, *Second Edition*

JOHNSON, KOTZ, and BALAKRISHNAN · Continuous Univariate Distributions, Volume 2, *Second Edition*

JOHNSON, KOTZ, and BALAKRISHNAN · Discrete Multivariate Distributions

JOHNSON, KOTZ, and KEMP · Univariate Discrete Distributions, *Second Edition*

JUREČKOVÁ and SEN · Robust Statistical Procedures: Aymptotics and Interrelations

KADANE · Bayesian Methods and Ethics in a Clinical Trial Design

KADANE AND SCHUM · A Probabilistic Analysis of the Sacco and Vanzetti Evidence

KALBFLEISCH and PRENTICE · The Statistical Analysis of Failure Time Data

KELLY · Reversability and Stochastic Networks

KHURI, MATHEW, and SINHA · Statistical Tests for Mixed Linear Models

KLUGMAN, PANJER, and WILLMOT · Loss Models: From Data to Decisions

KLUGMAN, PANJER, and WILLMOT · Solutions Manual to Accompany Loss Models: From Data to Decisions

KOVALENKO, KUZNETZOV, and PEGG · Mathematical Theory of Reliability of Time-Dependent Systems with Practical Applications

LAD · Operational Subjective Statistical Methods: A Mathematical, Philosophical, and Historical Introduction

LANGE, RYAN, BILLARD, BRILLINGER, CONQUEST, and GREENHOUSE · Case Studies in Biometry

LAWLESS · Statistical Models and Methods for Lifetime Data

LEE · Statistical Methods for Survival Data Analysis, *Second Edition*

LePAGE and BILLARD · Exploring the Limits of Bootstrap

LINHART and ZUCCHINI · Model Selection

LITTLE and RUBIN · Statistical Analysis with Missing Data

MAGNUS and NEUDECKER · Matrix Differential Calculus with Applications in Statistics and Econometrics

MALLER and ZHOU · Survival Analysis with Long Term Survivors

MANN, SCHAFER, and SINGPURWALLA · Methods for Statistical Analysis of Reliability and Life Data

*Now available in a lower priced paperback edition in the Wiley Classics Library.

Applied Probability and Statistics (Continued)

McLACHLAN and KRISHNAN · The EM Algorithm and Extensions

McLACHLAN · Discriminant Analysis and Statistical Pattern Recognition

McNEIL · Epidemiological Research Methods

MILLER · Survival Analysis

MONTGOMERY and PECK · Introduction to Linear Regression Analysis, *Second Edition*

MYERS and MONTGOMERY · Response Surface Methodology: Process and Product
in Optimization Using Designed Experiments

NELSON · Accelerated Testing, Statistical Models, Test Plans, and Data Analyses

NELSON · Applied Life Data Analysis

OCHI · Applied Probability and Stochastic Processes in Engineering and Physical
Sciences

OKABE, BOOTS, and SUGIHARA · Spatial Tesselations: Concepts and Applications
of Voronoi Diagrams

PANKRATZ · Forecasting with Dynamic Regression Models

PANKRATZ · Forecasting with Univariate Box-Jenkins Models: Concepts and Cases

PIANTADOSI · Clinical Trials: A Methodologic Perspective

PORT · Theoretical Probability for Applications

PUTERMAN · Markov Decision Processes: Discrete Stochastic Dynamic Programming

RACHEV · Probability Metrics and the Stability of Stochastic Models

RÉNYI · A Diary on Information Theory

RIPLEY · Spatial Statistics

RIPLEY · Stochastic Simulation

ROUSSEEUW and LEROY · Robust Regression and Outlier Detection

RUBIN · Multiple Imputation for Nonresponse in Surveys

RUBINSTEIN · Simulation and the Monte Carlo Method

RUBINSTEIN and MELAMED · Modern Simulation and Modeling

RYAN · Statistical Methods for Quality Improvement

SCHUSS · Theory and Applications of Stochastic Differential Equations

SCOTT · Multivariate Density Estimation: Theory, Practice, and Visualization

*SEARLE · Linear Models

SEARLE · Linear Models for Unbalanced Data

SEARLE, CASELLA, and McCULLOCH · Variance Components

STOYAN, KENDALL, and MECKE · Stochastic Geometry and Its Applications, *Second
Edition*

STOYAN and STOYAN · Fractals, Random Shapes and Point Fields: Methods of
Geometrical Statistics

THOMPSON · Empirical Model Building

THOMPSON · Sampling

TIJMS · Stochastic Modeling and Analysis: A Computational Approach

TIJMS · Stochastic Models: An Algorithmic Approach

TITTERINGTON, SMITH, and MAKOV · Statistical Analysis of Finite Mixture
Distributions

UPTON and FINGLETON · Spatial Data Analysis by Example, Volume 1: Point
Pattern and Quantitative Data

UPTON and FINGLETON · Spatial Data Analysis by Example, Volume II:
Categorical and Directional Data

VAN RIJCKEVORSEL and DE LEEUW · Component and Correspondence Analysis

WEISBERG · Applied Linear Regression, *Second Edition*

WESTFALL and YOUNG · Resampling-Based Multiple Testing: Examples and
Methods for p-Value Adjustment

WHITTLE · Systems in Stochastic Equilibrium

WOODING · Planning Pharmaceutical Clinical Trials: Basic Statistical Principles

WOOLSON · Statistical Methods for the Analysis of Biomedical Data

*ZELLNER · An Introduction to Bayesian Inference in Econometrics

*Now available in a lower priced paperback edition in the Wiley Classics Library.

Texts and References Section

AGRESTI · An Introduction to Categorical Data Analysis

ANDERSON · An Introduction to Multivariate Statistical Analysis, *Second Edition*

ANDERSON and LOYNES · The Teaching of Practical Statistics

ARMITAGE and COLTON · Encyclopedia of Biostatistics: Volumes 1 to 6 with Index

BARTOSZYNSKI and NIEWIADOMSKA-BUGAJ · Probability and Statistical Inference

BERRY, CHALONER, and GEWEKE · Bayesian Analysis in Statistics and Econometrics: Essays in Honor of Arnold Zellner

BHATTACHARYA and JOHNSON · Statistical Concepts and Methods

BILLINGSLEY · Probability and Measure, *Second Edition*

BOX · R. A. Fisher, the Life of a Scientist

BOX, HUNTER, and HUNTER · Statistics for Experimenters: An Introduction to Design, Data Analysis, and Model Building

BOX and LUCEÑO · Statistical Control by Monitoring and Feedback Adjustment

BROWN and HOLLANDER · Statistics: A Biomedical Introduction

CHATTERJEE and PRICE · Regression Analysis by Example, *Second Edition*

COOK and WEISBERG · An Introduction to Regression Graphics

COX · A Handbook of Introductory Statistical Methods

DILLON and GOLDSTEIN · Multivariate Analysis: Methods and Applications

DODGE and ROMIG · Sampling Inspection Tables, *Second Edition*

DRAPER and SMITH · Applied Regression Analysis, *Third Edition*

DUDEWICZ and MISHRA · Modern Mathematical Statistics

DUNN · Basic Statistics: A Primer for the Biomedical Sciences, *Second Edition*

FISHER and VAN BELLE · Biostatistics: A Methodology for the Health Sciences

FREEMAN and SMITH · Aspects of Uncertainty: A Tribute to D. V. Lindley

GROSS and HARRIS · Fundamentals of Queueing Theory, *Third Edition*

HALD · A History of Probability and Statistics and their Applications Before 1750

HALD · A History of Mathematical Statistics from 1750 to 1930

HELLER · MACSYMA for Statisticians

HOEL · Introduction to Mathematical Statistics, *Fifth Edition*

JOHNSON and BALAKRISHNAN · Advances in the Theory and Practice of Statistics: A Volume in Honor of Samuel Kotz

JOHNSON and KOTZ (editors) · Leading Personalities in Statistical Sciences: From the Seventeenth Century to the Present

JUDGE, GRIFFITHS, HILL, LÜTKEPOHL, and LEE · The Theory and Practice of Econometrics, *Second Edition*

KHURI · Advanced Calculus with Applications in Statistics

KOTZ and JOHNSON (editors) · Encyclopedia of Statistical Sciences: Volumes 1 to 9 wtih Index

KOTZ and JOHNSON (editors) · Encyclopedia of Statistical Sciences: Supplement Volume

KOTZ, REED, and BANKS (editors) · Encyclopedia of Statistical Sciences: Update Volume 1

KOTZ, REED, and BANKS (editors) · Encyclopedia of Statistical Sciences: Update Volume 2

LAMPERTI · Probability: A Survey of the Mathematical Theory, *Second Edition*

LARSON · Introduction to Probability Theory and Statistical Inference, *Third Edition*

LE · Applied Survival Analysis

MALLOWS · Design, Data, and Analysis by Some Friends of Cuthbert Daniel

MARDIA · The Art of Statistical Science: A Tribute to G. S. Watson

MASON, GUNST, and HESS · Statistical Design and Analysis of Experiments with Applications to Engineering and Science

MURRAY · X-STAT 2.0 Statistical Experimentation, Design Data Analysis, and Nonlinear Optimization

*Now available in a lower priced paperback edition in the Wiley Classics Library.

Texts and References (Continued)

PURI, VILAPLANA, and WERTZ · New Perspectives in Theoretical and Applied
 Statistics
RENCHER · Methods of Multivariate Analysis
RENCHER · Multivariate Statistical Inference with Applications
ROSS · Introduction to Probability and Statistics for Engineers and Scientists
ROHATGI · An Introduction to Probability Theory and Mathematical Statistics
RYAN · Modern Regression Methods
SCHOTT · Matrix Analysis for Statistics
SEARLE · Matrix Algebra Useful for Statistics
STYAN · The Collected Papers of T. W. Anderson: 1943–1985
TIERNEY · LISP-STAT: An Object-Oriented Environment for Statistical Computing
 and Dynamic Graphics
WONNACOTT and WONNACOTT · Econometrics, *Second Edition*

WILEY SERIES IN PROBABILITY AND STATISTICS
ESTABLISHED BY WALTER A. SHEWHART AND SAMUEL S. WILKS

Editors
Robert M. Groves, Graham Kalton, J. N. K. Rao, Norbert Schwarz,
Christopher Skinner

Survey Methodology Section

BIEMER, GROVES, LYBERG, MATHIOWETZ, and SUDMAN · Measurement
 Errors in Surveys
COCHRAN · Sampling Techniques, *Third Edition*
COX, BINDER, CHINNAPPA, CHRISTIANSON, COLLEDGE, and KOTT (editors) ·
 Business Survey Methods
*DEMING · Sample Design in Business Research
DILLMAN · Mail and Telephone Surveys: The Total Design Method
GROVES and COUPER · Nonresponse in Household Interview Surveys
GROVES · Survey Errors and Survey Costs
GROVES, BIEMER, LYBERG, MASSEY, NICHOLLS, and WAKSBERG ·
 Telephone Survey Methodology
*HANSEN, HURWITZ, and MADOW · Sample Survey Methods and Theory,
 Volume 1: Methods and Applications
*HANSEN, HURWITZ, and MADOW · Sample Survey Methods and Theory,
 Volume II: Theory
KASPRZYK, DUNCAN, KALTON, and SINGH · Panel Surveys
KISH · Statistical Design for Research
*KISH · Survey Sampling
LESSLER and KALSBEEK · Nonsampling Error in Surveys
LEVY and LEMESHOW · Sampling of Populations: Methods and Applications
LYBERG, BIEMER, COLLINS, de LEEUW, DIPPO, SCHWARZ, TREWIN (editors) ·
 Survey Measurement and Process Quality
SKINNER, HOLT, and SMITH · Analysis of Complex Surveys

*Now available in a lower priced paperback edition in the Wiley Classics Library.